Contents

Introduction

This series by Peter Brett is intended to form a complete reference work for a wide range of woodworking occupations including site carpenters, bench joiners, formworkers, maintenance carpenters, shop fitters, wood machinists and both carpenters and joiners working in the heritage skills sector.

These versatile resources have been written to provide an in-depth knowledge of the basic methods, tools and materials that are essential for students, plus extended material and specialist topics for those working within the industry.

- *Book 1: Job Knowledge* covers the introductory skills and an appreciation of the materials used in the woodworking industry. This is followed up with coverage of the background knowledge that is required to undertake a wide range of practical activities. The final chapters give an overview of the building industry and its controls, which are desirable for understanding and progression.

- *Book 2: Practical Activities* is divided into the specific areas of work, that may be undertaken by carpenters and joiners in a wide range of situations. The introductory topics in each chapter are covered in depth, supported by bulleted step-by-step text and illustrations, which are intended to guide the reader through the task. This is reinforced by a wider coverage of each topic area to aid understanding and enable the transfer/application of skills to other tasks.

- *Carpentry and Joinery Online* kerboodle! resources provide comprehensive, online teaching, learning and assessment resources that work hand-in-hand with the books to help every student reach their full potential. Resources include videos, animations, case studies, multiple choice question practice and plenty of worksheets and handouts. These online resources are available on kerboodle! which can be accessed via the internet at **www.kerboodle.com/live**, anytime, anywhere. If your college subscribes to kerboodle! you will be provided with your own personal login details. Once logged in, access your course and locate the required activity. Please visit **kerboodle.helpserve.com** if you would like more information and help on how to use kerboodle!.

This series will prove to be invaluable to both the new entrant undergoing training and qualified craftspeople that are undertaking a task for the first time or wish to extend their knowledge of craft processes and the interrelationship with the activities of the building industry as a whole. They are also ideal background reading for people studying or employed at all levels in the building industry and advanced DIY enthusiasts.

Peter Brett has written numerous books on carpentry and joinery and has extensive experience in both the industry and its education and training systems.

He has had over 20 years' experience as a tutor in a variety of establishments and has worked with both the City and Guilds of London Institute and Construction Skills, in the development and assessment of awards in carpentry and joinery.

Peter is now back in industry, undertaking design, value engineering, systems analysis and consultancy for construction projects on a freelance basis.

Acknowledgements

The author and publisher would like to thank the following for providing photographs:

Art Directors & Trip: fig 7.60 top, fig 9.52 top; Beaman Antiques / Lynford Beaman: page 267; Birmingham Picture Library: page 224; EDIFICE / Alamy: page 238; fotofacade.com / Alamy: page 197; H K Thorburn & Son: fig 4.7; HSP Garden Buildings www.birstall.com: page 79; Iconpix / Alamy: fig 7.19 right; iStockphoto: fig 6.2, fig 6.103, fig 12.4, page 98; Jacqueline Banerjee, PhD, Victorian Web (www.victorianweb.org): page 65; Jinny Goodman / Alamy: fig 12.1; Joe Fairs / Alamy: fig 12.9; Michael Holmes: page 45; Philip Pound / Alamy: page 430; Photolibrary.com/ Chris Laurens: fig 7.19 left; Rodger Tamblyn / Alamy: fig 7.60 bottom; Ron Chapple Stock / Alamy: fig 6.1; Reproduced by kind permission of The National Trust for Scotland Photo Library: page 122; The Tool Shop (www.antiquetools.co.uk): page 252; Wayne Boucher: page 198; Wellstead Woodworking/ Doug Johnston: fig 9.52 (bottom).

Chapter One

Timber ground and upper floor construction

This chapter covers the work of the site carpenter. It is concerned with principles of ground and upper floor construction and includes the following:

- An introduction to floors.
- Timber floor construction.
- Decking, boards and sheets.
- Insulation and finishing.

Floors are primary elements of a building; they form the lower horizontal surfaces of a room or area. The actual laying of timber joists, which form the structure of a floor, is termed carcassing. While the laying of joist covering and insulation, which should not be undertaken until after the building has been made weather tight, is termed first fixing.

Types of floor

Floors may be divided into two groups: ground floors and upper floors.

The main functions of a floor are to:
- provide an acceptable surface for walking, living and working
- provide acceptable levels of thermal and sound insulation
- support and transfer any loads imposed upon it.

Ground floors are also required to prevent moisture penetration and weed growth.

Ground floors

There are two types of ground floor construction: solid and suspended or hollow. A solid ground floor is in direct contact with the ground and a suspended floor spans from wall to wall.

Solid ground floors

These are built up from a number of layers, see Figure 1.1. Hardcore provides a suitable base for the construction. A thin blinding layer of sand or weak concrete is often used to fill voids or level out the rough edges of

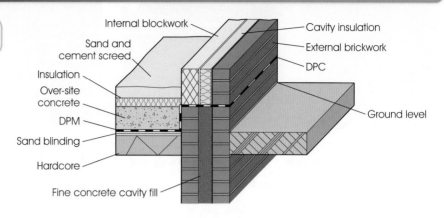

▶ **Figure 1.1** Solid ground floor construction

Internal blockwork — Cavity insulation
Sand and cement screed — External brickwork
Insulation — DPC
Over-site concrete
DPM — Ground level
Sand blinding
Hardcore
Fine concrete cavity fill

the hardcore in order to reduce grout loss or the risk of puncturing the damp-proof membrane (DPM). The DPM is placed between the hardcore and concrete slab to prevent ground moisture rising through the floor. The wall should include a damp-proof course (DPC) at least 150 mm above the external ground level to prevent ground moisture or rain splashes bouncing off the ground level and rising up the wall. It is essential that the DPM is lapped into the walls' DPC in order to prevent any possibility of ground moisture by passing the DPM. The concrete floor slab (also known as the over-site concrete, since it covers the whole site of the building) is not built into the walls but acts as an independent raft transferring its imposed loading directly to the ground and not the foundations.

For industrial use, warehouses, garages and so on, the over-site concrete may be power floated or grinded to form the actual floor finish. In other buildings, it is normal to lay a cement and sand screed. This screed levels the slab, takes out any irregularities and provides a smooth surface to receive the final floor finish (e.g. carpet, lino, tiles, woodblocks or strips etc.). Alternatively, a floating timber floor can be laid over the concrete subfloor to provide the final finish. Insulation is included between the oversite and screed to provide the required thermal standards.

Suspended ground floors

Traditionally, these were constructed using timber joists and floorboading, although they are rarely used today. The modern method is to use pre-cast concrete beams, in-filled with concrete blocks, known as beam and block floors.

Timber ground floors. In common with solid ground floors, these still require the hardcore and over-site concrete layers. The dwarf or sleeper walls, which support the timber floor construction, are built on top of the over-site at about 1.8 m centres using honeycomb bond (a bond that leaves half-brick voids in the wall to allow air circulation). Airbricks must also be included in the outside walls at ground level to enable through ventilation in the under-floor space. This prevents the moisture content of the timber rising too high. (Dry-rot fungi will almost inevitably attack timber in poorly ventilated areas with moisture content above 20 per cent.) The timber floor construction consists of floorboards or sheets, supported by joists that bear on wall plates that in turn spread the load evenly along the sleeper walls. A DPC is placed between the wall plate and sleeper walls to prevent the rise of ground moisture into the timberwork. See Figure 1.2.

Beam and block ground floors. The ends of the inverted 'tee' section pre-cast concrete beams are supported on either the inner skin of the cavity wall or by separate sleeper walls. Intermediate sleeper walls may also be required, to provide mid-span support or to enable beams to be joined in length. The beams should be evenly spaced to receive the standard size (440 mm × 215 mm × 100 mm) concrete infill blocks. These are laid edge to edge and are supported on the beam's protruding bottom edge. See Figure 1.3.

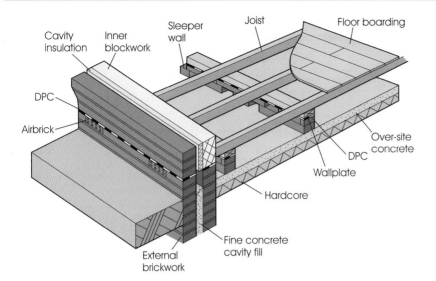

◄ Figure 1.2 Timber suspended ground floor

Cavity insulation
Inner blockwork
Sleeper wall
Joist
Floor boarding
DPC
Airbrick
Over-site concrete
DPC
Wallplate
Hardcore
Fine concrete cavity fill
External brickwork

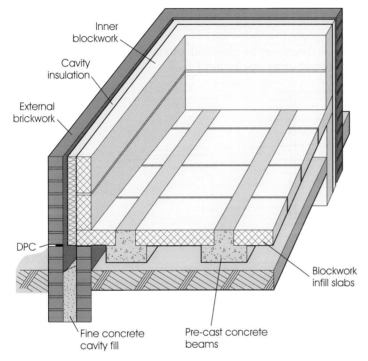

◄ Figure 1.3 Beam and block suspended ground floor

Inner blockwork
Cavity insulation
External brickwork
DPC
Fine concrete cavity fill
Pre-cast concrete beams
Blockwork infill slabs

After laying, any gaps between the blocks and beams can be filled by brushing in a sand and cement grout. The actual finished floor surface can be formed using either a sand and cement screed or a timber floating floor.

Upper floors

These are also known as suspended floors. Timber construction is mainly used for house construction and concrete for other works.

Timber suspended upper floors. These consist of a number of bridging joists supported at either end by load-bearing walls. The joists are covered on their top by floorboarding or sheeting to provide the floor surface and on their bottom with plasterboard to form the ceiling. The end of the joists can either be supported by building them into the internal leaf of the cavity wall or, alternatively, metal joist hangers may be used; these have a flange that locates in the joint of the wall. Where openings in the floor are required for stairs, chimney breasts and so on, the joists must be framed or trimmed around the opening. This entails cutting short a number of bridging joists so that they do not protrude into the opening. These are called trimmed joists. See Figure 1.4.

► **Figure 1.4** Timber suspended upper floor

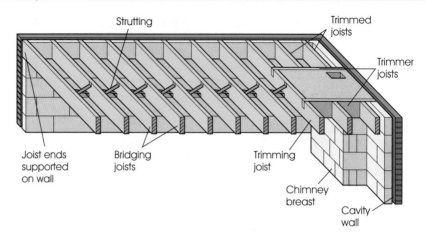

Trimmer and trimming joists, which are thicker than the bridging, are framed around the opening to provide support for the ends of the trimmed joists. In order to prevent joists longer than 2 m buckling when under load, they must be strutted at their centre. This strutting has the effect of stiffening the whole floor.

Timber upper floors may be classified into three groups according to their method of construction, see Figure 1.5.

► **Figure 1.5** Types of timber upper floor construction

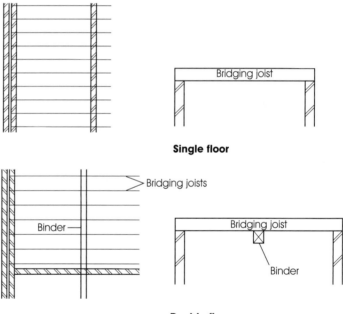

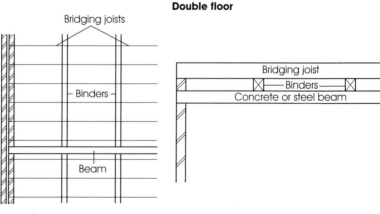

- *Single floors:* These are the most common type of upper floor for domestic house construction, where common or bridging joists span the shortest dimension of an area from wall to wall.

- *Double floors:* These are used where the shortest span of the floor is over about 4.8 m or the bridging joists run in the longest dimension. In this case the area can be sub-divided into a number of bays, each of which is not more than 4.8 m long, by binders, which provide intermediate support for the bridging joists.

- *Triple or framed floors:* These are now obsolete in new construction, as floors where the shortest span is sufficiently long to require them will almost certainly be constructed using other forms of floor construction. However, triple floors may still be encountered by carpenters and joiners when undertaking maintenance or renovation work. The area is divided up into bays by very heavy, large-sectioned, steel universal beams or concrete beams. These beams provide support for the binders, which in turn provide support for the bridging joists.

Concrete suspended upper floors. These are supported on the walls or structural framework. They may be formed using either reinforced *in situ* cast concrete or some form of pre-cast units. See Figure 1.6. In both cases, the top surface will require screeding to receive the floor finish or a timber floating floor can be used. The bottom surface will need plastering or another form of finishing to provide the ceiling surface. *In situ* concrete floors will require the erection of formwork to cast and support the wet concrete. In general, concrete floors are more sound resistant to airborne noise and have better fire-resistant properties than timber floors.

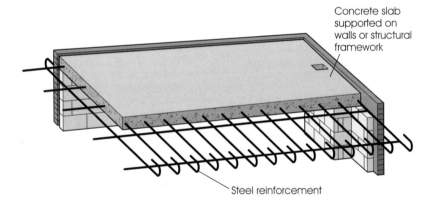

Concrete slab supported on walls or structural framework

Steel reinforcement

◀ **Figure 1.6** Concrete suspended upper floor

Timber floor construction

Joists

These are of a series of parallel timber beams used to span the gap between walls and directly support a floor surface and the ceiling surface below. Traditionally, solid timber is used for joists, however, manufactured alternatives are also available including: glulam, which is seen as environmentally friendly because it utilises smaller sections and shorter lengths, often off-cuts or recycled timber; box and 'I' joists are lighter than solid or glulam sections and are also seen as environmentally friendly because they use small solid sections for their top and bottom flanges often with orientated strand board (OSB) for the outside or central webs, which can be manufactured from recycled timber. Composite I joists use vee-shaped galvanised steel webs nailed to either side of the timber top and bottom flanges. See Figure 1.7.

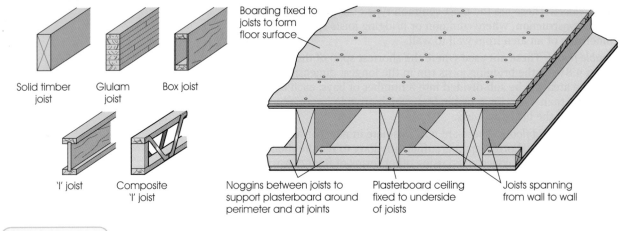

Boarding fixed to joists to form floor surface

Noggins between joists to support plasterboard around perimeter and at joints

Plasterboard ceiling fixed to underside of joists

Joists spanning from wall to wall

Solid timber joist

Glulam joist

Box joist

'I' joist

Composite 'I' joist

▲ **Figure 1.7** Joists

Sectional size of joists

The sectional size of a joist (breadth and depth) depends on its span, spacing, the weight or loading placed upon it and the strength class of the timber used. For new work, the sectional size will always be specified. The Timber Research and Development Association (TRADA) and other bodies produce tables of suitable sectional joist size for use in different situations, see Table 1.1 for an example.

Table 1.1 Maximum clear span of floor joists (m)

Breadth × Depth	Spacing (centre to centre)			Based on using C16 grade softwood with a dead load excluding weight of joist between 0.25 and 0.5 kN/m²
Size of joist (mm)	400 mm	450 mm	600 mm	Example
38 × 97	1.72	1.56	1.21	To determine size of joists, spaced at 600 mm centres and having a clear span of 3.75 m:
38 × 122	2.37	2.22	1.76	
38 × 147	2.85	2.71	2.33	
38 × 170	3.28	3.10	2.69	1. Look down the 600 mm column to get the nearest value above 3.75 m.
38 × 195	3.72	3.52	3.06	
38 × 220	4.16	3.93	3.42	
50 × 97	1.98	1.87	1.54	2. This is 3.90 m. Move to the left hand column to get the required joist size.
50 × 122	2.60	2.50	2.19	
50 × 147	3.13	3.01	2.69	
50 × 170	3.61	3.47	3.08	3. 50 mm × 220 mm joists are required in this case.
50 × 195	4.13	3.97	3.50	
50 × 220	4.64	4.47	3.90	4. As an alternative, 75 mm × 195 mm joists could be used, but would not be economical.
75 × 122	2.97	2.86	2.60	
75 × 147	3.56	3.43	3.13	
75 × 170	4.11	3.96	3.61	
75 × 195	4.68	4.52	4.13	
75 × 220	5.11	4.97	4.64	

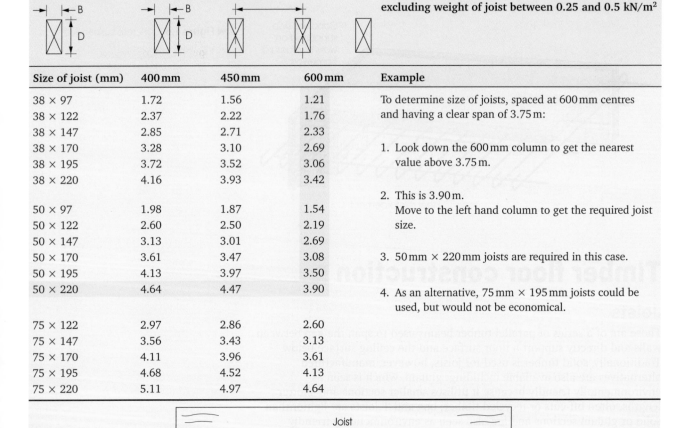

Joist

Clear span

Distance between supports

The approximate depth of 50 mm breadth joists spaced at 400 mm centres can be found by using the following rule-of-thumb formula:

$$\text{Depth of joist} = \frac{\text{span of joist in millimetres}}{20} + 20$$

Example

for joists spanning 4000 mm

$$\text{Depth of joist} = \frac{4000}{20} + 20 = 220\,\text{mm}$$

The nearest commercial size of timber to use would be 50 mm × 225 mm or 50 mm × 220 mm using machined timber. These are suitable for normal domestic use, although on the generous side, as 50 mm × 195 mm joists could be used for spans up to 4130 mm (see Table 1.1).

Span of joists

- *Clear span* is the distance between joist supports.
- *Effective span* is the distance between the centres of the joist bearings.

The bearing itself is the length of the end of a joist that rests on the support. The overall length of a joist is its clear span plus the length of its end bearings (e.g. a joist with a 3400 mm clear span and 100 mm end bearings will have an effective span of 3500 mm and an overall joist length of 3600 mm). See Figure 1.8.

Joists are commonly laid out to span the shortest distance between the supporting walls of a room or other area. This keeps to a minimum the size of joist required. Once the depth of the joist is determined for the longest span it is normal practice to keep all other joists the same. The shorter span joists will be oversize, but all joist coverings and ceiling surfaces will be level.

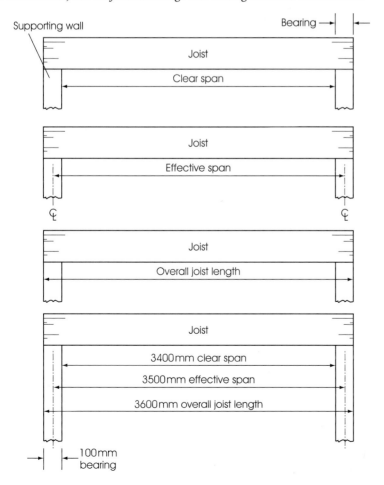

◄ Figure 1.8 Span of joists

Spacing of joists

Joist spacing is the distance between the centres of adjacent joists. Commonly called joist centres or c/c (centre to centre), see Figure 1.9. They range between 400 mm to 600 mm, depending on the joist covering material. Joists should be spaced to accommodate surface dimensions of their covering material, see Table 1.2 for maximum joist spacings for a range of joist covering materials.

▶ **Figure 1.9** Spacing of joists

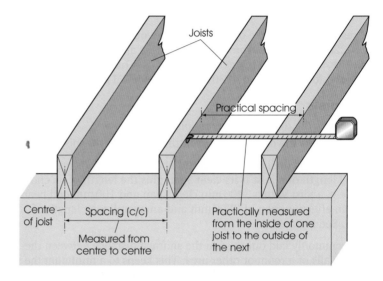

Table 1.2 Guide to maximum span of decking (joist covering materials)

Decking material	Finished thickness	Maximum span
		Span = joist spacing
Softwood planed, tongued and grooved (PTG)	16 mm	450 mm
Softwood planed, tongued and grooved (PTG)	19 mm	600 mm
Flooring grade chipboard	18 mm	450 mm
Flooring grade chipboard	22 mm	600 mm
Flooring grade plywood	16 mm	400 mm
Flooring grade plywood	19 mm	600 mm

Related information

Unventilated and damp timber is open to an attack of rot (dry rot or wet rot), which leads to a loss of strength and possible collapse. For detailed information on timber rots, their prevention and remedial treatment see Book 1, Chapter 1, pages 32–6.

End joists adjacent to walls should be kept 50 mm away from the wall surface, in order to allow air circulation and prevent dampness being transferred from wall to joist and thus reduce the risk of rot. In addition, this gap helps reduce the transmission of noise at party walls.

To determine the number of joists required and their centres for a particular area the following procedure can be used, see Figure 1.10:

1. Measure the distance between adjacent walls, for example 3150 mm. The first and last joist would be positioned 50 mm away from the walls. The centres of 50 mm breadth joists would be 75 mm away from the wall. The total distance between end joists centres would be 3000 mm.

2. Divide the distance between end joist centres by specified joist spacing for example 400 mm. This gives the number of spaces between joists. Where a whole number is not achieved, round up to the nearest whole number. There will always be one more joist than the number of spaces, so remember to add one to this figure to determine the number of joists.

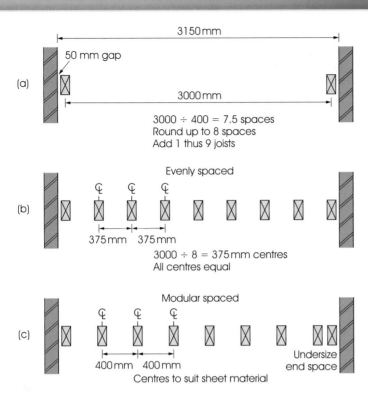

(a)

3150 mm

50 mm gap

3000 mm

3000 ÷ 400 = 7.5 spaces
Round up to 8 spaces
Add 1 thus 9 joists

Evenly spaced

(b)

375 mm 375 mm

3000 ÷ 8 = 375 mm centres
All centres equal

Modular spaced

(c)

400 mm 400 mm

Undersize
end space

Centres to suit sheet material

Where tongued and grooved (T&G) boarding is used as a floor covering, the joist centres may be spaced out evenly (i.e. divide the distance between end joist centres by the number of spaces).

Where sheet material is used as a joist covering to form a floor, ceiling or roof surface, the joist centres are normally maintained at a 400 mm or 600 mm modular spacing to coincide with sheet sizes. This would leave an undersized space between the last two joists.

Joist sections

The strength of a joist varies in direct proportion to changes in its breadth and in proportion to the square of its depth. For example, doubling the breadth of a section doubles its strength (see Figure 1.11):

A 100 mm × 100 mm joist has double the strength of a 50 mm × 100 mm joist.

Doubling the depth of a section increases its strength by four times:

A 50 mm × 200 mm joist has four times the strength of a 50 mm × 100 mm joist.

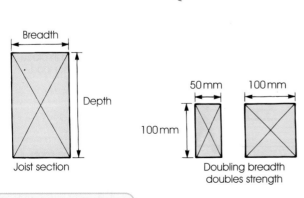

Breadth

Depth

Joist section

50 mm 100 mm

100 mm

Doubling breadth
doubles strength

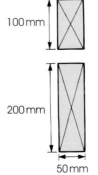

100 mm

200 mm

50 mm

Doubling depth
increases strength
by four times

$100^2 = 100 \times 100 = 10\,000$
$200^2 = 200 \times 200 = 40\,000$

▲ **Figure 1.11** Joist sections

Less material is required for the same strength when the greatest sectional dimension is placed vertically rather than horizontally:

> A 100 mm × 100 mm joist has the same sectional area as a 50 mm × 200 mm joist, but the deeper joist would be twice as strong.

Therefore, joists are normally placed so that the depth is the greatest sectional dimension. Joists of the same sectional size and span would clearly have different strengths if their breadths and depths were reversed. Those with the smaller depth would sag under load, possibly leading to structural collapse.

When stating the sectional size of a joist the first measurement given is the breadth and the second the depth.

Machined timber (formally termed regularised timber) is preferred for joists as all timber will be a consistent depth. This aids floor levelling and ensures a flat fixing surface for joist coverings and ceilings. Joists are machined to a consistent depth by re-sawing or planing one or both edges. A reduction in size of 3 mm must be allowed for timber up to 150 mm in depth and 5 mm over this depth. See Figure 1.12. Thus, 50 mm × 150 mm may be machined to 50 mm × 147 mm and 50 mm × 200 mm may be machined to 50 mm × 195 mm.

Positioning of joist defects. Any joists that are not straight should be positioned with their camber or crown (curved edge) upwards. When loaded these joists will tend to straighten out, rather than sag further if laid the other way. Joists with edge knots should be positioned with the knots on their upper edge. When loaded, the knots will be held in position as the joists sag, rather than fall out, weaken the joist and possibly lead to structural collapse, if laid the other way. See Figure 1.13.

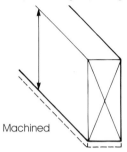

50 mm × 200 mm joist machined to 50 mm × 195 mm by re-sawing or planing

Machined

▲ **Figure 1.12** Machined joist

▶ **Figure 1.13** Positioning of joists with camber or edge knots

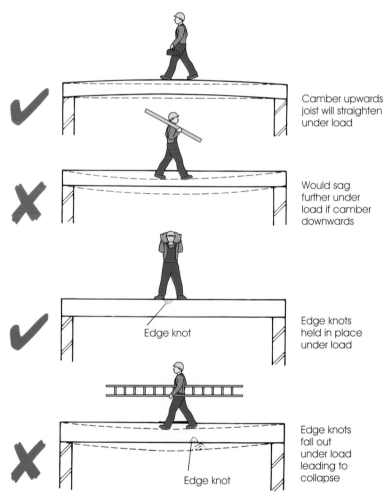

Camber upwards joist will straighten under load

Would sag further under load if camber downwards

Edge knot

Edge knots held in place under load

Edge knot

Edge knots fall out under load leading to collapse

Joist supports and restraints

Joist supports

The ends of joists may be supported:

- by building in
- on hangers
- on wall plates
- on binders.

Building in the inner leaf of a cavity wall. The minimum bearing in a wall is normally 90 mm, see Figure 1.14. Shorter bearings do not tie in the wall sufficiently and can lead to a crushing of the joist end possibly leading to collapse. A steel bearing bar may be incorporated into the mortar joint where lightweight blocks are used to reduce the risk of the blocks crumbling. The ends of the joists are often splayed but they must not project into the cavity where they could possibly catch mortar droppings during building, leading to dampness in the joist and rotting. The ends of joists that are in contact with the external wall should be treated with a timber preservative to protect them from dampness and subsequent rot.

Hangers. Ends of joists may be supported on galvanised steel joist hangers, which are themselves built into or bear on a wall (see Figure 1.15). Double hangers are available, which saddle internal walls to provide a support for joists on both sides. An advantage of this method is that the joist can be positioned independently of the building process. Hangers are useful when forming extensions as they are simply inserted into a raked-out mortar joint. The bottom edge or bearing surface of the hanger must be recessed into the joist to prevent the hangers obstructing any ceiling covering. Joists should be secured into the hanger using 32 mm galvanised clout nails in each hole provided.

In modern construction, the preferred method of supporting the ends of upper floor joists in external walls is to use joist hangers, as any subsequent timber shrinkage or warping may result in gaps in the wall around a built-in end leading to a reduction in both sound and thermal insulation.

Wall plates. Ends of joists may be supported by a wall plate bedded on the top of a wall. This is normally used for ground floor construction and over internal load-bearing partitions for upper floors (see Figure 1.16). The minimum bearing recommended on a wall plate and thus the minimum width is 75 mm, although 100 mm wide wall plates are normally used for ground floor sleeper walls. Often joists from either side meet over a wall plate; it is usual to nail them together side by side, both overlapping the wall plate by about 300 mm.

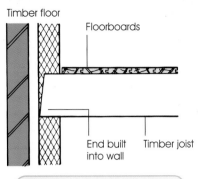

▲ **Figure 1.14** Building in a joist

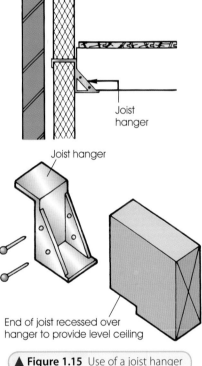

▲ **Figure 1.15** Use of a joist hanger

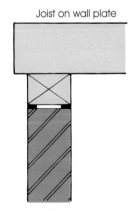

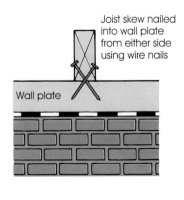

▲ **Figure 1.16** Joist supported on a wall plate

The use of wall plates provides a means of securing joists by skew nailing with 75 mm or 100 mm wire nails. In addition, wall plates also spread the loading of a joisted surface evenly over a wide area rather than a point load. Cambered joists may be straightened over a wall plate by partly sawing through and nailing down. Wall plates are not suitable for use in external walls of upper floor construction. This is due to shrinkage movement and the likelihood of rot.

The carpenter uses halving joints, see Figure 1.17, to join wall plates at corners and where required in length. The plates are then normally bedded and levelled in position by the bricklayer using bedding mortar.

▶ **Figure 1.17** Wall plate joint

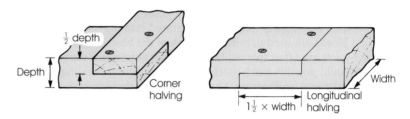

Intermediate support to joists

Binders. These are introduced into a floor structure in order to provide an intermediate support for large span joists used in double and triple floors. These binders may be made of:

- steel (when they are known as a universal beam (UB))
- timber (either of solid section, glue laminated section (glulam) or an OSB plywood box beam)
- concrete.

Depending on the space available, binders may be positioned below the joists, or accommodated partly within the joist depth projecting above or below as required, see Figure 1.18. Where joists are fitted to a steel universal beam, a plywood template may be cut to speed the marking out of the joist end and ensure a consistent, accurate fit, see Figure 1.19.

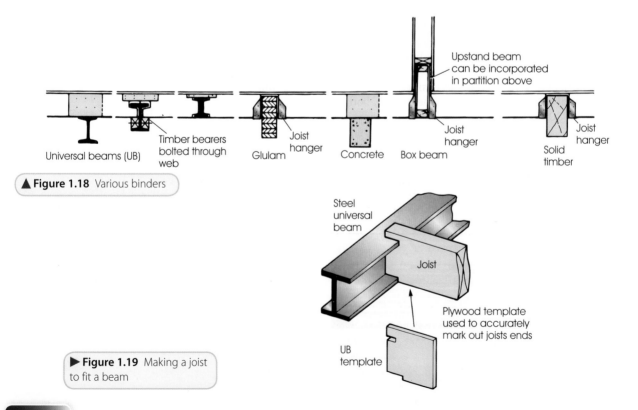

▲ **Figure 1.18** Various binders

▶ **Figure 1.19** Making a joist to fit a beam

Fire protection of steel binders. Where universal beams are used in floor construction they must be protected as a fire precaution. This protection is required in order to prevent early structural collapse resulting from the fact that the strength of steel rapidly decreases at temperatures above 300 °C. Various methods of protection include:

▪ a sprayed coat of non-combustible insulating material

▪ a solid casing of concrete

▪ encasing in expanded metal lathing and plastered

▪ encasing with plasterboard or other non-combustible material.

Figure 1.20 shows a typical method of encasing a steel beam using plasterboard and a skim coat of finishing plaster. The plasterboard has been fixed to timber cradling members nailed to each joist.

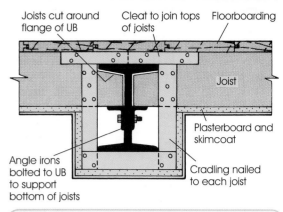

▲ **Figure 1.20** Plasterboard and cradling to a universal beam

Joist restraints

Restraint straps. With modern lightweight structures, walls and joisted areas require positive tying together for strength, to ensure wall stability in windy conditions. This is a requirement of the Building Regulations. Galvanised mild steel restraint straps, also known as lateral restraint straps, must be used at not more than 2 m intervals and carried over at least three joists. Noggins should be used under the straps where the joists run parallel to the wall; these should be a minimum of 38 mm thick and extend at least half the depth of the joist. For example, with 200 mm deep joists at least 100 mm deep noggins are required. See Figure 1.21.

◀ **Figure 1.21** Restraint strap

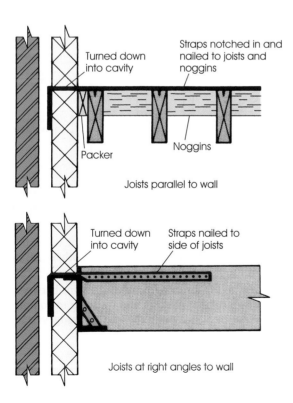

Strutting. Where deep joists exceed a span of 2.5 m, they tend to buckle or sag under loads unless restrained by strutting. Joists spanning between 2.5 and 4.5 m should be restrained by strutting at their mid-span; larger span joists over 4.5 m will require two rows of strutting at one third and two thirds of their span. Strutting is a requirement of the Building Regulations.

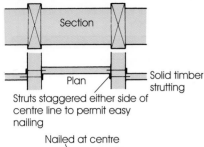

Struts staggered either side of centre line to permit easy nailing

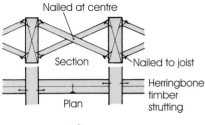

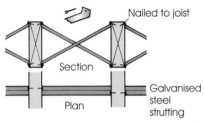

▲ **Figure 1.22** Strutting to joists

There are three main types of strutting in use, see Figure 1.22:

- *Solid strutting* is quick to install but is considered inferior, as it tends to loosen and become ineffective when joists shrink. It should be at least 38 mm thick and extend to at least ¾ of the joist's depth (e.g. a 200 mm deep joist will require at least 150 mm deep struts).

- *Herringbone timber strutting* is considered the most effective as it actually tightens when joists shrink. However, it takes longer to install and so is more expensive in terms of labour costs. The minimum sectional size is 38 mm × 38 mm. Timber herringbone struts can only be used where the spacing between the joists is less than three times the depth of the joist (e.g. machined for 150 mm deep joists herringbone struts are suitable up to 450 mm spacing, but for 200 mm deep joists a spacing of up to 600 mm is permitted).

- *Galvanised steel herringbone strutting* is quick to install as no cutting is required, but has a disadvantage in that the depth and centres of the joist must be specified when ordering; different depths and spacings will require different-sized struts.

Herringbone timber struts can be marked out using the following procedure, see Figure 1.23:

1. Mark across the joists the centre line of the strutting.
2. Mark a second line across the joist so that the distance between the lines is 10 mm less than the depth of the joist.
3. Place the length of strutting on top of the joists as shown and mark underneath against the joists at A and B.
4. Cut two struts to these marks. If all joists are spaced evenly, all the strutting will be the same size and can be cut using the first one as a template. If not, each set of struts will have to be marked individually.
5. Fix struts on either side of the centre line using wire nails, one in the top and bottom of each strut and one through the centre.

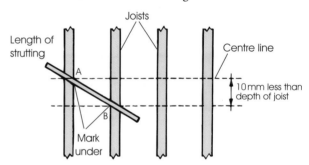

▶ **Figure 1.23** Marking out strutting

The gaps between the end joists will require packing and wedging to complete the system. See Figure 1.24.

Whichever method of strutting is used, care should be taken to ensure that they are clear of the tops and bottoms of the joists, otherwise they may subsequently distort the joist covering or ceiling surface.

Again, whichever method of strutting is used, the gaps between the end joists and the walls will require packing and wedging to complete the system. Care must be taken not to over tighten the folding wedges as it is possible to dislodge the blockwork.

▶ **Figure 1.24** Packing and wedging

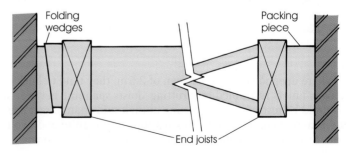

Jointing and positioning joists

Trimming

Where openings are required in joisted areas or where projections occur in supporting walls, joists must be framed or trimmed around them (see Figure 1.25). Members used for trimming each have their own function and are named accordingly:

- ■ *Bridging joist:* a joist spanning from support to support, also known as a common joist.
- ■ *Trimmed joist:* a bridging joist that has been cut short (trimmed) to form an opening in the floor.
- ■ *Trimmer joist:* a joist placed at right angles to the bridging joist in order to support the cut ends of the trimmed joists.
- ■ *Trimming joist:* a joist with a span the same as the bridging joist but supporting the end of a trimmer joist.

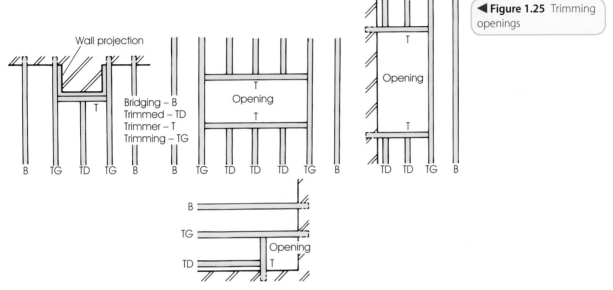

▶ **Figure 1.25** Trimming openings

As both the trimmer and the trimming joists take a greater load, they are usually 25 mm thicker in breadth than the bridging joists. For example, use 75 mm × 200 mm trimmer and trimming joists with 50 mm × 200 mm bridging joists.

Trimming joints. Traditionally, tusk mortise and tenon joints were used between the trimmer and trimming joists (see Figure 1.26), while housing joints were used between the trimmed joists and the trimmer. Once wedged, the tusk mortise and tenon joint requires no further fixing. The housing joints will require securing with 100 mm wire nails. The proportions of these joints must be followed. They are based on the fact that there are neutral stress areas in joists (see Figure 1.27). If any cutting is restricted mainly to this area then the reduction in the joist strength will be kept to a minimum.

Joist hangers (see Figure 1.28), which are a quicker, modern alternative to the traditional joints, are now used almost exclusively. As these hangers are made from thin galvanised steel, they do not require recessing in as do the thicker wall hangers, but they must be securely nailed in each hole provided with 32 mm galvanised clout nails.

To provide sufficient headroom for quarter-turn stairs, an L-shaped stairwell opening is often required. Figure 1.29 shows how the landing is formed using cantilever joists, which must be carried through the wall and across the adjacent room. Solid strutting should be fixed diagonally across the landing for strength. Where the joists meet at the external angle a dovetail joint can be used.

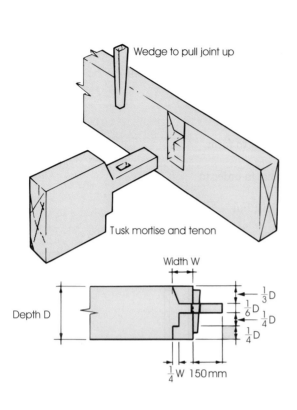

Wedge to pull joint up

Tusk mortise and tenon

Width W

Depth D

$\frac{1}{3}$D
$\frac{1}{6}$D
$\frac{1}{4}$D
$\frac{1}{4}$D

$\frac{1}{4}$W   150 mm

Bevelled housing

$\frac{1}{2}$D

$\frac{1}{3}$W

Square housing

$\frac{1}{2}$D

$\frac{1}{4}$W

▲ **Figure 1.26** Traditional trimming joints

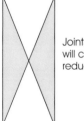

Joints cut in shaded areas will cause the minimum reduction in strength

▲ **Figure 1.27** Neutral stress areas

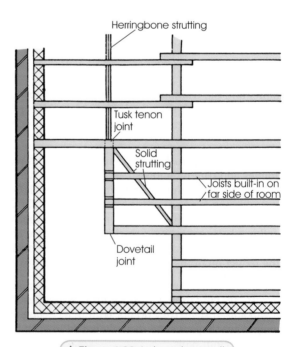

Herringbone strutting

Tusk tenon joint

Solid strutting

Joists built-in on far side of room

Dovetail joint

▲ **Figure 1.29** L-shaped stairwell

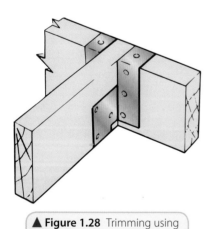

▲ **Figure 1.28** Trimming using a joist hanger

Positioning joists

Vertical position of joists. This will normally have been established by the bricklayers. They will have built up to the required brickwork course for the underside of built-in floor joists or bedded a wall plate on at this position. When joist hangers are being used the brickwork is built up to the required course to take the lug of the hanger. This will be near the top edge of the joist. See Figure 1.30. Alternatively, where straight lugged hangers rather than the turned down restraint hangers are used, the wall may be built beyond the floor level and the bed joint is raked out to receive the hanger.

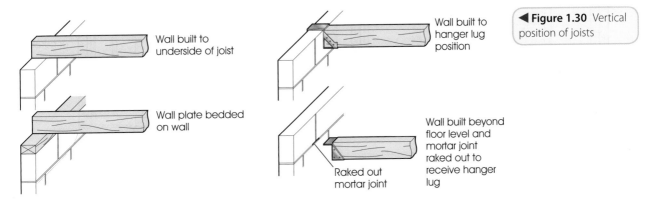

Wall built to underside of joist

Wall plate bedded on wall

Wall built to hanger lug position

Raked out mortar joint

Wall built beyond floor level and mortar joint raked out to receive hanger lug

◀ **Figure 1.30** Vertical position of joists

Laying out joists. Joists are first cut to length, see Figure 1.31. Ground floor joists should be cut between the walls to give a gap of about 20 mm at both ends. Upper floor joists that are built in should be cut with a splayed end to avoid any possibility of the end protruding into the cavity. Joists in hangers must be cut to the exact length between hangers to keep the hangers tight against the wall. Take care not to cut them too long, as forcing them in can easily push the freshly laid walls out of position.

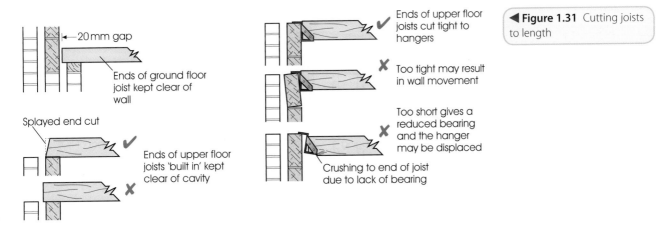

20 mm gap

Ends of ground floor joist kept clear of wall

Splayed end cut

Ends of upper floor joists 'built in' kept clear of cavity

Ends of upper floor joists cut tight to hangers

Too tight may result in wall movement

Too short gives a reduced bearing and the hanger may be displaced

Crushing to end of joist due to lack of bearing

◀ **Figure 1.31** Cutting joists to length

The outside joists are placed in position first, leaving a 50 mm gap between them and the wall. The other joists are then spaced out to the required centres in the remaining area.

In situations where openings are required in a joisted area for stairwells, flues, fireplaces or so on, these govern the layout of the joists. The outside bridging, trimmings and trimmer joists are the first to be positioned. Followed by the remaining bridging joists and trimmed joists, which are spaced out at the required centres in the respective areas, see Figure 1.32.

Change of joist direction. In order to span the shortest direction it is sometimes necessary to change the direction of joists in a particular joisted area. This is done over a load-bearing wall where the bridging joists from one area are allowed to overhang by 50 mm; the end-bridging joist in the other area is simply nailed to the overhanging ends, see Figure 1.33.

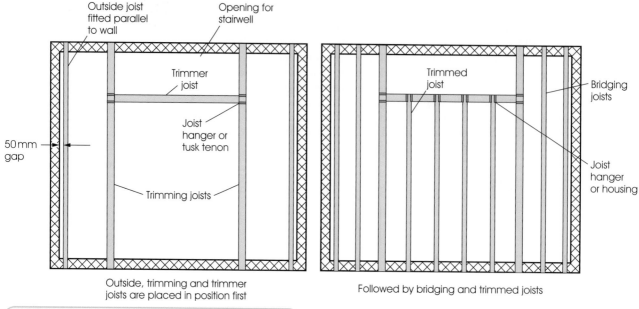

▲ **Figure 1.32** Layout of joists around opening for stairwell

▶ **Figure 1.33** Typical joist layout showing change of direction

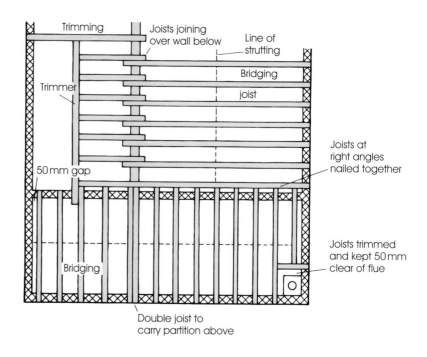

Figure 1.34 Double joists required to support blockwork above

Double joists are required where block work partition walls are to be built on a joisted area. These double joists are either nailed or bolted together (depending on specification) and positioned under the intended wall to take the additional load, see Figure 1.34. When laying out a joisted area, double joists are positioned along with the trimming and end bridging joists, before the other joists are spaced.

Timber near sources of heat

The Building Regulations restrict the positioning and trimming of timber near sources of heat. They state that the construction around heat-producing appliances should minimise the risk of the building catching fire. In order to comply with this requirement the use of combustible material in the vicinity of a fireplace, chimney or flue is restricted, see Figure 1.35.

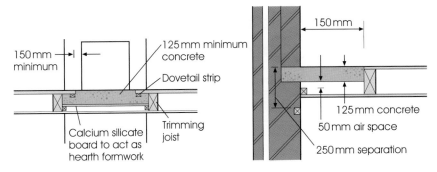

Flooring, skirting and other trim do not require 40 mm separation

Metal fixings at least 50 mm

Joist

Minimum of 40 mm

200 mm for structural timber

Flue wall

Flue lining

150 mm minimum

125 mm minimum concrete

Dovetail strip

Calcium silicate board to act as hearth formwork

Trimming joist

150 mm

125 mm concrete

50 mm air space

250 mm separation

- No timber is to be built into the flue, or be within 200 mm of the flue lining, or be nearer than 40 mm to the outer surface of a chimney or fireplace recess. (A 50 mm air space between any joist and wall is standard practice.) The exceptions to this requirement are: flooring; skirting; architrave; mantelshelf and other trim. However, any metal fixings associated with these must be at least 50 mm from the flue.

- There must be a 125 mm minimum thickness solid non-combustible hearth, which extends at least 500 mm in front of the fireplace recess and 150 mm on both sides.

- Combustible material used underneath the hearth is to be separated from the hearth by air space of at least 50 mm or be at least 250 mm from the hearth's top surface. Combustible material supporting the edge of the hearth is permitted.

Ground floors. The layout of joists around a fireplace is shown in Figure 1.36. The ends of the joists in front of the fireplace are supported by a wall plate on a fender wall. The fire hearth is framed out with lengths of joist that sit on the fender wall. This framing performs two functions:

- It acts as formwork when the fire hearth is concreted.
- It supports the ends of the floorboards around the hearth.

Upper floors. A typical method of trimming joists around a projecting flue, either from a ground floor fireplace or other heat-producing appliance situated on the lower floor is shown in Figure 1.37. Fireplace openings in upper floors are virtually obsolete in new construction, however, they are covered in this chapter as they may be encountered in maintenance or rehabilitation work. The layout of joists for a single upper floor with a fireplace opening is shown in Figure 1.38. The dovetail strips are bedded into the top of the concrete hearth to provide a fixing for the ends of the floorboards.

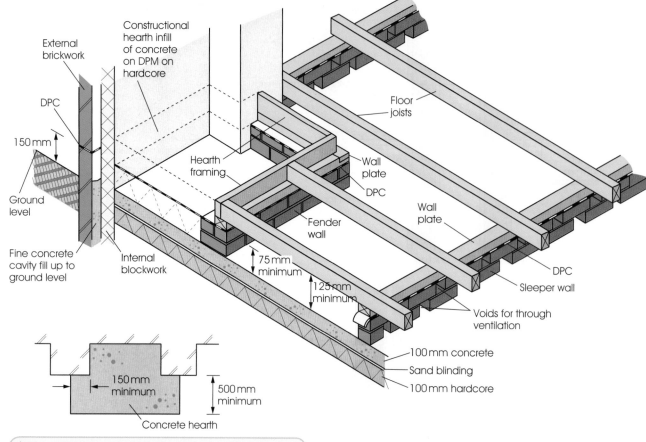

External brickwork

Constructional hearth infill of concrete on DPM on hardcore

DPC

150 mm

Ground level

Fine concrete cavity fill up to ground level

Internal blockwork

Hearth framing

Wall plate

DPC

Fender wall

75 mm minimum

125 mm minimum

Floor joists

Wall plate

DPC

Sleeper wall

Voids for through ventilation

100 mm concrete

Sand blinding

100 mm hardcore

150 mm minimum

500 mm minimum

Concrete hearth

▲ **Figure 1.36** Layout of joists around ground floor fireplace

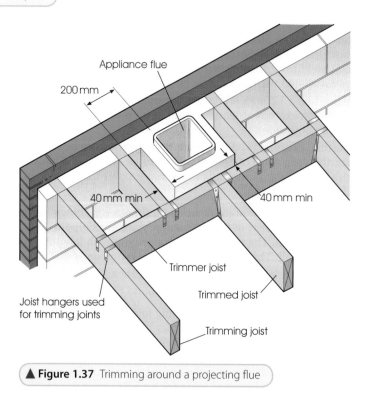

Appliance flue

200 mm

40 mm min

40 mm min

Trimmer joist

Trimmed joist

Joist hangers used for trimming joints

Trimming joist

▲ **Figure 1.37** Trimming around a projecting flue

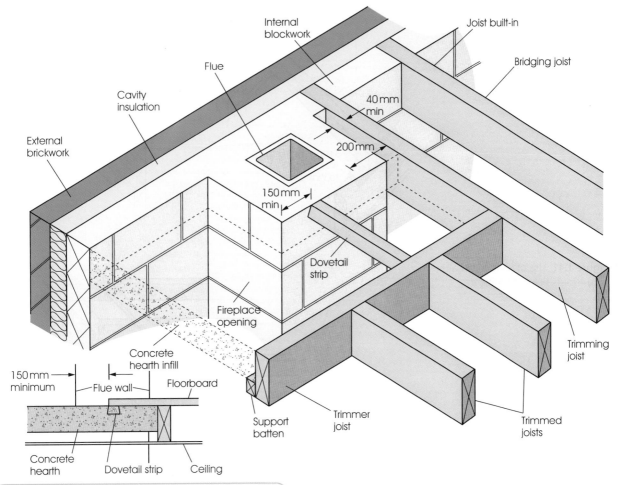

150 mm
minimum

Concrete
hearth

Flue wall

Dovetail strip

Floorboard

Concrete
hearth infill

Ceiling

▲ Figure 1.38 Layout of joists around upper floor fireplace

Levelling joists

After positioning, the joists should be checked for line and level. End joists are set using a spirit level; intermediate joists are lined through with a straightedge and spirit level (see Figure 1.39). Where the brickwork or blockwork course has been finished to a level line by the bricklayer or where wall plates are used and have been accurately bedded level, and where regularised joists have been used, the bearings of the joist should not require any adjustment to bring them into line and level.

However, minor adjustments may be required, see Figure 1.40. Joists may be housed into or packed off wall plates. Where packings are required for built-in joists, these should be slate or another durable material. Do not use timber packings as these may shrink and work loose and in any case are susceptible to rot. Joists may be recessed, to lower their bearing, providing the reduced joist depth is still sufficient for its span.

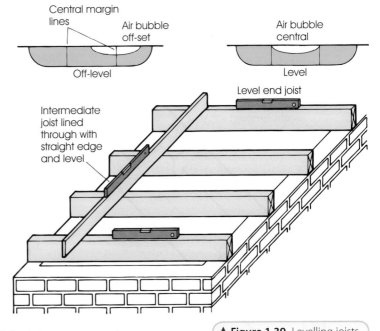

▲ Figure 1.39 Levelling joists

Temporary battens can be nailed across the top of the joists to ensure that their spacing remains constant before and during their building in, see Figure 1.41. Joists fixed to wall plates can be skew nailed to them using 75 mm or 100 mm wire nails.

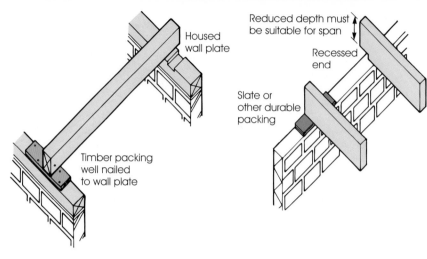

▶ **Figure 1.40** Packing, housing or recessing joists to maintain level

Housed wall plate

Timber packing well nailed to wall plate

Reduced depth must be suitable for span

Recessed end

Slate or other durable packing

▶ **Figure 1.41** Use of temporary batten to secure joists before building in

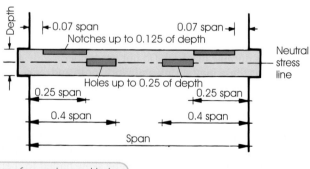

Temporary batten secures joists before building in

Notching and drilling joists for services

The position of notches for pipes and holes for cables in joists should have been determined at the building design stage and indicated on the drawings. Both reduce the joist strength. Notches and holes in joists should, therefore, be kept to a minimum and conform to the following (see Figure 1.42):

■ Notches on the joist's top edge of up to one eighth (0.125) of the joist's depth located between 7 per cent and 25 per cent (0.07 and 0.25) of the span from either support are permissible.

■ Holes of up to one quarter (0.25) of the joist's depth drilled on the neutral stress line (centre line) and located between 25 per cent and 40 per cent (0.25 and 0.4) of the span from either support are permissible. Adjacent holes should be separated by at least three times their diameter measured centre to centre.

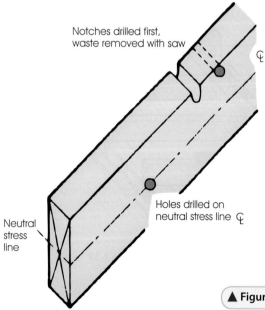

Notches drilled first, waste removed with saw

Holes drilled on neutral stress line ℄

Neutral stress line

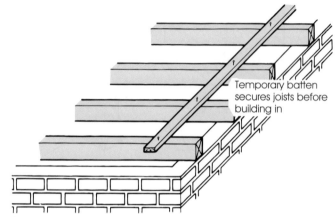

Depth

0.07 span

Notches up to 0.125 of depth

0.07 span

Neutral stress line

Holes up to 0.25 of depth

0.25 span

0.25 span

0.4 span

0.4 span

Span

▲ **Figure 1.42** Positions for notches and holes

Example

The position and sizes for notches and holes in a 200 mm depth joist spanning 4000 mm are:

- *Notches* between 280 mm and 1000 mm in from each end of the joist and up to 25 mm deep.
- *Holes* between 1000 mm and 1600 mm in from each end of the joist and up to 50 mm diameter.

Excessive notching and drilling of holes even within the permissible limits will seriously weaken the joist and may lead to structural failure.

Preservative treatment of joists

It is recommended that all timber used for structural purposes is preservative treated before use. Any preservative treated timber cut to size on site will require retreatment on the freshly cut edges and ends. Applying two brush flood coats of preservative will achieve this. Timber preservatives prevent rot by poisoning the food supply on which fungi feed and grow.

Floor covering and finishing

Floor decking is also variously termed as flooring, floorboarding or joist covering. The main materials in common use for this are shown in Figure 1.43:

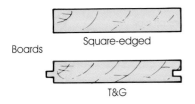

Boards
Square-edged

T&G

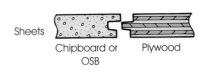

Sheets
Chipboard or OSB
Plywood

◀ **Figure 1.43** Joist coverings

- Timber floorboards, mainly PTG (planed, tongued and grooved), but square-edged boarding may be found in older buildings.
- Flooring grade particleboard, mainly T&G (tongued and grooved) or square-edged chipboard, but OSB (orientated strand board) is also used.
- Flooring grade plywood either T&G or square-edged.

The laying of floor decking is normally carried out after plumbing and electrical under-floor carcassing work and preferably also after the window glazing and roof tiling is complete so that it is not exposed to the weather. Where sheet material is used in kitchens and bathrooms, a moisture resisting grade (green tinted) should be used.

Timber flooring boards

Hardwood strip flooring. This is used where the floor is to be a decorative feature. It is made from narrow strips of T&G hardwood. These are usually random lengths up to 1.8 m long and 50 mm to 100 mm in width. The heading joints are also usually T&G. Figure 1.44 shows a section through a piece of hardwood strip floorboard. It has a splayed and stepped tongue to ease secret nailing and also to reduce any likelihood of the tongue splitting. The purpose of the recessed back is to ensure positive contact between the board and joist.

It is essential that the newly laid hardwood strip flooring is protected from possible damage by completely covering it with building paper or a polythene sheet, which should remain in position until the building is ready for occupation.

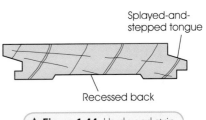

Splayed-and-stepped tongue

Recessed back

▲ **Figure 1.44** Hardwood strip floorboards

► **Figure 1.45** Fixing softwood flooring

Floor brad nailed though face and punched in

Lost heads used to secret nail through tongue

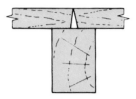

Square heading

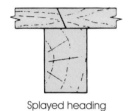

Splayed heading

▲ **Figure 1.46** Heading joints for softwood flooring

Softwood flooring. This usually consists of ex 25 mm × 150 mm T&G boarding. A standard floorboard section has the tongue and groove offset away from the board's face. This identifies the upper face and also provides an increased wearing surface before exposing the tongue and groove. Boards can be fixed either by floor brads nailed through the surface of the boards and punched in, or by lost-head nails secret-fixed through the tongue, see Figure 1.45. Nails should be approximately 2½ times the thickness of the floorboard in length.

Heading joints. Square-butted or splayed heading joints are introduced as required to utilise off-cuts of board and avoid wastage, see Figure 1.46. Splayed heading joints are preferred, as there is less risk of the board end splitting and they don't show a wide open gap if shrinkage takes place. Heading joints should be staggered evenly throughout the floor for strength; these should never be placed next to each other, as the joists and covering would not be tied together properly. See Figure 1.47.

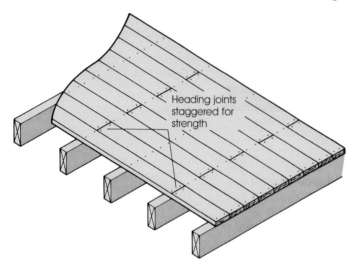

Heading joints staggered for strength

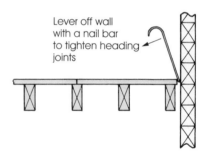

Lever off wall with a nail bar to tighten heading joints

▲ **Figure 1.47** Positioning of heading joints

Laying timber flooring boards

Surface fixing. Boards are laid at right angles to the joists. The first board should be fixed at least 10 mm away from the outside wall. This gap, which is later covered by the skirting, helps to prevent dampness being absorbed through direct wall contact. In addition, the gap also allows the covering material to expand without either causing pressure on the wall or a bulging of the floor surface. The remainder of the boards are laid four to six at a time, cramped up with floorboard cramps (see Figure 1.48) and surface nailed to the joists. The final nailing of a floor is often termed 'bumping', which should be followed by punching the nail head just below the surface. Boards up to 100 mm in width require one nail to each joist while boards over this require two nails.

Figure 1.49 shows two alternatives to the use of floorboard cramps – folding or wedging – although neither of these is as quick or efficient. Folding a floor entails fixing two boards spaced apart 10 mm less than the width of five boards. The five boards can then be placed with their tongues and grooves engaged. A short board is laid across the centre and 'jumped on' to press the boards in position. This process is then repeated across the rest of the floor. Alternatively, the boards may be cramped, four to six at a time using dogs

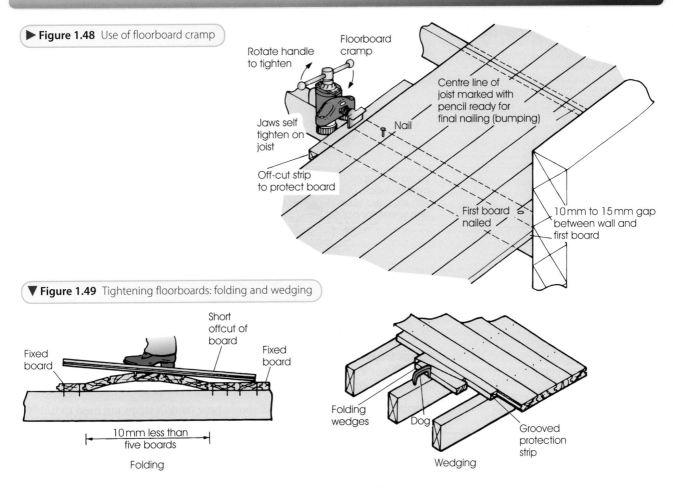

► **Figure 1.48** Use of floorboard cramp

Rotate handle to tighten

Floorboard cramp

Jaws self tighten on joist

Nail

Centre line of joist marked with pencil ready for final nailing (bumping)

Off-cut strip to protect board

First board nailed

10mm to 15mm gap between wall and first board

▼ **Figure 1.49** Tightening floorboards: folding and wedging

Fixed board

Short offcut of board

Fixed board

10mm less than five boards

Folding

Folding wedges

Dog

Grooved protection strip

Wedging

and wedges. Whichever method is used the final board will invariably require cutting to width; rip it about 10 mm narrower than the gap and lever off the wall to give a tight joint before fixing.

Secret fixing. Secret-fixed boards must be laid and tightened individually and so cramping is not practical. Figure 1.50 shows how they may be tightened by levering them forward with a firmer chisel driven into the top of the joist, or with the aid of a floorboard nailer. This tightens the boards and drives the nail when the plunger is struck with a hard mallet. Secret fixing is normally only used on high-class work or hardwood flooring, as the increased laying time makes it considerably more expensive.

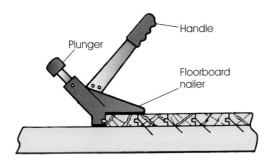

Handle

Plunger

Floorboard nailer

Firmer chisel

Short grooved protection strip

▲ **Figure 1.50** Tightening secret-fixed boards

Services

Where services such as water and gas pipes or electric cables are run within the floor, there is a danger of driving nails into them. They should be marked on, saying in chalk or pencil 'PIPES NO FIXING', so that the danger area is kept clear when nailing. The floorboards over services can be fixed with recessed cups and screws to permit easy removal for subsequent access, and also to provide easy recognition of location, see Figure 1.51.

► **Figure 1.51** Marking position of services

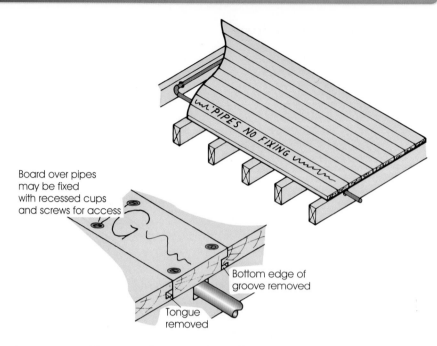

Board over pipes may be fixed with recessed cups and screws for access

Bottom edge of groove removed

Tongue removed

Access traps. These may be required in a floor over areas where water stopcocks or electrical junction boxes are located, see Figure 1.52. Again these can be fixed with recessed cups and screws to permit easy removal.

Fireplace openings. Figure 1.53 shows how margin strips are used to finish the ends of the floor boarding around a fire hearth. The actual arrangement will vary depending on whether the boards run into or across the hearth. In either case the layout of boarding should be planned before any fixing takes place, to ensure an even margin around the hearth.

► **Figure 1.52** Provision of access traps for services

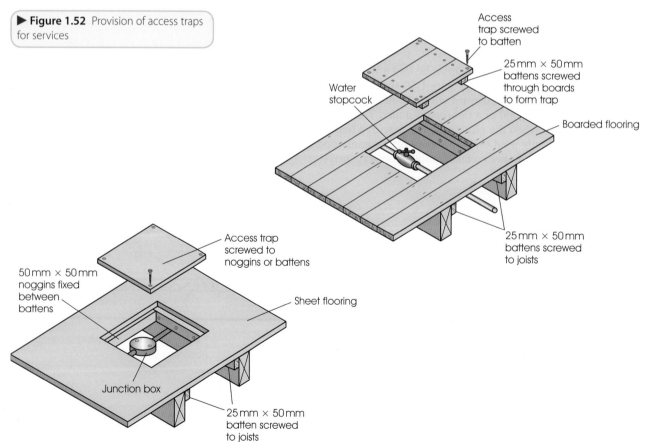

Access trap screwed to batten

25 mm × 50 mm battens screwed through boards to form trap

Water stopcock

Boarded flooring

25 mm × 50 mm battens screwed to joists

Access trap screwed to noggins or battens

50 mm × 50 mm noggins fixed between battens

Sheet flooring

Junction box

25 mm × 50 mm batten screwed to joists

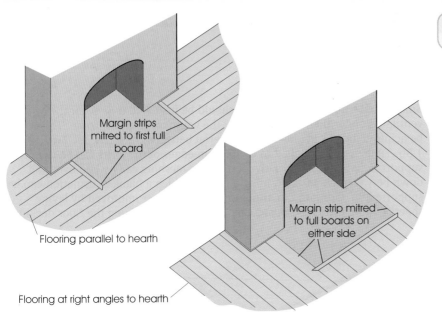

Margin strips mitred to first full board

Flooring parallel to hearth

Margin strip mitred to full boards on either side

Flooring at right angles to hearth

Sheet floor decking

This is now being increasingly used for domestic flooring. Flooring grade chipboard is available with square edges in 1220 mm × 2440 mm sheets and with T&G edges in 600 mm × 2440 mm sheets. Square-edged sheets are normally laid with their long edges over a joist. Noggins must be fixed between the joists to support the short ends. T&G sheets are usually laid with their long edges at right angles to the joists and their short edges joining over the joist. Both types require noggins between the joists where the sheet abuts a wall. See Figure 1.54. Joists should be spaced to accommodate the dimensions of the sheet flooring.

▼ **Figure 1.54** Layout of chipboard floors

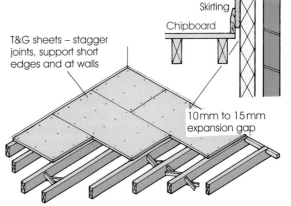

Plaster

Skirting

Chipboard

Square-edged sheets – stagger joints, support all edges

10 mm to 15 mm expansion gap around each wall

Support noggins

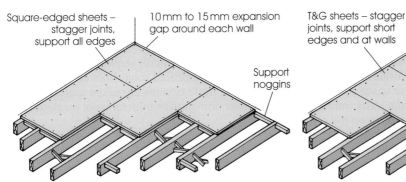

T&G sheets – stagger joints, support short edges and at walls

10 mm to 15 mm expansion gap

Fixing sheet flooring

Sheets are laid staggered and fixed at 200 mm to 300 mm centres with 50 mm lost-head nails or, for additional strength, annular ring shanked or serrated nails. A gap of 10 mm must be left along each wall to allow for expansion and prevent absorption of dampness from the wall. Manufacturers of chipboard often recommend the gluing of the tongues and grooves with a PVA adhesive to prevent joint movement and to stiffen the floor. See Figure 1.55.

For protection, it is recommended that the floor be covered with building paper after laying and that this is left in position until the building is occupied.

Plywood and OSB sheet floor decking are laid using the same procedures as chipboard flooring.

Apply PVA adhesive to the groove

▲ **Figure 1.55** Gluing joints in chipboard flooring

Insulation and finishing

Insulation of floors

Thermal insulation

Suspended timber ground floors should be insulated to prevent the cool ventilated under floor space drawing the warm air from the room. This can be achieved by placing insulating material between the joists, see Figure 1.56. Timber bearers nailed to the bottom edges of the joists can be used to support rigid panels. Alternatively, a plastic meshing can by laid over and down between the joists, forming a tray on which a blanket insulation can be supported. In addition, the insulation can be further improved by laying sheets of fibre insulation board over the joists before fixing the floor decking.

▶ **Figure 1.56** Thermal insulation to suspended timber ground floors

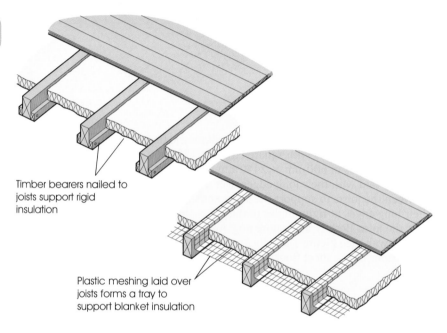

Timber bearers nailed to joists support rigid insulation

Plastic meshing laid over joists forms a tray to support blanket insulation

Suspended timber upper floors do not normally require added thermal insulation except where they separate a heated room from an unheated room or space below, see Figure 1.57. In these circumstances the space between the joists can be filled with either fibreglass or mineral wool insulation resting on the ceiling.

▶ **Figure 1.57** Thermal insulation to suspended timber upper floors

Related information

Also see Book 1, Chapter 11, pages 471–2, which give details of thermal insulation materials.

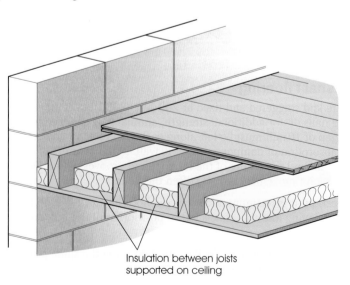

Insulation between joists supported on ceiling

Sound insulation

The need for sound insulation in domestic floors is fairly limited. Where required in upper floors to separate say a children's noisy bedroom from the lounge below, the details shown in Figure 1.58 can be used:

■ Filling the space between the joists with a sound insulating slag wool pugging or blanket insulation will reduce the passage of airborne sound.

■ A floating floor will insulate against impact or structural-borne sound. This uses a resilient quilt to separate the floor decking from the floor structure. To be effective the battens that support the flooring should be loose laid over the joists and not fixed to them.

■ A combination of the previous two methods can be used to reduce both airborne and structural-borne sound.

The passage of sound will find the point of least resistance; therefore do not forget to insulate the 50 mm gap between the outside joists and the wall. In addition a resilient seal should be used at the skirting to floor intersection.

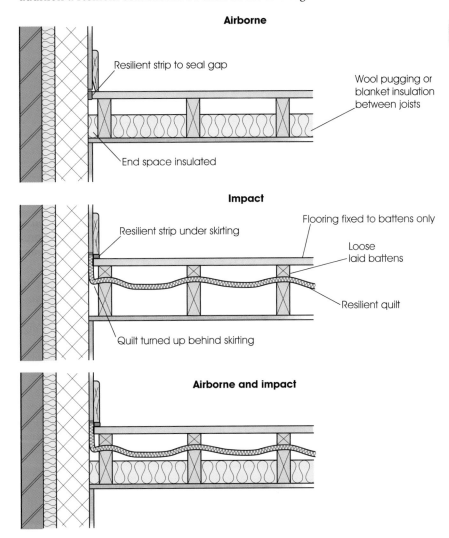

Figure 1.58 Sound insulation to suspended timber upper floors

Timber finishes to solid floors

There are three main methods that can be used to provide a timber finish to solid floors:

■ Timber floor boarding or sheets fixed to timber battens.

■ Sheet floor panels laid on rigid insulation.

■ Wood blocks bonded directly to the concrete.

Timber floor boarding or sheets fixed to timber battens

The timber bearers or battens may be bedded in, clipped to, fixed directly to, or left floating on the solid floor, see Figure 1.59. These bearers, which are spaced at about 400 mm or 600 mm centres, act as floor joists and so provide a means of fixing the floorboards.

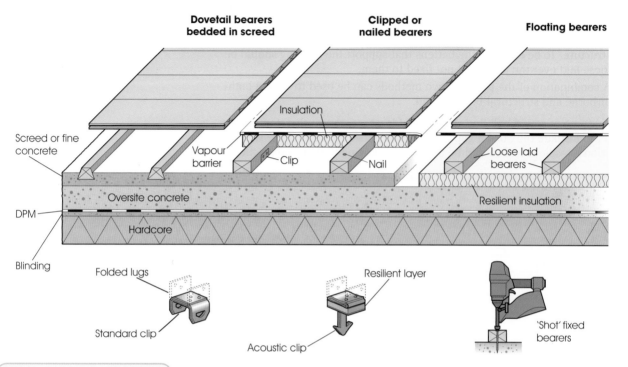

▲ **Figure 1.59** Timber floor finish

■ *Dovetail-shaped bearers* are used for bedding in. Small mounds of screed or concrete are deposited along the bearer line into which the bearers are tapped down to a level position. When set, further screed or concrete can be laid using the bearers as a guide for levelling off.

■ *Floor clips* are made from folded galvanised mild steel and are available in either a standard pattern or with a separating resilient pad, known as an acoustic pattern (used to increase the impact sound insulation of solid upper floors). These are pushed into freshly laid screed or concrete, along the bearer lines at about 600 mm centres. When set, the lugs of the clips are opened up to form a cradle in which the bearers are located. Special improved nails are used to secure the bearers to the clips.

■ *Directly fixed bearers* are normally shot fixed to the concrete at about 600 mm centres using a cartridge-operated fixing tool.

■ *Floating bearers* are laid unfixed on a layer of resilient insulating material. When the floor boarding is fixed to the bearers the whole floor floats on the insulation and is held down by its own weight and the skirting around the perimeter of the room.

Insulation. In order to provide thermal insulation the space between the bearers can be filled with fibreglass or mineral wool. Alternatively, a layer of insulating fibreboard can be used between the bearers and the floor boarding.

Vapour barrier. A polythene vapour barrier should be laid over the bearers or fibreboard before fixing the floor boarding in order to prevent condensation forming in the floor space. The vapour barrier should be turned up the walls at its edges and any joints overlapped and sealed with a waterproof tape.

Sheet floor panels laid on rigid insulation

T&G plywood or chipboard flooring panels are laid directly and floating on a rigid layer of insulation, normally polystyrene. See Figure 1.60. The joints of the sheeting should be glued as they are laid, to provide an insulating floor finish that is held down by its own weight and the skirting around the perimeter of the room. Composite flooring panels are also available for this purpose, with the insulation being pre-bonded to the back.

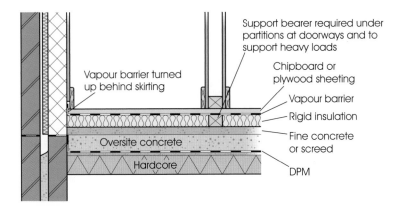

◀ **Figure 1.60** Sheet flooring on rigid insulation

Labels in figure:
- Support bearer required under partitions at doorways and to support heavy loads
- Vapour barrier turned up behind skirting
- Chipboard or plywood sheeting
- Vapour barrier
- Rigid insulation
- Oversite concrete
- Fine concrete or screed
- Hardcore
- DPM

Wood blocks bonded directly to the concrete

This method is mostly used for laying wood block or parquet floor finishes using a bituminous mastic compound as the adhesive. See Figure 1.61. An expansion gap around the perimeter of the floor is required; this should be filled with either polystyrene or cork so that the blocks are allowed an amount of moisture movement but are prevented from 'creeping' and closing the gap.

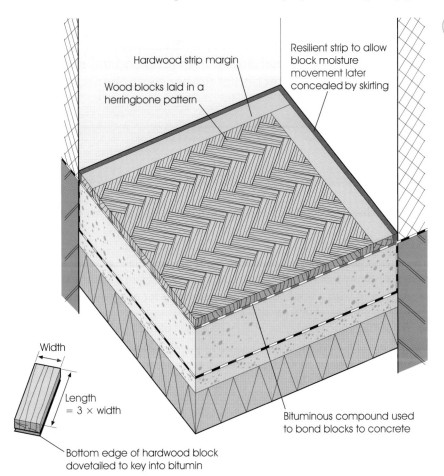

◀ **Figure 1.61** Wood block floor

Labels in figure:
- Hardwood strip margin
- Wood blocks laid in a herringbone pattern
- Resilient strip to allow block moisture movement later concealed by skirting
- Width
- Length = 3 × width
- Bituminous compound used to bond blocks to concrete
- Bottom edge of hardwood block dovetailed to key into bitumin

2
Chapter Two

Flat and pitched roof construction

This chapter covers the work of the site carpenter. It is concerned with principles of flat and pitched roof construction and includes the following:

- An introduction to roof types.
- Timber flat roof construction.
- Traditional pitched roof construction.
- Pitched roof construction using trussed rafters.
- Verges and eaves.
- Dormer windows and ornamental roofs.
- Roof erection.
- Prefabricated rafters.

Roofs are part of the external envelope that spans the building at high level and has structural, weathering and insulation functions. The cutting and erection of timber components, which form the structure of a roof, is termed carcassing.

Roof types

Roofs are classified according to their pitch, shape and also type of structure.

Roof pitch

The pitch or slope of a roof surface to the horizontal may be expressed either in degrees or as a ratio of the rise to the span, see Figure 2.1. For example:

- A flat roof with a slope of 1 : 40 will rise 1 unit for every 40 units run, that is a rise of 25 mm for every metre run.
- A one third pitch roof ($\frac{1}{3}$ or 1 : 3), with a span of 3 m will have a rise of 1 m.

Pitch is dependent on the type and size of the materials used for the weatherproof covering and the building's exposure to windy conditions. They may be classified as either:

- flat roofs, where the pitch of the roof surface does not exceed 10°
- pitched roofs, where the pitch of the roof surface exceeds 10°.

Flat roofs are normally weathered using either rolls of bituminous felt built up in layers and bonded to the decking with cold or hot applied bitumen; or mastic asphalt applied hot in layers on a sheathing felt that covers the

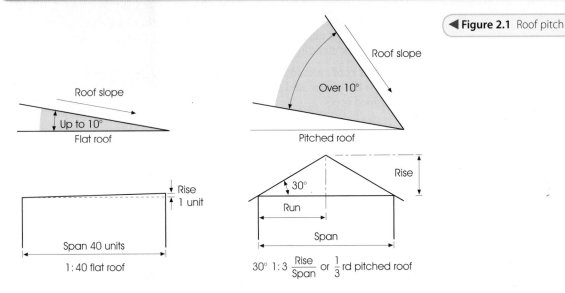

◀ **Figure 2.1** Roof pitch

decking. In order to reflect some of the sun's heat, both types of weathering may be protected with a heat reflecting paint or covered with a layer of pale coloured stone chippings.

Pitched roofs can be covered using clay or concrete tiles; quarried or man-made slates; wood shingles (normally cedar); thatch (reeds); and various profiled sheet materials manufactured from protected metals, plastics and cement/mineral fibre mix. In general, as the pitch of the roof lowers, the unit size of the covering material must increase.

Roof shape

Roofs may be further divided into a number of basic forms by their shape, the most common of which are shown in Figure 2.2.

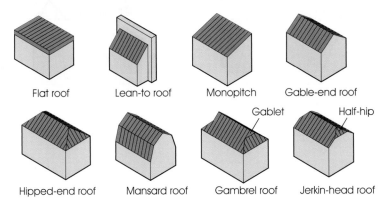

◀ **Figure 2.2** Roof shapes

Lean-to and mono-pitch roofs. Both have a single sloping surface. The lean-to roof abuts a higher wall or building and is often used for extensions, whereas mono-pitched roofs are normally free-standing.

Gable-end roofs have two sloping surfaces (double pitched) terminating at one or both ends with a triangular section of brickwork.

Hipped-end roofs are again double-pitched, but in these cases the roof slope is returned around one or both the shorter sides of the building to form a sloping triangular end.

Mansard roofs are double-pitched roofs, where each slope of the roof has two pitches. The lower part has a steep pitch (to act as walls) and often incorporates dormer windows for room in the roof applications, while the upper part (acting as the main weathering) rarely exceeds 30°. The shorter sides of the building may be finished with either gable or hipped ends.

Gambrel roofs are double-pitched roofs incorporating a small gable or gablet at the ridge and a half-hip below. The gablet may be finished with tile hanging, timber cladding or a louvred ventilator.

Jerkin-head roofs are double-pitched roofs, which are hipped from the ridge part way to the eaves, and the remainder gabled. These are also termed as Dutch hipped roofs.

Roof elements

See Figure 2.3.

- *Abutment.* The intersection where a lean-to or flat roof meets the main wall structure.
- *Eaves.* The lower edge of a roof, which overhangs the walls and where rainwater is discharged, normally into a gutter. May be finished with a fascia board and soffit.
- *Gable.* The triangular upper section of a wall that closes the end of a building with a pitched roof.
- *Hip.* The line between the ridge and eaves of a pitched roof, where the two sloping surfaces meet at an external angle.
- *Parapet.* The section of a wall that projects above and is terminated some way beyond the roof surface.
- *Ridge.* The horizontal line of intersection between two sloping roof surfaces at their highest point or apex.
- *Valley.* The line between the ridge and eaves of a pitched roof, where the two sloping surfaces meet at an internal angle.
- *Verge.* The termination or edge of a pitched roof at the gable or the sloping edge (non-drained) of a flat roof. May be finished with a bargeboard and soffit.

▶ **Figure 2.3** Roof elements

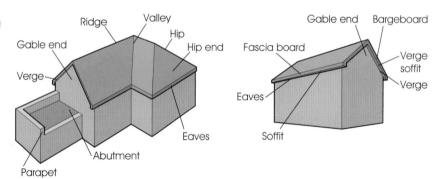

Flat roof construction

Timber flat roofs are similar in construction to that of upper floors. Their means of support; tying in and down; stiffening with the use of strutting and the decking all follow on similar lines.

Although the surface of the roof is flat, it is not horizontal. It should have a slope or fall on its top surface to ensure that rainwater will be quickly cleared and not accumulate on the roof. There are three ways of forming the slope as shown in Figure 2.4:

1. *Level joists:* The joists are laid level with the slope being formed by long tapering wedged shaped pieces of timber (firring pieces) being nailed to their upper edges. This is the most common method as it forms a level undersurface for fixing the ceiling plasterboard.

2. *Sloping joists:* The joists themselves are laid to the required slope and so no firrings are required. However, it has the disadvantage of forming a sloping ceiling soffit.

3. *Diminishing firrings:* The joists are laid level, at right angles to the slope, and reducing or diminishing sections of timber are nailed to their top edge to create the fall. This method is rarely seen, but it is a useful method when timber boards are used for the roof decking, as these should be fixed parallel to the roof slope.

Flat roof carcass

Joist size and spacing. The size of flat roof joists can be found in the same way as floor joists. For new work, the sectional size and spacing will always be specified. In common with floor joists, the Timber Research and Development Association (TRADA) and other bodies produce tables of suitable sectional sizes for use in different situations. Alternatively, they may be determined by calculation. Slightly longer spans are permitted for flat roof joists than floor joists of the same sectional size, as flat roofs are only normally walked upon for maintenance purposes. The actual spacing of the joists must be related to the type of material used for the decking. Table 2.1 can be used as a guide.

> #### Example
> To determine the sectional size of a flat roof joist having a clear span of 3.25 m using 18 mm roofing grade chipboard as the decking material.
> Using Table 2.1:
> - Roofing grade chipboard is suitable for spans of up to 500 mm; this will be the maximum joist spacing, the nearest below this spacing on the table is 450 mm.
> - Look down the 450 mm joist spacing column to get the nearest span above 3.25 m.
> - This is 3.34 m.
> - Look left to get the suitable sectional joist size, 50 × 147 mm in this case.

Layout of joists. The positioning and levelling of flat roof joists is similar to the positioning and levelling of floor joists, except that flat roofs only occasionally have to be trimmed around openings (e.g. roof lights, chimney stacks, etc.) Flat roofs, like floors, also require some form of strutting at their mid-span or at 1.8 m centres, see Figure 2.5.

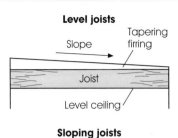

Level joists

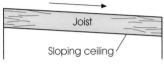

Sloping joists

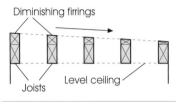

Diminishing firrings

▲ **Figure 2.4** Methods of forming flat roof slopes

◀ **Figure 2.5** Flat roof carcass

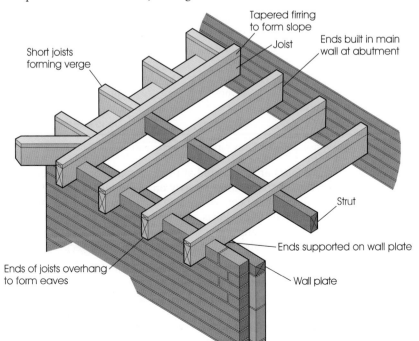

Table 2.1 Maximum clear span of flat roof joists and decking (m)

Based on the use of C16 grade softwood with dead load of up to 0.5 kN/m² with access only for maintenance and repair purposes.

Flat roof joist

Up to 10° pitch

Clear span

Size of joist (mm) B × D	Joist spacing (c/c)		
	400 mm	450 mm	600 mm
38 × 97	1.74	1.72	1.67
38 × 122	2.37	2.34	2.25
38 × 147	3.02	2.97	2.85
38 × 170	3.63	3.57	3.37
38 × 195	4.30	4.23	3.86
38 × 220	4.94	4.76	4.34
50 × 97	1.97	1.95	1.89
50 × 122	2.67	2.64	2.53
50 × 147	3.39	3.34	3.19
50 × 170	4.06	3.99	3.69
50 × 195	4.79	4.62	4.22
50 × 220	5.38	5.19	4.74
75 × 122	3.17	3.12	3.00
75 × 147	3.98	3.92	3.64
75 × 170	4.74	4.58	4.19
75 × 195	5.42	5.23	4.79
75 × 220	6.07	5.87	5.38

Decking material	Finished thickness	Span (joist spacing)
Softwood T&G	16 mm	up to 500 mm
Softwood T&G	19 mm	up to 600 mm
Roofing grade chipboard	18 mm	up to 500 mm
Roofing grade chipboard	22 mm	up to 600 mm
OSB	15 mm	up to 500 mm
OSB	18 mm	up to 600 mm
External plywood WBP (weather and boil proof)	16 mm	up to 500 mm
External plywood WBP (weather and boil proof)	19 mm	up to 600 mm
Wood wool slabs	50 mm	up to 600 mm

Note: Recommended span of decking may vary between manufacturers of panel products.

Joist supports. The joists are supported at the eaves and/or verge depending on the joist's direction of span, either directly on the inner leaf of the cavity wall or a timber wall that has been secured on top of the inner leaf, see Figure 2.6. Where the ends of the joists abut a wall, either at the main building or at a parapet, various methods can be used for support including building in, joist hangers, timber wall pieces and steel angle sections.

Anchoring joists. In order to prevent strong winds lifting the roof, the joists must be anchored to the walls at a maximum of 2 m centres. This anchoring must be carried out to the joists that run both parallel with or at right angles to the wall. Figure 2.7 shows how a flat roof joist can be anchored to the wall plate when used with a framing anchor or truss clip, nailed to both the joist and wall plate. The wall plate or actual joist must also be tied down using tie

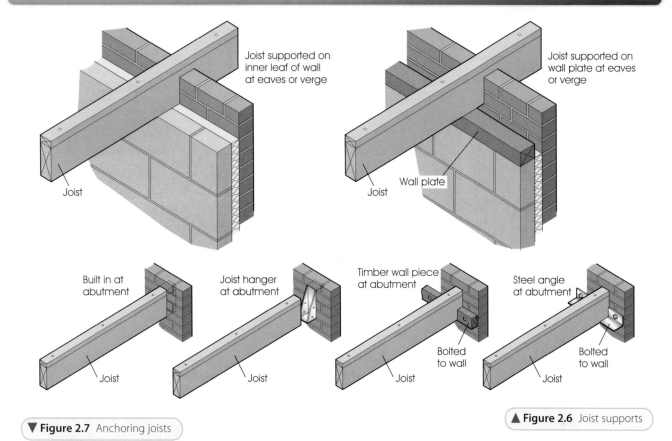

Joist supported on inner leaf of wall at eaves or verge

Joist

Joist supported on wall plate at eaves or verge

Wall plate

Joist

Built in at abutment

Joist

Joist hanger at abutment

Joist

Timber wall piece at abutment

Joist

Bolted to wall

Steel angle at abutment

Joist

Bolted to wall

▲ **Figure 2.6** Joist supports

▼ **Figure 2.7** Anchoring joists

Joist

Wall plate

Framing anchor or truss clip

Tie down strap over wall plate and screwed to wall

Joist

Twisted tie down strap screw-fixed to joist and wall

down straps. The strap is nailed to the wall plate or the side of the joist and fixed to the wall by either nailing or plugging and screwing.

Single or double flat roofs. Flat roofs, in common with floors, are termed as single where the joists span from wall to wall. Binders may be incorporated in large flat roofs, either to provide intermediate support to long joists or to enable the roof to slope or fall in both directions, see Figure 2.8. The use of a binder creates what is termed as a double flat roof. This reduces the effective span of the joists and enables the use of smaller sectional sizes. In addition, the required width of verge and eave finishes are reduced.

Flat roof decking. Boarding or decking to flat roofs should be carried out just before they are to be waterproofed. Where timber boards are used, they should be laid either with, or at a diagonal to, the fall of the roof's surface. This is because any cupping or distortion of the boards can lead to pools of water being trapped in hollows formed on the roof's surface if boards are laid at right angles to the fall. See Figure 2.9. Sheet materials should be of a moisture-resisting grade and sold as suitable for use as flat roof decking. Decking sheets may be either square-edged or T&G and may also be available pre-covered with a layer of felt to provide some measure of initial weather protection. The methods of laying, edge supporting and fixing of both board and sheet flat roof decking materials are the same as those used for floor decking.

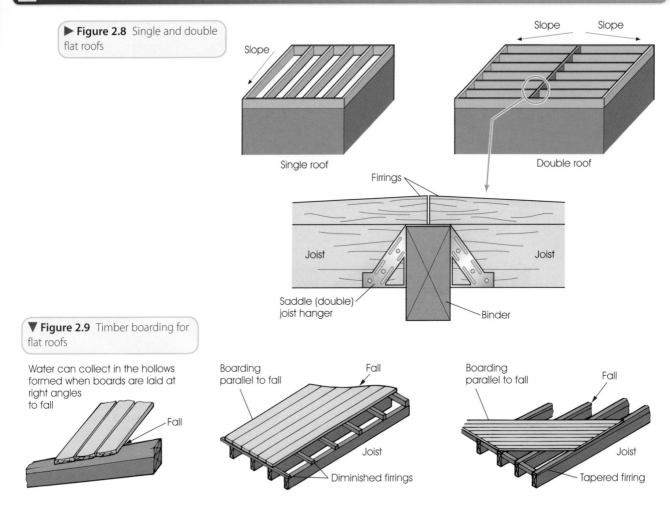

▶ **Figure 2.8** Single and double flat roofs

Single roof

Double roof

Slope

Slope Slope

Firrings

Joist

Joist

Saddle (double) joist hanger

Binder

▼ **Figure 2.9** Timber boarding for flat roofs

Water can collect in the hollows formed when boards are laid at right angles to fall

Fall

Boarding parallel to fall

Fall

Joist

Diminished firrings

Boarding parallel to fall

Fall

Joist

Tapered firring

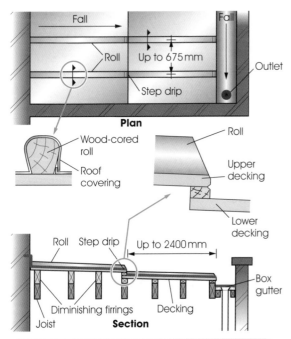

Fall

Fall

Roll Up to 675 mm

Step drip

Outlet

Plan

Wood-cored roll

Roof covering

Roll

Upper decking

Lower decking

Roll Step drip Up to 2400 mm

Box gutter

Diminishing firrings Decking

Joist **Section**

▲ **Figure 2.10** Section through metal-covered flat roof

Flat roof weatherproofing. Timber boarded/sheeted flat roofs are normally weathered using bitumen felt or mastic asphalt as explained previously. When rolled sheet metal such as lead, copper or zinc is used as a flat roof covering material, wood-cored rolls in the direction of the fall and stepped drips across the fall must be incorporated in the roof construction. This is to enable sheets to be joined and also limit the size of the sheets to combat the effects of expansion in warm weather. Figure 2.10 shows a part plan and section through a lead flat roof with rolls, drips and a box gutter.

Other weatherproofing materials include various profiled sheet materials manufactured in protected metals, translucent plastics and cement/mineral fibre and are supplied as a finished product along with a range of fixings and flashings. Profiled sheets are most often seen on steel framed structures and pitched timber roofs, but can also be used on timber joisted flat roofs. They do not require a decked surface as they can be fixed either directly to the joists or cross bearers in accordance with the manufacturer's instructions. See Figure 2.11.

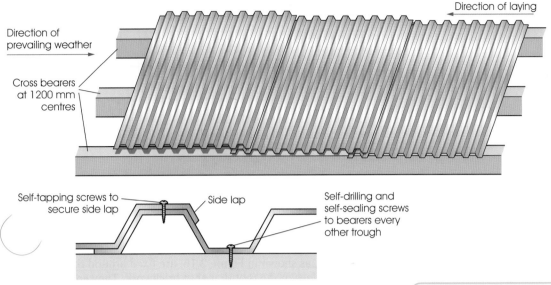

Direction of prevailing weather

Direction of laying

Cross bearers at 1200 mm centres

Self-tapping screws to secure side lap

Side lap

Self-drilling and self-sealing screws to bearers every other trough

▲ **Figure 2.11** Use of profiled sheets

Eaves, verge and abutment finishes

Both the eaves and verge of a flat roof may be finished as either flush or overhanging, see Figure 2.12.

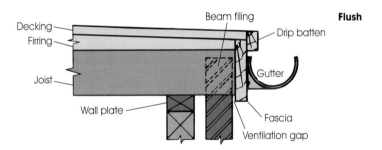

Flush

Decking
Firring
Joist
Wall plate

Beam filing
Drip batten
Gutter
Fascia
Ventilation gap

◀ **Figure 2.12** Flush and overhanging eaves

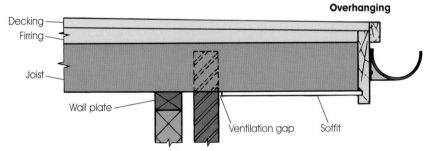

Overhanging

Decking
Firring
Joist
Wall plate
Ventilation gap
Soffit

A drip batten is fixed at the lower eaves to the top of the fascia to extend the roof edge into the gutter, see Figure 2.13. This extension enables the roofing felt to be turned around it. Rainwater flows off the drip batten into the centre of the gutter, so ensuring efficient discharge and reducing the risk of rot damage to timber. A drip batten is also fixed at higher eaves and verge edges, again to enable roofing felt to be turned around it. Any moisture is then allowed to drip clear of the fascia and not creep back under the felt by capillary action, with the subsequent likelihood of rot.

Due to the depth of joists, deep fascia boards are often required. These may be formed from solid timber, plywood (WBP) or matchboarding. An alternative is to reduce the depth of the joists at the ends by either a splay or square cut, see Figure 2.14.

▼ **Figure 2.13** Use of drip batten to eaves of flat roof

Rainwater runs over edge

Felt turned over drip batten

Rainwater drips off

► **Figure 2.14** Reducing the depth of the joist

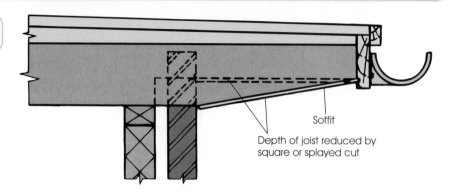

Soffit

Depth of joist reduced by square or splayed cut

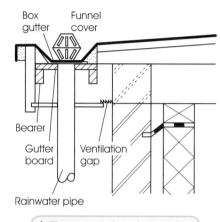

Box gutter

Funnel cover

Bearer

Gutter board

Ventilation gap

Rainwater pipe

▲ **Figure 2.15** Boxed gutter to a flat roof

Figure 2.15 shows how a boxed or internal gutter can be formed at the eaves as an alternative to external gutters. The gutter fall (1:60) can be achieved by progressively increasing the depth of cut-out in the joist ends towards the outlet.

Angle fillets, as shown in Figure 2.16, are fixed around the upper eaves, verge and edges of the roof that abut brickwork. They enable the roofing felt to be gently turned at the junction; sharp turns on felt lead to cracking, subsequent leaking and risk of rot. In addition, the use of angle fillets also prevents rainwater from dripping or being blown over the edges of the roof.

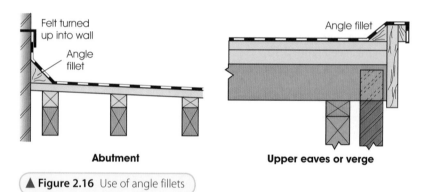

Felt turned up into wall

Angle fillet

Angle fillet

Abutment

Upper eaves or verge

▲ **Figure 2.16** Use of angle fillets

Overhanging verges, returned eaves and upper eaves that are positioned at right angles to the main joist run can be formed using either: short return joists fixed at right angles to the last main joist with hangers; or by forming a ladder frame from two joists and noggins, again fixed to the last main joist, see Figure 2.17.

Thermal insulation. There are two methods by which a flat roof can be insulated, see Figure 2.18.

■ *Insulation in the roof space* known as cold roof construction, this has thermal insulation and a vapour check at ceiling level. The roof space itself is cold and must be vented to the outside air to prevent interstitial condensation.

▼ **Figure 2.17** Arrangement of joists at verge and returned eaves

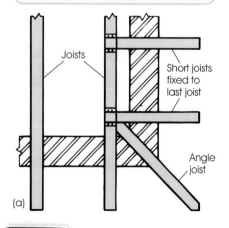

Joists

Short joists fixed to last joist

Angle joist

(a)

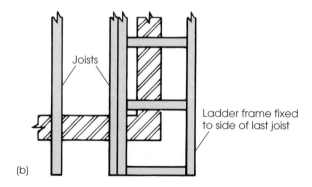

Joists

Ladder frame fixed to side of last joist

(b)

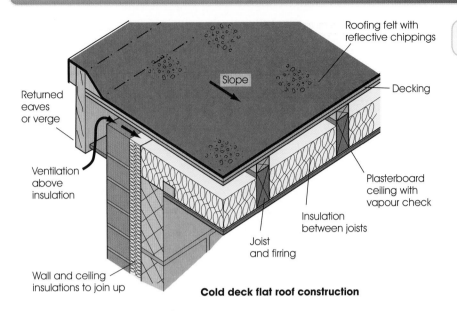

Roofing felt with reflective chippings

Slope

Decking

Returned eaves or verge

Ventilation above insulation

Plasterboard ceiling with vapour check

Insulation between joists

Joist and firring

Wall and ceiling insulations to join up

Cold deck flat roof construction

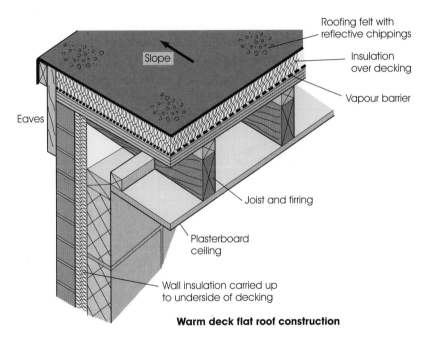

Roofing felt with reflective chippings

Slope

Insulation over decking

Vapour barrier

Eaves

Joist and firring

Plasterboard ceiling

Wall insulation carried up to underside of decking

Warm deck flat roof construction

- *Insulation above the roof space* known as warm roof construction, this has its thermal insulation and vapour barrier placed over the roof decking. The roof space is kept warmer than the outside air temperature and does not require ventilation.

The actual thickness of insulation used will be dependent on its type, positioning and the use of the building. The Building Regulations should be consulted for further information.

Ventilation. To reduce the likelihood of condensation within the roof space, ventilation is required by the Building Regulations. All cold flat roofs must be cross-ventilated at verge or eaves level by permanent vents. These must have an equivalent area equal to a continuous gap along both sides of the roof of 25 mm. The void in the roof between the insulation and the underside of the decking should be a minimum of 50 mm to allow free air movement. See Figure 2.19. Joists that run at right angles to the airflow should be counter-battened before decking to provide the free air space.

Related information

Also see Book 1, Chapter 10, Building control, which gives further details of Building Regulations. You may also view documents on the following website:

www.planningportal.gov.uk

▶ **Figure 2.19** Flat roof ventilation

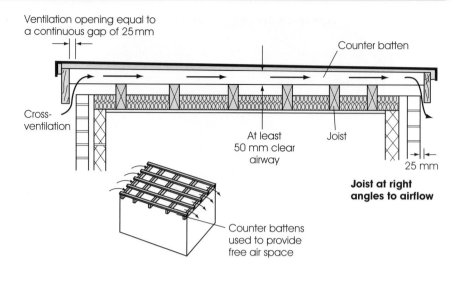

Ventilation opening equal to a continuous gap of 25 mm

Counter batten

Cross-ventilation

At least 50 mm clear airway

Joist

25 mm

Joist at right angles to airflow

Counter battens used to provide free air space

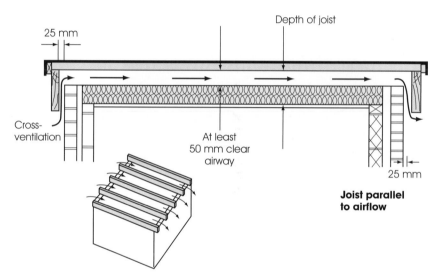

Depth of joist

25 mm

Cross-ventilation

At least 50 mm clear airway

25 mm

Joist parallel to airflow

This ventilation requirement can be achieved by either:

■ leaving a gap between the wall and soffit (this may be covered with a wire mesh to prevent access by birds, rodents and insects, etc.)

■ the use of proprietary ventilation systems in the soffit.

Where cross-ventilation is not possible due to abutting walls, a warm roof construction should be used.

Pitched roof construction

Timber pitched roofs may be divided into two broad but distinct categories:

■ *Traditional framed cut roofs.* Almost entirely constructed onsite from loose timber sections and utilising simple jointing methods.

■ *Prefabricated trussed rafters.* Normally manufactured under factory conditions.

Traditional roof component terminology

The main structural component parts of a roof are shown in Figure 2.20.

■ *Common rafters.* The main load-bearing timbers in a roof, which are cut to fit the ridge and birdsmouthed over the wall plate.

- *Ridge.* The backbone of the roof, which provides a fixing point for the tops of the rafters, keeping them in line.
- *Wall plate.* Transfers the loads imposed on the roof, uniformly over the supporting brickwork. It also provides a bearing and fixing point for the feet of the rafters.
- *Hip rafter.* Used where two sloping roof surfaces meet at an external angle. It provides a fixing point for the jack rafters and transfers their loads to the wall.
- *Jack rafters.* Span from the wall plate to the hip rafter, like common rafters that have had their tops shortened.
- *Crown rafter.* A common rafter that is used in the centre of a hip end.
- *Cripple rafters.* Span from the ridge to the valley, like common rafters that have had their feet shortened (the reverse of jack rafters).
- *Valley rafter.* Like a hip rafter, but used at an internal angle. It provides a fixing point for the cripple rafters and transfers their loads to the wall.
- *Purlin.* A beam that provides support for the rafters in their mid span. Like a binder used in upper floor construction.
- *Ladder frame.* Also known as a *gable ladder* and is fixed to the last common rafter to form the overhanging verge on a gable roof. It consists of two rafters with noggins nailed between them.
- *Ceiling joists.* On which the ceiling plasterboard is fixed. They also act as ties for each pair of rafters at wall plate level.
- *Binders and hangers.* Stiffen and support the ceiling joists in their mid span to prevent them from sagging and distorting the ceiling.
- *Struts.* Transfer the loads imposed by the purlins onto a load-bearing partition wall.
- *Spreader plate.* Provides a suitable bearing for the struts at ceiling level.
- *Collar tie.* Used to prevent the spread of rafters in the same way as ceiling joists and may also be used to provide some support for the purlins.

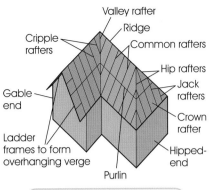

▲ **Figure 2.20** Traditional roof terminology

Traditional roof structures

Traditional framed cut roofs may be constructed as single, double or triple roofs, according to their span.

Single roofs

The rafters of all forms of single roof span from wall plate to ridge; they do not have any intermediate support (see Figure 2.21). Single roofs are only suitable for spans up to 5.5 m, as it is not economically viable to use increased sectional sizes of timber required for larger spans.

- *Couple roofs* consist of pairs of rafters fixed at one end to the wall plates and at the other to the ridge board. The span is restricted as, without a tie to the feet, the forces that act on the roof have a tendency to spread the walls.
- *Close couple roofs* are similar to the couple roofs, but with the feet of the rafters closed with a tie. This enables the span to be increased. The tie also acts as the ceiling joist. In addition, a central binder may be fixed to the ceiling joist, hung from the ridge at every third or forth rafter spacing, to bind together and prevent any sagging in the ceiling joists.
- *Collar tie roofs* are similar to the close couple roofs, but in order to increase the span the tie is moved up the rafter a maximum of one third of the rise.

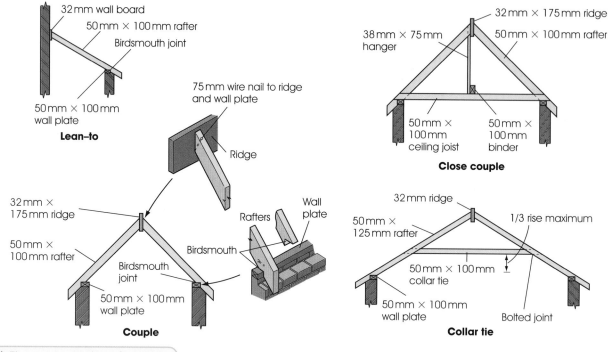

▲ **Figure 2.21** Single roof structures

Double roofs

The rafters of double roofs are of such a length that they require an intermediate support by purlins in their mid span, see Figure 2.22. This effectively reduces the span of the rafters and limits their size to economical sections. Struts, collars and hangers are spaced evenly along the roof at every third or fourth pair of rafters, to provide intermediate support for the purlins. Binders may also be fixed to the ceiling joist, hung from above to bind together and prevent any sagging in the ceiling joists.

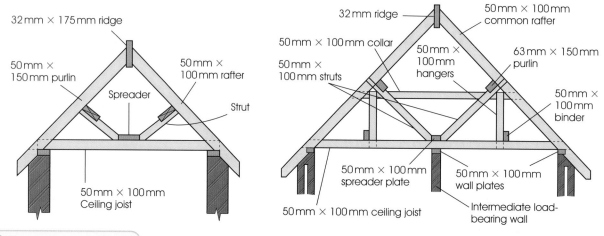

▲ **Figure 2.22** Double roofs

Triple roofs

These use fabricated roof trusses spaced at intervals along a roof to provide intermediate support for the purlins, which in turn provide intermediate support for the common rafters. In common with triple floors they have been largely obsolete for many years, but you may still come across them in maintenance or renovation work (see Figure 2.23). Early trusses were mortised and tenoned together from fairly large section timbers. Later, bolted trusses were developed; these were pre-fabricated in a workshop using simple overlapping joints, which were secured with bolts and timber connectors.

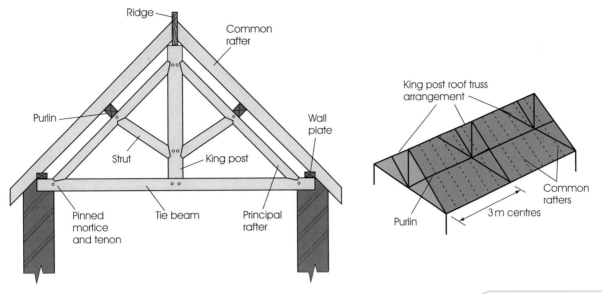

▲ **Figure 2.23** Triple roof

Pitched roof ends and intersections

Gables and hips

The ends of pitched roofs may be terminated with a gable wall, a sloping hip-end roof or a combination of both, as in gambrel and jerkin roofs. Figures 2.24 and 2.25 show typical details of gable and hip ends.

A **saddle board** is sometimes fixed to the common rafters as a bearing for the crown and hip rafters, see Figure 2.26. The order of fixing can vary: either the crown is fixed first and the hips fitted in between; or the hips are fitted first and the crown is fitted in between.

An **angle tie** (see Figure 2.27), traditionally termed as a **dragon tie**, may be fitted across the wall plate to strengthen the joint and provide an additional bearing to take the thrust of the hip.

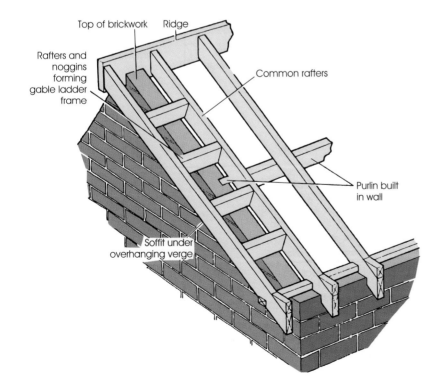

◄ **Figure 2.24** Gable-end roof detail

Heritage Link

Examples of early roof trusses used in triple roofs can be seen in many churches. The photo below shows hammer beam trusses or principal rafters used to support purlins. These date from the Gothic period circa 1400 to 1500, where to give greater height in the centre the tie beam was cut through, leaving short portions at either end, termed hammer beams, which are supported by curved braces back to the walls. The trusses shown are richly decorated with tracery infills and fine carvings.

▶ **Figure 2.25** Hip-end roof detail

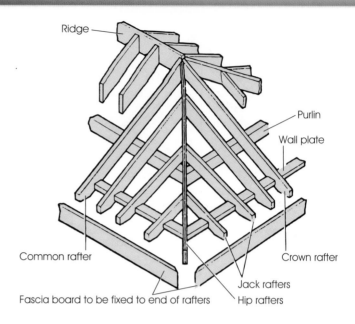

Ridge

Purlin

Wall plate

Common rafter

Crown rafter

Jack rafters

Fascia board to be fixed to end of rafters

Hip rafters

▼ **Figure 2.26** Use of a saddle board at hip/ridge intersection

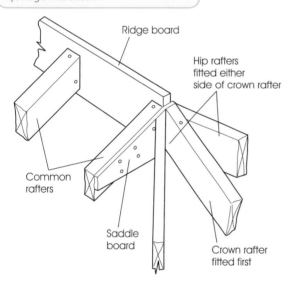

Ridge board

Hip rafters fitted either side of crown rafter

Common rafters

Saddle board

Crown rafter fitted first

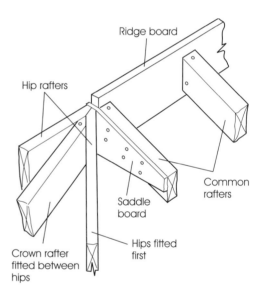

Ridge board

Hip rafters

Common rafters

Saddle board

Crown rafter fitted between hips

Hips fitted first

▶ **Figure 2.27** Use of angle tie at hip across wall plates

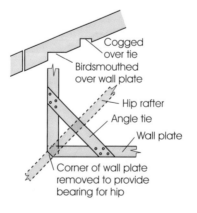

Cogged over tie

Birdsmouthed over wall plate

Hip rafter

Angle tie

Wall plate

Corner of wall plate removed to provide bearing for hip

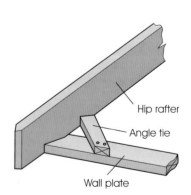

Hip rafter

Angle tie

Wall plate

Valleys

Where two pitched roofs intersect at an internal angle, a valley is formed between the two sloping surfaces. This valley may be constructed in one of two ways. Either a valley rafter is used and the feet of the common rafters (cripple rafters) of both roofs are trimmed into it, or the rafters of one roof are run through and lay boards are used to take the feet of the cripple rafters of the other roof. See Figure 2.28.

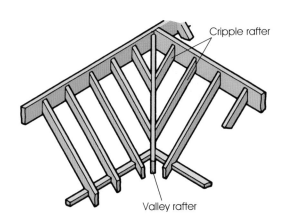

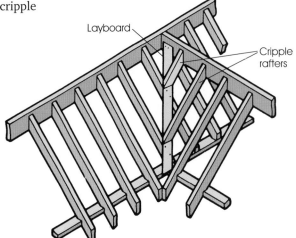

▲ **Figure 2.28** Valley designs

Parapet gutter is where a pitched roof terminates at the eaves by a parapet wall. Support and framework for a lead- or zinc-lined parapet gutter must be formed. See Figure 2.29.

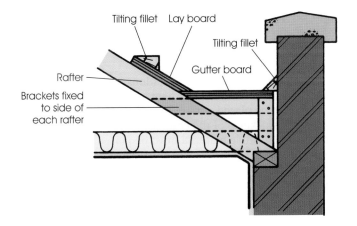

◀ **Figure 2.29** Parapet gutter

Related information

Also see Book 1, Chapter 10, Building control, which gives further details of Building Regulations. You may also view documents on the following website:

www.planningportal.gov.uk

Trimming to openings

Where openings occur in roofs for roof lights, chimneys or loft access hatches, the rafters and or the ceiling joists will have to be trimmed to create the required opening size. Framing anchors/joist hangers or well-nailed housing joints are used to join the rafters, joists, trimmers and trimmed components together, see Figure 2.30. Openings in the roof slope for roof lights and in ceilings for loft access hatches must be trimmed to accommodate the window or hatch lining size, so consult with the manufacturer's details before setting out. Openings in the roof slope and ceiling for chimneys should be of sufficient size to allow for a 50 mm gap between the structural timber and brickwork. This allows for ventilation and also reduces the possibility of the timber catching fire due to the hot brickwork. To weatherproof the roof around a chimney stack and prevent the build-up of rainwater behind the chimney, a supporting framework is required at the upper slope and layboards to the lower slope and sides for the lead or zinc-lined back gutter and flashings, see Figure 2.31.

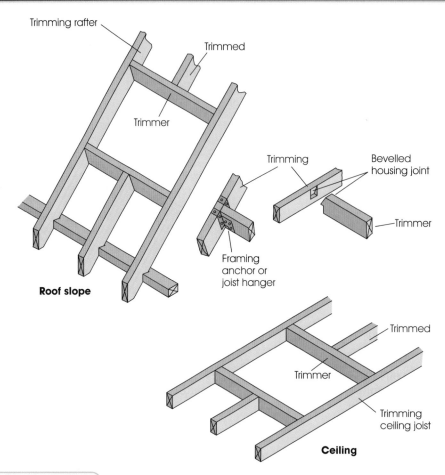

Trimming rafter

Trimmed

Trimmer

Trimming

Bevelled housing joint

Trimmer

Framing anchor or joist hanger

Roof slope

Trimmed

Trimmer

Trimming ceiling joist

Ceiling

▲ **Figure 2.30** Trimming to openings

In common with floors, the Building Regulations places restrictions on the use of structural timber near sources of heat, no combustible material, including timber, is to be placed within 200 mm of the inside of the flue lining; or, where the thickness of the chimney surrounding the flue is less than 200 mm, no combustible material must be placed within 40 mm of the chimney, see Figure 2.32. However, a 50 mm air gap between brickwork and structural timber is best practice, providing the timber is still 200 mm away from the flue lining.

▼ **Figure 2.31** Trimming to chimney stack

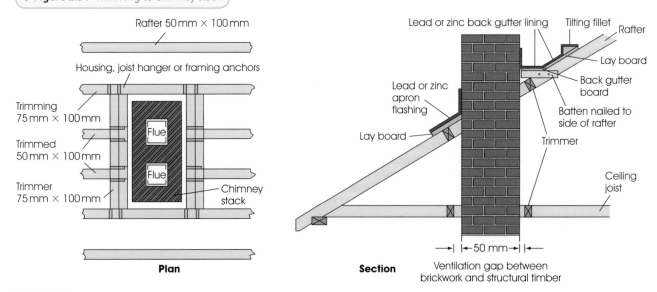

Rafter 50 mm × 100 mm

Housing, joist hanger or framing anchors

Trimming 75 mm × 100 mm

Trimmed 50 mm × 100 mm

Trimmer 75 mm × 100 mm

Flue

Flue

Chimney stack

Plan

Lead or zinc back gutter lining

Tilting fillet

Rafter

Lay board

Back gutter board

Lead or zinc apron flashing

Lay board

Batten nailed to side of rafter

Trimmer

Ceiling joist

|←50 mm→|

Section

Ventilation gap between brickwork and structural timber

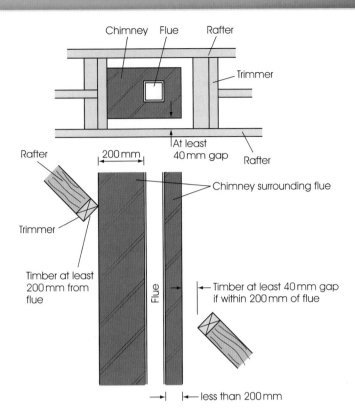

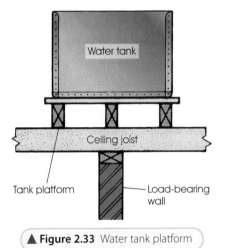

Chimney Flue Rafter

Trimmer

At least
40 mm gap

Rafter Rafter

200 mm

Chimney surrounding flue

Rafter

Trimmer

Flue

Timber at least
200 mm from
flue

Timber at least 40 mm gap
if within 200 mm of flue

less than 200 mm

◀ **Figure 2.32** Separation of flue and combustible materials

Water tank platforms

These are ideally situated centrally over a load-bearing wall. The platform should consist of boarding supported by joists laid on top of the ceiling joists and at right angles to them. Figure 2.33 shows a typical water tank platform.

Anchoring roofs

Rafters can be skew nailed to the wall plates or, in areas noted for high winds, framing anchors or truss clips can be used where appropriate. See Figure 2.34. Wall plates must be secured to the wall with straps at 2 m centres. The rafters and ceiling joists adjacent to the gable end should also be tied into the wall with metal restraint straps at 2 m centres.

Water tank

Ceiling joist

Tank platform Load-bearing
wall

▲ **Figure 2.33** Water tank platform

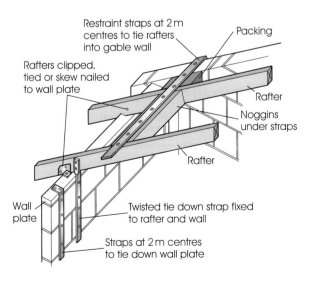

Restraint straps at 2 m
centres to tie rafters
into gable wall

Packing

Rafters clipped,
tied or skew nailed
to wall plate

Rafter

Noggins
under straps

Rafter

Wall
plate

Twisted tie down strap fixed
to rafter and wall

Straps at 2 m centres
to tie down wall plate

◀ **Figure 2.34** Anchoring of rafters and wall plates

Size and spacing of pitched roof members

The size and spacing of pitched roof members is dependent on the roof's location, span and pitch. In most situations, see Table 2.2 as a guide. Roofs in exposed locations or subject to extreme weather conditions, such as strong winds and heavy prolonged snow, are subject to additional load. This results in the need to use larger sectional sizes, shorter spans and closer spacing. TRADA and other bodies produce tables of suitable sectional sizes for use in a wide range of situations.

Table 2.2 Maximum clear span of pitched roof rafters and ceiling joists (m)

Based on using C16 grade softwood rafters for roofs over 30° and up to 45° pitch, with a dead load of up to 0.5 kN/m², imposed loading of 0.75 kN/m² suitable for most sites located up to 100 m above ordnance datum with access only for maintenance and repair. Ceiling joists are for dead loads between 0.25 and 0.50 kN/m².

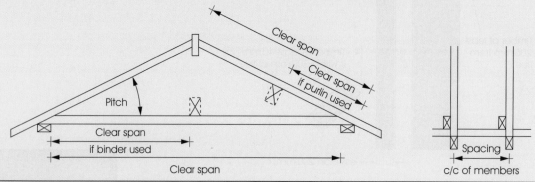

Size of rafter (mm) B × D	Spacing (centre to centre)		
	400 mm	450 mm	600 mm
38 × 100	2.28	2.23	2.10
38 × 125	3.07	2.95	2.69
38 × 150	3.67	3.53	3.22
50 × 100	2.69	2.59	2.36
50 × 125	3.35	3.23	2.94
50 × 150	4.00	3.86	3.52
Size of ceiling joist (mm)			
38 × 72	1.11	1.10	1.06
38 × 97	1.67	1.64	1.58
38 × 122	2.25	2.21	2.11
38 × 147	2.85	2.80	2.66
50 × 72	1.27	1.25	1.21
50 × 97	1.89	1.86	1.78
50 × 122	2.53	2.49	2.37
50 × 147	3.19	3.13	2.97

Related information

Also see Book 1, Chapter 1, Strength grading, page 53.

Determining lengths and bevels

There are three main methods that can be used to determine the lengths and bevels required for a traditional cut roof:

- Full size setting out; only really suitable for fairly small rise/span gable end roofs.
- Scale drawings and geometrical development.
- Roofing square.

Other methods to determine lengths and bevels are: by calculation using Pythagoras' rule and trigonometry; and various roofing 'ready reckoners' are published, consisting of a series of tables that show angles and lengths of members based on the roof's span and pitch.

Full size setting out

This method involves setting out in full size on a suitable surface, such as a concrete base, a couple of sheets of plywood, a couple of scaffold boards or joists positioned at right angles to each other.

For practical purposes the initial setting out is done using single lines, without the width or thickness of the members being shown, see Figure 2.35. Lines to indicate the rise and run (half the roof span) of the rafter are drawn first, followed by the diagonal rafter line (pitch line). This will indicate the required plumb and seat cuts as well as the theoretical or true length of the rafter. The end sections of the wall plate and the ridge can be added and then the width of the rafter, two thirds above and one third below the pitch line.

Related information

Also see Book 1, Chapter 9, Applied geometry, page 384, which refers to the geometrical development of pitched roofs; Chapter 8, Applied calculations, which refers to the methods of calculating lengths and angles.

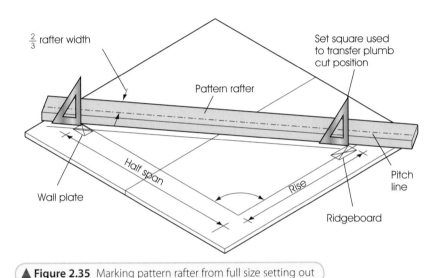

$\frac{2}{3}$ rafter width

Set square used to transfer plumb cut position

Pattern rafter

Half span

Rise

Pitch line

Wall plate

Ridgeboard

▲ **Figure 2.35** Marking pattern rafter from full size setting out

A pattern rafter can then be laid on top of this and marked out. The rest of the rafters can then be marked out and cut from the pattern.

Scale drawings and geometrical development

After having produced your own scale drawings and geometrical developments, or after a setter-out has produced them, either manually or with the aid of a computer-aided drawing programme (CAD), you will have to interpret the information in order to mark out the roof members.

Figure 2.36 shows a typical scale drawing of a part plan and section of a hipped-end roof with purlins. The drawing shows all the developments, angles and true lengths required to set out and construct the roof. The same plumb and seat cuts indicated for the common rafter are used for the crown, jack and cripple rafters. However, those of the hip rafter will be different due to its increased run.

An adjustable bevel square can be set up to the required angles by placing over the actual drawing, as shown in Figure 2.37.

The true lengths shown on the drawing are measured on the pitch line, which is a line marked up from the underside of the common rafter, one third of its depth. As the hip and valley rafters are usually of deeper section, the pitch line on these is marked down from the top edge at a distance equal to two thirds the depth of the common rafter as shown in Figure 2.38.

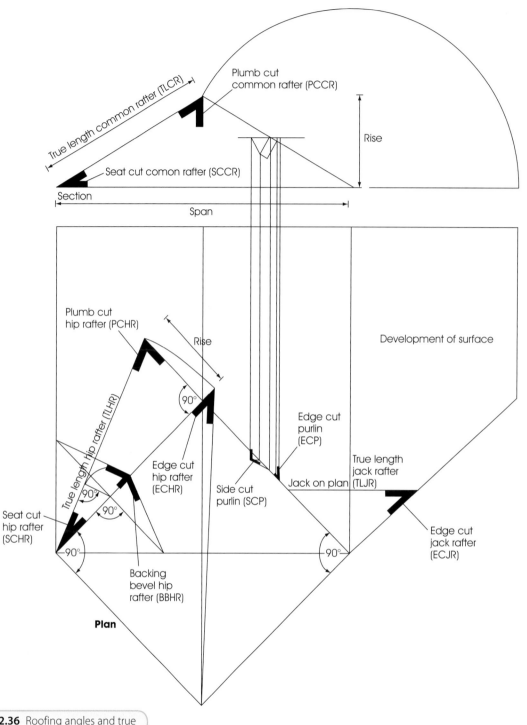

True length common rafter (TLCR)

Plumb cut common rafter (PCCR)

Rise

Seat cut comon rafter (SCCR)

Section

Span

Plumb cut hip rafter (PCHR)

Rise

Development of surface

90°

True length hip rafter (TLHR)

Edge cut hip rafter (ECHR)

Edge cut purlin (ECP)

90°

90°

Side cut purlin (SCP)

True length jack rafter

Jack on plan (TLJR)

Seat cut hip rafter (SCHR)

90°

90°

Edge cut jack rafter (ECJR)

Backing bevel hip rafter (BBHR)

Plan

▲ **Figure 2.36** Roofing angles and true lengths

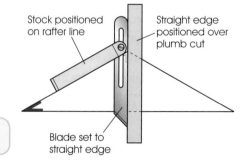

Stock positioned on rafter line

Straight edge positioned over plumb cut

▶ **Figure 2.37** Setting an adjustable bevel for a rafter plumb cut

Blade set to straight edge

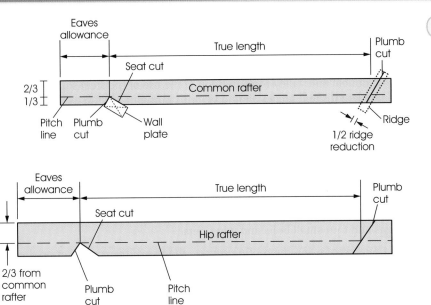

Take care when scaling the dimensions from the drawing as 1 mm difference on a 1:20 drawing is 20 mm difference at full size. It's best always to cut a pair of rafters and try them up on the roof for size, before progressing onto cutting the full requirement. See Figure 2.39 for the effect of any possible errors in the length of rafters.

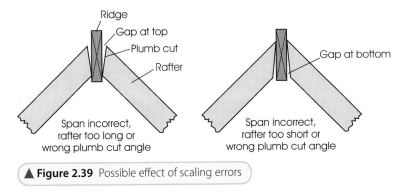

▲ **Figure 2.39** Possible effect of scaling errors

The true length of the common and hip rafters is measured on the pitch line from the centre line of the ridge to the outside edge of the wall plate. For jack rafters, it is from the centre line of the hip to the outside edge of the wall plate; for the cripple rafter it is from the centre line of the ridge to the centre line of the valley.

Therefore, when marking out the true lengths of the roofing components from the single line drawing, an allowance in measurement must be made. This allowance should be an addition for the eaves overhang and a reduction to allow for the thickness of the components. If this reduction is not apparent, it may be found by drawing full size the relevant intersecting components.

Figure 2.40 shows the intersection between the ridge, common, crown and hip rafters. The reduction of the hip rafters is shown. The reduction for the common rafters is always half the ridge thickness and for the crown rafter half the common rafter thickness. These reductions should be marked out at right angles to the plumb cut. When a saddle board is fitted, its thickness must be allowed for when making the reductions to the hips and crown. Figure 2.41 shows details at the corner intersection of the wall plates and the hip rafter, as the corner of the wall plates has been cut off to provide a flat surface for the hip plumb cut, the plumb cut itself must be moved forward to fit.

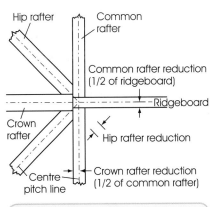

▲ **Figure 2.40** Reductions for material thickness

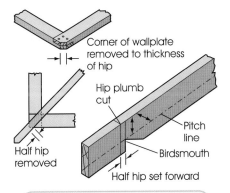

▲ **Figure 2.41** Hip rafter birdsmouth

Roofing square

Setting out rafters

After having studied geometrically the lengths and angles required for hipped, double and valley roofs in Book 1, the use of the steel square to find the same lengths and angles should be fairly straightforward, as it is merely the application of the geometric principles.

Most roofing squares contain sets of tables on them, which give rafter lengths per metre run for standard pitches, although in practice these tables are rarely used.

The wide part of a roofing square is the blade and the narrow part, the tongue (see Figure 2.42). Both the blade and the tongue are marked out in millimetres. Most carpenters will make a fence for themselves using two battens and four small bolts and wing nuts.

To set out a roof using the roofing square, the rise of the roof is set on the tongue and the run of the rafter is set on the blade (run of rafter = half of the span). In order to set the rise and run on the roofing square, these measurements must be scaled down and it is usual to divide them by 10.

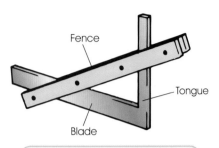

▲ **Figure 2.42** Roofing square and fence

> ■ *Example*
> For a roof with a rise of 2.5 m and a rafter run of 3.5 m, the scale lengths to set on the roofing square would be:
> Rise 2.5 m ÷ 10 = 250 mm; run 3.5 m ÷ 10 = 350 mm

▼ **Figure 2.43** Setting up a roofing square to obtain various scale lengths and angles

Figure 2.43 shows how to set up the roofing square and fence to obtain the required lengths and angles. The lengths will, however, be scale lengths.

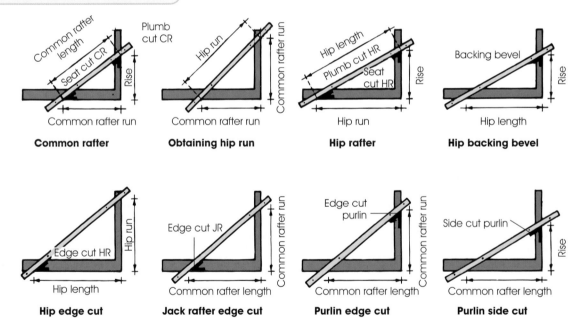

Figure 2.44 shows how the roofing square may be stepped down the rafter 10 times to obtain its actual length. Alternatively, the scale length can be measured off the roofing square and multiplied by 10 to give its actual length.

▼ **Figure 2.44** Using a roofing square

Allowances. The roofing square, like the geometrical method, gives the true lengths of members on the pitch line. Therefore, the same allowances in measurement as stated before must be made. The length for the shortest jack rafter can be found by dividing the length of the common rafter by one more than the number of jack rafters on each side of the hip. This measurement is then added to each successive jack rafter to obtain its length.

> ### Example
> For the roof shown in Figure 2.45 with three jack rafters on each side of the hip and a common rafter length of say 2.1 m:
> Length of short jack (1) = 2.1 m ÷ 4 = 525 mm
> Length of middle jack (2) = 525 mm + 525 mm = 1.05 m
> Length of long jack (3) = 1.05 m + 525 mm = 1.575 m

The true lengths and angles for valley and cripple rafters can be found by using the same methods as used for the hip and jack rafters.

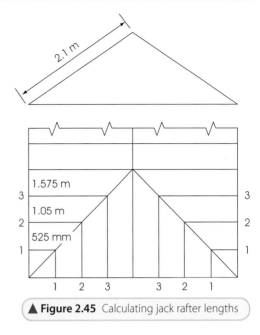

▲ **Figure 2.45** Calculating jack rafter lengths

Roof erection

The procedure used for erection is similar for most types of traditional cut roof.

Gable roof

The procedure of erection for a gable-end single roof would be as follows (see Figure 2.46):

1. The wall plate, having been bedded and levelled by the bricklayer, must be tied down.
2. Mark out the position of the rafters on the wall plate and transfer these onto the ridge.
3. Place ceiling joists in position and fix by nailing to the wall plate; a temporary working platform inside the roof space can be created by placing scaffold boards over the ceiling joists.
4. Make up two temporary A-frames. These each consist of two common rafters with a temporary tie joining them at the top, leaving a space for the ridge and a temporary tie nailed to them in the position of the ceiling joists for ease of handling.

▼ **Figure 2.46** Gable roof erection

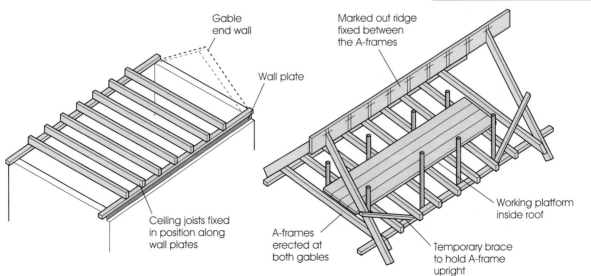

Gable end wall

Marked out ridge fixed between the A-frames

Wall plate

Working platform inside roof

Ceiling joists fixed in position along wall plates

A-frames erected at both gables

Temporary brace to hold A-frame upright

5. Stand up the A-frames at either end of the roof, fix in position by skew nailing to the wall plate and the ceiling joists.

6. Fix temporary braces to hold the A-frames upright.

7. Insert the marked out ridge board and nail in position through the top edge of the A-frame rafters.

8. Progressively position pairs of rafters along the roof, securing them by nailing them at both the ridge and wall plate. Diagonal rafter bracing may be required to stabilise the roof in windy locations.

9. Add noggins to form a gable ladder where an overhanging verge is required.

10. Allow bricklayers to build the gable end walls.

11. Finish the roof at the verge and eaves with bargeboard, fascia and soffit as required. (This stage is covered later in this chapter.)

The erection of double-cut roofs follows the same procedure, except for the positioning and fixing of the purlin and any associated struts and binders. These may be fitted either after the A-frames and a central pair of rafters have been fixed, or left until all the common rafters have been fixed. The former method is often preferred, especially for longer spans as an early fixed purlin provides a point on which to slide the intermediate common rafters up to the ridge and eases any potential problems associated with fixing to a run of possibly sagging rafters.

Hipped-end roof

The procedure of erection for a hipped-end roof would be as follows (see Figure 2.47):

1. The wall plate, having been bedded and levelled by the bricklayer, must be tied down.

2. Mark out the position of the rafters on the wall plate and transfer these onto the ridge.

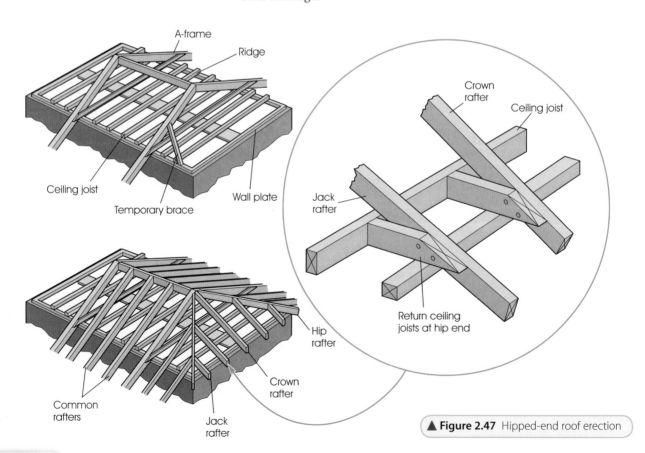

▲ **Figure 2.47** Hipped-end roof erection

3. Place the ceiling joists in position and fix by nailing to the external and any internal wall plates, a temporary working platform inside the roof space can be created by placing scaffold boards over the ceiling joist. Where additional height is required to reach the ridge, a temporary scaffold from the floor below may need erecting.

4. Make up two temporary A-frames. These each consist of two common rafters with a temporary tie joining them at the top, leaving a space for the ridge and a temporary tie nailed to them in the position of the ceiling joists.

5. Stand up the A-frames towards either end of the roof at the position of the last common rafter (half span of roof from corner of wall plate to centre line of rafter), fix in position by skew nailing to the wall plate and the ceiling joists.

6. Fix temporary braces to hold the A-frames upright.

7. Insert the marked-out ridge board and nail in position through the top edge of the A-frame rafters.

8. Fix the crown and hip rafters.

9. Fix the purlins, struts and binders (where required in double cut roofs).

10. Fix the remaining common rafters and jack rafters. Short return ceiling joists are fixed to the feet of the crown and jack rafters at the hipped ends.

11. Fix the collars and the hangers.

12. Finish the roof at the eaves with fascia and soffit as required.

Prefabricated trussed rafters

Prefabricated roof components were developed because of the need to:
- reduce the amount of timber in a roof
- simplify the often-complicated on-site cutting of a roof
- enable large spans without the need for intermediate supporting walls.

The earlier designs were bolted roof trusses, which developed from the traditional jointed king and queen post trusses. Then came nail plate trussed rafters, which are almost exclusively used today for new work. The essential difference between the two is that in a trussed rafter roof, every rafter is trussed and carries only its own proportion of the total load, whereas the older bolted roof trusses supported a purlin, which in turn carried the roof loads from a number of adjacent common rafters.

> **Related information**
>
> Also see in this chapter, triple roofs, page 44, and bolted roof trusses, page 76.

Nail plate trussed rafters

These are prefabricated by a number of specialist manufacturers in a wide range of shapes and sizes. Figure 2.48 shows a range of standard trussed rafter configurations. They consist of prepared timber laid out in one plane, with their butt joints fastened with nail plates. Figure 2.49 shows a typical design for a trussed rafter.

In use, the trusses are spaced along the roof at between 400 and 600 mm centres and fixed to the wall plate, preferably using truss clips or tie down straps, see Figure 2.50. Unlike common rafters the trusses simply rest on top of the wall plates and are not birdsmouthed over them.

In order to provide lateral stability, the roof requires binders at both ceiling and apex level and diagonal rafter bracing fixed to the underside of the rafters (see Figure 2.51). These must be fixed in accordance with the individual manufacturer's requirements.

Figure 2.52 shows how the gable wall must be tied back to the roof for support. This is done using lateral restraint straps at 2 m maximum centres both up the rafter slope and along the ceiling tie.

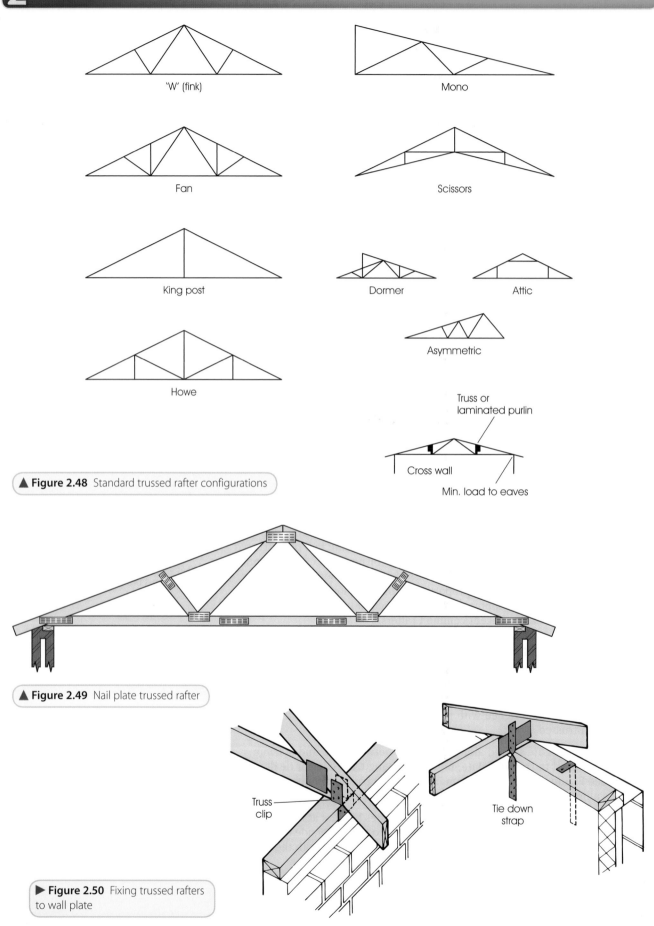

'W' (fink)

Mono

Fan

Scissors

King post

Dormer

Attic

Asymmetric

Howe

Truss or
laminated purlin

Cross wall

Min. load to eaves

▲ **Figure 2.48** Standard trussed rafter configurations

▲ **Figure 2.49** Nail plate trussed rafter

Truss
clip

Tie down
strap

▶ **Figure 2.50** Fixing trussed rafters
to wall plate

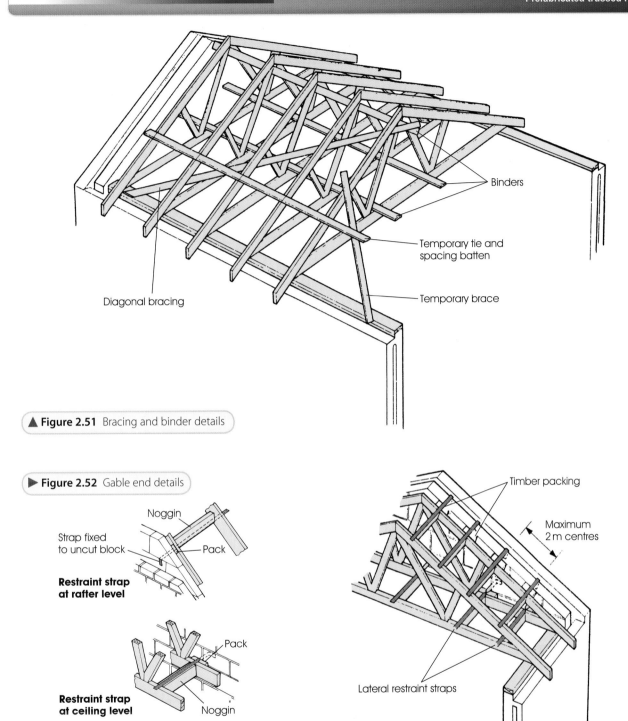

▲ Figure 2.51 Bracing and binder details

▶ Figure 2.52 Gable end details

Roof end and intersections using trussed rafters

Prefabricated gable ladders, see Figure 2.53, can be fixed to the last truss when an overhanging verge is required. Bargeboards and soffits are in turn nailed directly to the ladder.

Where hip ends and valleys occur in trussed rafter roofs, these may be formed either by using loose timber and cutting normal hip and valley rafters, and so on, in the traditional manner, or by using specially manufactured components.

Hip end. A hip end may be formed using cut timber in the traditional way, see Figure 2.54. These are pitched against a girder truss formed by nailing two or more trusses together.

▶ **Figure 2.53** Prefabricated gable ladder

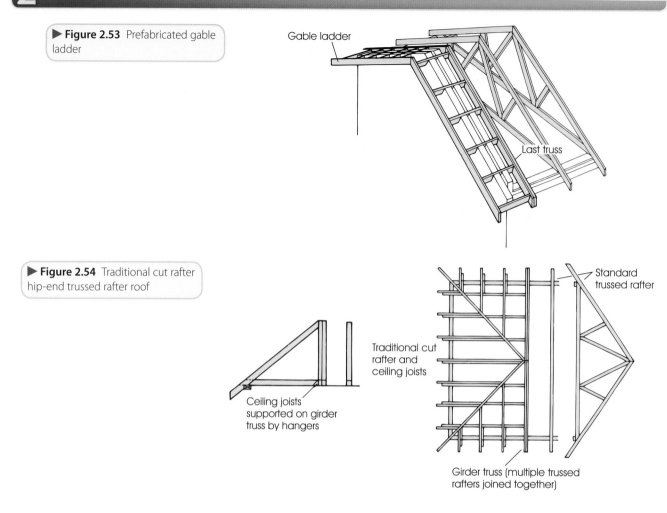

Gable ladder

Last truss

▶ **Figure 2.54** Traditional cut rafter hip-end trussed rafter roof

Standard trussed rafter

Traditional cut rafter and ceiling joists

Ceiling joists supported on girder truss by hangers

Girder truss (multiple trussed rafters joined together)

A hip end may be formed using a compound hip girder truss to support hip mono trusses, see Figure 2.55. Loose hip rafters, infill jack rafters and ceiling joists must still be cut and fixed in the normal way.

▼ **Figure 2.55** Hip-end detail using mono trusses

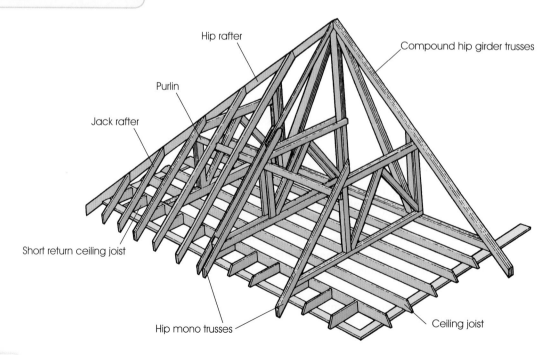

Hip rafter

Compound hip girder trusses

Purlin

Jack rafter

Short return ceiling joist

Hip mono trusses

Ceiling joist

Valley. A valley may be formed where two roofs intersect at a tee junction, using diminishing jack rafter frames nailed on to lay boards, see Figure 2.56. The ends of the rafters on the main roof are carried across the opening by suspending them from the compound girder truss using suitable joist hangers.

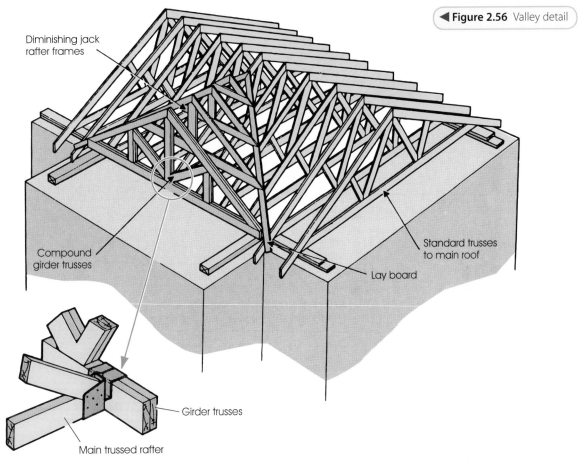

▲ **Figure 2.56** Valley detail

Diminishing jack rafter frames

Compound girder trusses

Standard trusses to main roof

Lay board

Girder trusses

Main trussed rafter

Support detail

Water tank platforms. These should be placed centrally in the roof with the load spread over at least three trussed rafters. The lower bearers of the platform should be positioned so that the load is transferred as near as possible to the mid-third points of the span, see Figure 2.57.

Trimming. Wherever possible, openings in roofs for chimney stacks or loft hatches should be accommodated within the trussed rafter spacing. If larger openings are required the method shown in Figure 2.58 can be used. This entails positioning a trussed rafter on either side of the opening and infilling the space between with normal rafters, purlins and ceiling joists. For safety reasons, on no account should trussed rafters be trimmed or otherwise modified without the structural designer's approval.

Water tank
Built up platform

▲ **Figure 2.57** Water tank platform

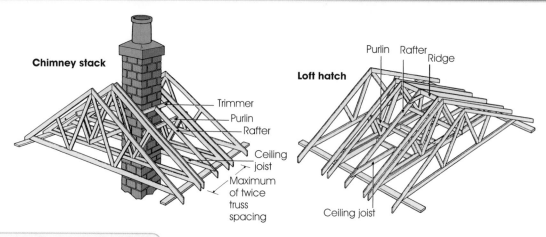

Purlin   Rafter Ridge

Trimmer
Purlin
Rafter

Ceiling joist

Maximum of twice truss spacing

Ceiling joist

▲ **Figure 2.58** Trimmings to openings

Attic-trussed rafters

These are available to accommodate rooms in the roof space and are available in two types, each requiring a different approach, see Figure 2.59.

■ Trusses that include a large sectioned bottom tie that acts as the floor joist. The roof is normally erected in the same way as standard trussed rafters, with the centre section then being boarded over to form the floor. However, you may come across roofs that have been assembled and boarded out at ground level or even off-site in a factory and simply lifted in place.

■ The alternative is attic trusses that require an upper suspended floor to be formed at the eaves level first. The trusses are then erected and fixed to the floor joists, again with the centre section boarded over to form the floor.

▶ **Figure 2.59** Attic-trussed rafters

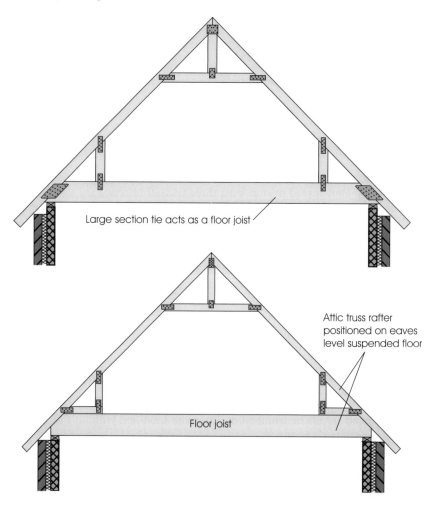

Large section tie acts as a floor joist

Attic truss rafter positioned on eaves level suspended floor

Floor joist

Standard windows in the gable-end wall can be used to provide much of the light and ventilation requirements, which can be supplemented by sloping roof lights positioned at intervals along the roof between the standard truss spacing.

Where **dormer windows** are required, an opening larger than the truss spacing will be required. This can be achieved using a similar method to that of other trimmed openings, see Figure 2.60.

▼ **Figure 2.60** Attic-trussed rafter roof dormer window details

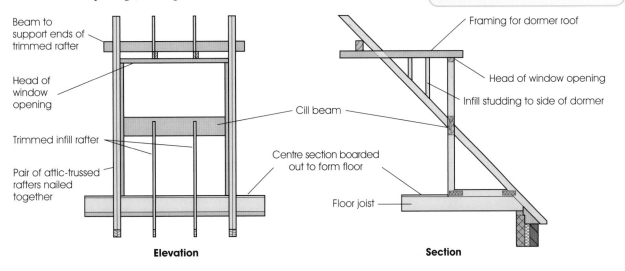

Elevation

Section

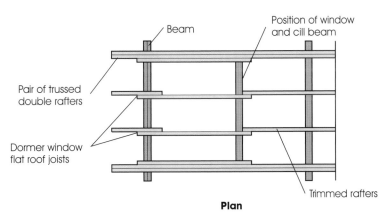

Plan

Pairs of attic trusses are nailed together and positioned at both ends of the required opening. The space in between is in-filled with either special trimmed trusses or normal cut rafters and joists. Again for safety reasons, on no account should trussed rafters be trimmed or otherwise modified without the structural designer's approval.

Erection of trussed rafters

The main problem encountered with the erection of trussed rafters is handling. In order not to strain the joints of the trussed rafters they should be lifted from the eaves, keeping the rafter in a vertical plane with its apex uppermost. Inadequate labour or care will lead to truss damage. Trussed rafters may be lifted into position with the aid of a crane, either singly from the node joints using a spreader bar and slings, or in banded sets. In both cases these should be controlled from the ground using a guide rope to prevent swinging. In addition, where large cranes are available, the entire roof may be assembled, braced, felted and battened at ground level and then lifted into position as one whole weatherproof unit, requiring only tiling at a later stage. This method is particularly suited to the rapid erection of timber frame buildings. Figure 2.61 shows the handling of trussed rafters.

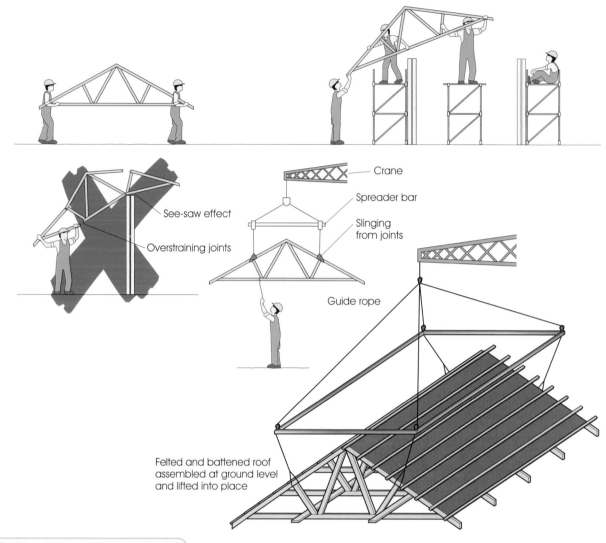

Crane

Spreader bar

Slinging from joints

See-saw effect

Overstraining joints

Guide rope

Felted and battened roof assembled at ground level and lifted into place

▲ **Figure 2.61** Handling trussed rafters

The erection procedure for a gable-end roof using trussed rafters is as follows (see Figure 2.62):

1. Mark the position of the trusses along the wall plates.
2. Once up on the roof, the first trussed rafter can be placed in position at the end of the under-rafter diagonal bracing. It can then be fixed at the eaves, plumbed and temporarily braced.
3. Fix the remaining trussed rafters in position one at a time up to the gable end, temporarily tying each rafter to the preceding one with a batten.
4. Fix diagonal bracing and binders.
5. Repeat the previous procedure at the other end of the roof.
6. Position and fix the trusses between the two braced ends one at a time, and fix binders.
7. Fix ladder frames and restraint straps.
8. Finish the roof at the eaves and verge with fascia, bargeboard and soffit as required.

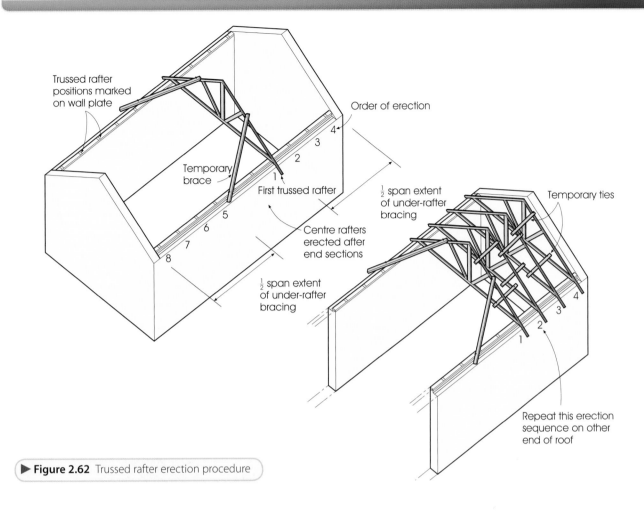

▶ **Figure 2.62** Trussed rafter erection procedure

Verge and eaves finishing

Verges

The termination or edge of a pitched roof at the gable end often overhangs the wall and is finished with a bargeboard and soffit, see Figure 2.63.

To finish the overhanging verge of a gable-end roof, the ridge and wall plate are extended past the gable-end wall, and an additional rafter is pitched to give the required gable overhang. Noggins are fixed between the last two rafters to form a gable ladder. This provides a fixing for the bargeboard, soffit and tile battens.

Bargeboard

The continuation of a fascia board around the verge or sloping edge of the roof (typically from 25 mm × 150 mm planed-all-round PAR softwood) provides a finish to the verge. The lower end of the bargeboard is usually built up to box in the wall plate and eaves. A template may be cut for the bargeboard eaves shaping in order to speed up the marking out where a number of roofs are to be cut and also to ensure that each is the same shape (especially where more elaborate designs are concerned).

Two methods may be used for determining the bevels at the apex (top) and foot (bottom) of the bargeboard:

■ *Marking in position.* The board is temporarily fixed in position, so a spirit level can then be used to mark the plumb cut (vertical) and seat cut (horizontal) in the required positions (see Figure 2.64).

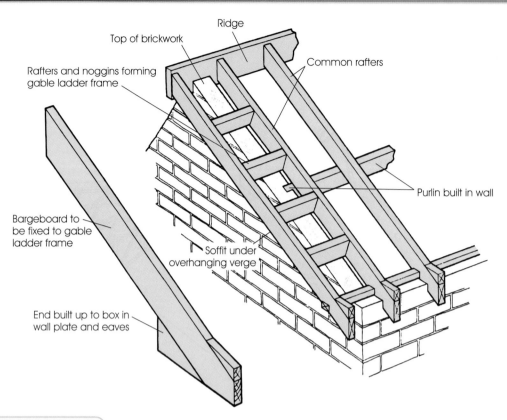

▲ **Figure 2.63** Gable-end detail

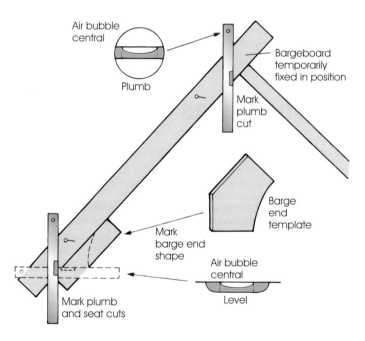

▶ **Figure 2.64** Marking out a bargeboard

■ *Determining bevels.* Adjustable bevel squares may be set to the required angles for the plumb and seat cuts, using a protractor (see Figure 2.65) or ones previously set from the scale geometrical drawings. These angles will be related to the pitch of the roof.

The sum of all three angles in a triangle will always be 180°. Therefore, in a 30° pitched roof, the apex angle will be 120° (180° – 2 × pitch), making the plumb cut for each barge board 60° (half of apex angle). The seat cut is at right angles to the plumb cut and is at the same angle as the roof pitch, see Figure 2.66.

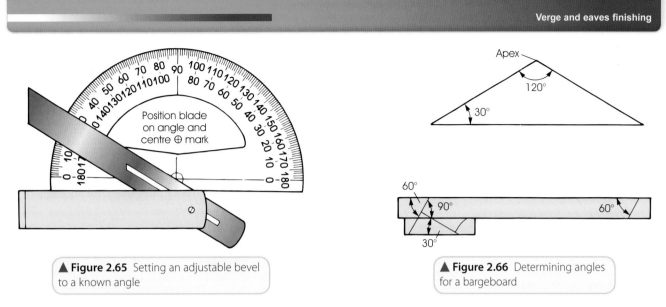

▲ **Figure 2.65** Setting an adjustable bevel to a known angle

▲ **Figure 2.66** Determining angles for a bargeboard

Fixing the bargeboard

The foot of a bargeboard may be either mitred to the fascia board, butted and finished flush with the fascia board or butted and extended slightly in front of the fascia board (see Figure 2.67). The actual method used will depend on the specification and/or supervisor's instructions.

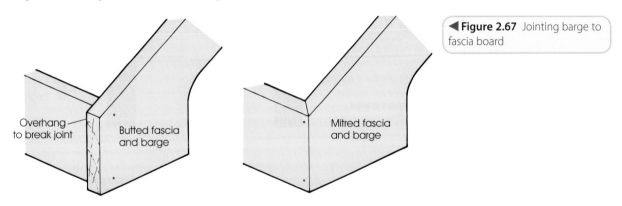

◀ **Figure 2.67** Jointing barge to fascia board

The mitred joint is preferred for high-quality work. The angle of the mitre for the bargeboard and fascia is best marked in position. Temporarily fix each in position, one at a time. Use a piece of timber of the same thickness to mark two lines across the edge of the board and join the opposite corners to form the mitre, see Figure 2.68. The face angle will be 90° for the fascia board and

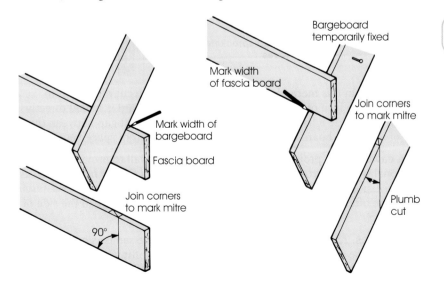

◀ **Figure 2.68** Marking barge to fascia board mitre

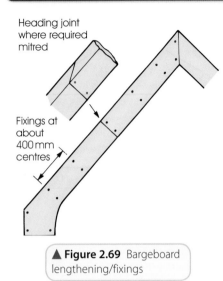

Heading joint where required mitred

Fixings at about 400 mm centres

▲ **Figure 2.69** Bargeboard lengthening/fixings

a plumb cut for the bargeboard. Where timber of sufficient length is not available for a continuous barge, splayed heading joints may be used, see Figure 2.69.

After marking, cutting to shape, mitring and fitting, the bargeboard can be fixed to its gable ladder by double nailing at approximately 400 mm centres. Use either oval nails, wire nails, lost-head nails or cut nails. These should be at least $2\frac{1}{2}$ times the thickness of the bargeboard in length in order to provide a sufficiently strong fixing. For example, an 18 mm thick bargeboard would require nails of at least 45 mm long (50 mm being the nearest standard length). All nails should be punched below the surface, ready for subsequent filling by the painter.

Eaves

The lowest part of a pitched roof slope, where the ends of the rafters terminate, usually overhangs the wall and is finished with a fascia board and soffit.

Eaves may be finished in one of three ways (see Figure 2.70):
■ Flush.
■ Overhanging, open or closed.
■ Sprocketed.

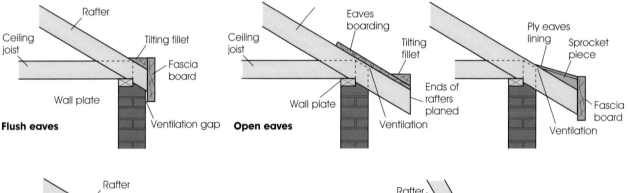

Flush eaves

Open eaves

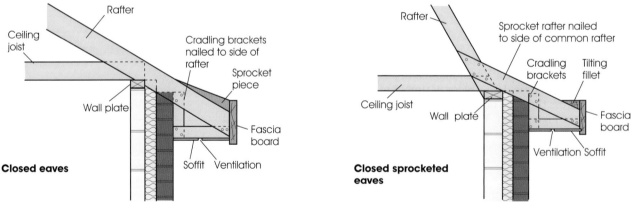

Closed eaves

Closed sprocketed eaves

▲ **Figure 2.70** Eaves details

Flush eaves. In this method the ends of the rafters are cut off 10 mm to 15 mm past the face of the brickwork and the fascia board is nailed directly to them to provide a fixing point for the gutter. The small gap between the back of the fascia board and the brickwork allows for roof space ventilation.

Open eaves. These project well past the face of the wall to provide additional weather protection. The ends of the rafters should be prepared as they are exposed to view from the ground. In cheaper quality work the fascia boards are often omitted and the gutter brackets fixed directly to the side of the rafter.

Closed eaves. These overhang the face of the wall the same as open eaves except that the ends of the rafters are closed with a soffit. Cradling brackets are nailed to the sides of the rafters to support the soffit at the wall edge.

Sprocket piece. Flush, open and closed eaves often use sprocket pieces nailed to the top of each rafter to reduce the pitch of the roof at the eaves. This has the effect of easing the fast-flowing rainwater under storm conditions into the gutter.

Sprocketed eaves. On steeply pitched roofs the flow of rainwater off the roof surface has a tendency to overshoot the gutter. Sprockets can be nailed to the side of each rafter to lower the pitch and slow down the rainwater before it reaches the eaves. This reduces the likelihood of rainwater overshooting the eaves and/or hitting the front of the gutter and splashing back soaking the eaves timbers, with the subsequent risk of rot. In addition, the use of sprockets also enhances the appearance of a roof giving it a distinctive 'bell-cast' appearance.

Fascia board

The fascia board is the horizontal board (typically ex 25 mm × 150 mm PAR softwood) that is fixed to the ends of the rafters to provide a finish to the eaves and a fixing for the guttering. Before fixing the fascia board the rafter feet will require marking and cutting to plumb and line, see Figure 2.71. A seat cut may also be required depending on the assembly detail.

- Measure out from brickwork the required soffit width and mark on the last rafter at either end of the roof.
- Mark the plumb cut and the seat cut if required using a spirit level.
- Stretch a string line between the end two rafters and over the tops of the other rafters. Use a spirit level to mark each individual plumb cut.
- Cut the plumb cuts using either a handsaw or portable circular saw.
- Where a seat cut is required, move the line down to the seat cut position on the end rafters. Use a spirit level to mark each individual seat cut.
- Cut the seat cuts using either a hand or portable circular saw.

Where timber of sufficient length is not available for a continuous fascia board, splayed heading joints may be used. These should be positioned centrally over a rafter end, see Figure 2.72.

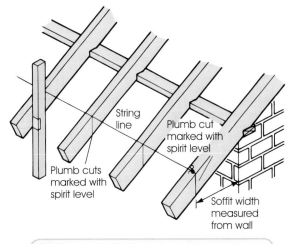

▲ **Figure 2.71** Marking out plumb cut at eaves

◀ **Figure 2.72** Jointing a fascia board over rafter end

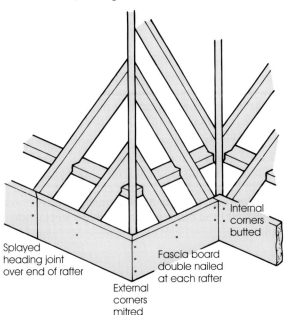

Where level fascia boards are returned around corners it is standard practice to use a mitre at the external corners and butt at the internal corners. Both of these joints should be secured by nailing (50 mm ovals).

Fascia boards are fixed to the end of each rafter using two nails. These are normally either oval nails, wire nails, lost-head nails or cut nails, at least $2\frac{1}{2}$ times the thickness of the fascia board in length. Typically 50 mm or 62 mm nails provide a sufficiently secure fixing. All nails should be punched below the surface ready for subsequent filling by the painter.

Prior to final fixing, the fascia should be checked for line, see Figure 2.73:

- Drive nails on the face of the fascia at each end of the roof. Strain a line between them.
- Cut three identical pieces of packing, place one at each end under the line and use the third to check the distance between the fascia and the line at each rafter position.
- Pack out or use a saw to ease ends of rafters as appropriate, so that the packing piece just fits between the fascia and the line.

▶ **Figure 2.73** Checking fascia board for line

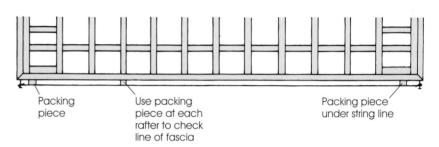

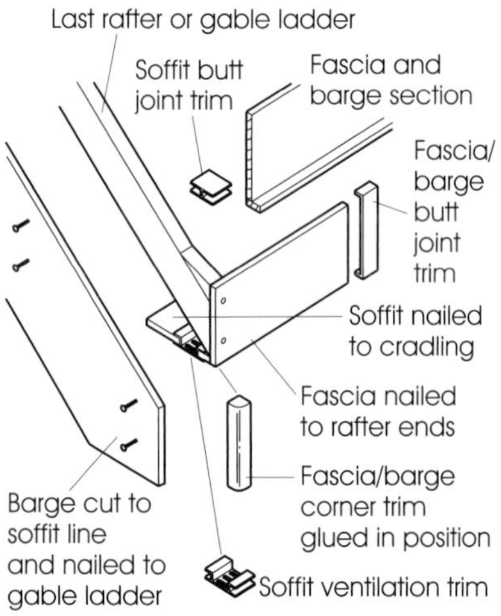

Packing piece

Use packing piece at each rafter to check line of fascia

Packing piece under string line

Soffit boards

Soffit boards are used to close the gap between the fascia and the wall of the building. They are normally strips of non-combustible sheet material, hardboard or formed from T&G matching.

Soffits normally tongue into the fascia board and are fixed to the underside of cleats or L-shaped brackets (cradling), which are themselves fixed to the sides of each rafter at the seat cut line or required soffit line.

Sheet material soffits are typically fixed using two 25 mm galvanised wire nails at each cleat or cradle position. Matchboarded soffits may be either surface nailed or secret nailed through the tongue using 38 mm oval nails at each cleat or cradle position.

Verge, eaves and soffit finishings in uPVC

Increasingly, extruded uPVC (unplasticised polyvinyl-chloride) sections are being used for new-build and refurbishment works. They have the advantage over timber in that they require very little in the way of ongoing maintenance, whereas timber can suffer from the effects of weather and requires regular repainting.

Standard profiles are available for fascias, soffits and bargeboards as shown in Figure 2.74. Corner and butt joint trims are also available, as are matching soffit ventilation sections. Profiles are easily cut to length using a fine tooth panel or dovetail saw. Fixing is normally directly into the rafter or cradling at about 600 mm centres, using stainless steel white plastic dome-headed pins.

Roof coverings

After the erection of the roof it should be covered without delay, both to protect the timber from the elements and make the building watertight so that internal work can commence. Sarking felt is laid over the rafters.

Last rafter or gable ladder

Soffit butt joint trim

Fascia and barge section

Fascia/ barge butt joint trim

Soffit nailed to cradling

Fascia nailed to rafter ends

Barge cut to soffit line and nailed to gable ladder

Fascia/barge corner trim glued in position

Soffit ventilation trim

▲ **Figure 2.74** Use of uPVC profiles for eaves and verge finishing

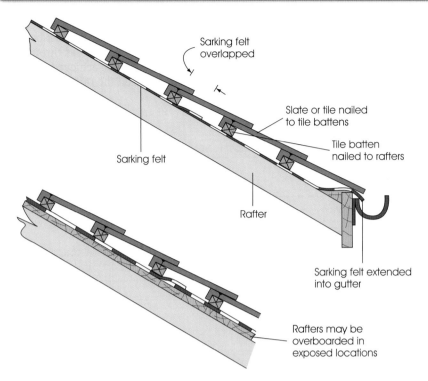

It should be rolled out along the eaves leaving sufficient to overhang into the gutter, overlapping each subsequent length as they progress up the slope. This is to prevent moisture entering the building should the covering tiles or slates get damaged, or the joints get bypassed by wind-driven rain or capillary action. Tile battens fixed at right angles to the rafters provide a fixing point for the tiles or slates and also hold-down the sarking felt, see Figure 2.75. Sarking felt and tile battens are normally fixed by the roof tiler. In exposed locations rafters may be overboarded before the sarking felt is laid. This provides additional lateral stability to the roof in severe wind conditions and also additional strength to support heavy snow loads for extended periods. Traditionally, solid timber boards were used for overboarding, however, in modern construction WBP sheathing plywood or moisture resistant OSB sheets are likely to be specified.

Preservation, insulation and ventilation

Preservative treatment

It is recommended that all timber used for verge and eaves finishes is preservative treated before use. Any preservative-treated timber cut to size on site will require retreatment on the freshly cut edges/ends. Applying two brush flood coats of preservative will achieve this.

Thermal insulation

This was normally done by placing rockwool or glassfibre insulation between the ceiling joists. A vapour check, such as foil-backed plasterboard, can be incorporated at ceiling level below the insulation. Vapour checks help reduce the amount of water vapour, which otherwise would pass from the room below, through the ceiling and into the cold roof void. However, with the increased levels of thermal insulation now required for new buildings, a second layer of insulation is often laid over the ceiling joists, giving a total insulating layer of around 300 mm. This is carried over the inner leaf of the cavity wall and well into the eaves. Care must be taken not to block the eaves with the insulating material, as the roof space must be well ventilated, in order to prevent any warm moist air that passes through the ceiling from condensating in the roof space. This can be achieved using a proprietary ventilator or timber boards nailed between the rafters, see Figure 2.76.

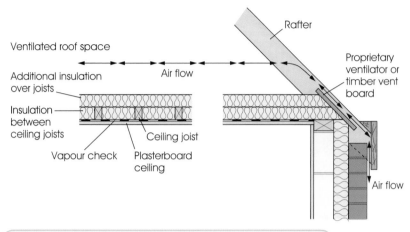

▲ **Figure 2.76** Pitch roof insulation between and over ceiling joists

Roof ventilation

To reduce the likelihood of condensation within the roof space, ventilation is required by the Building Regulations. All roofs must be cross-ventilated at eaves level by permanent vents. These must have an equivalent area equal to a continuous gap along both sides of the roof of 10 mm or 25 mm where the pitch of the roof is less than 15° or the insulation follows the pitch of the roof. See Figure 2.77. This ventilation requirement can be achieved in the following ways:

- Leaving a gap between the wall and soffit (this may be covered with a wire mesh to prevent access by birds, rodents, insects and so on).
- Using a proprietary ventilation strip fixed to the back of the fascia.
- Using proprietary circular soffit ventilators let into the soffit at about 400 mm centres.
- Lean-to roofs must have high-level ventilation in the roof slope close to the abutting wall. Proprietary vents are available for this purpose, these must have an equivalent area equal to a continuous gap along the abutment of 5 mm.
- Where the insulation follows the pitch of the roof, there must be at least a 50 mm air space between the insulation and the underside of the roof structure, and high-level ventilation at the ridge equal to a continuous gap of at least 5 mm. Proprietary ridge vents are available for this purpose.
- Roofs where the span exceeds 10 m or non-rectangular roofs may require additional ventilation, totalling 0.6 per cent of the roof area.

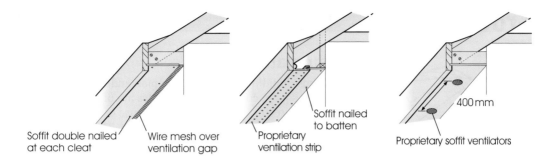

Soffit double nailed at each cleat

Wire mesh over ventilation gap

Soffit nailed to batten

Proprietary ventilation strip

400 mm

Proprietary soffit ventilators

Lean-to roof

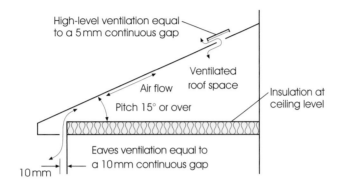

High-level ventilation equal to a 5 mm continuous gap

Ventilated roof space

Air flow

Insulation at ceiling level

Pitch 15° or over

Eaves ventilation equal to a 10 mm continuous gap

10 mm

Pitched roof

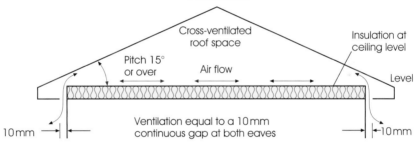

Cross-ventilated roof space

Insulation at ceiling level

Pitch 15° or over

Air flow

Level

Ventilation equal to a 10 mm continuous gap at both eaves

10 mm

10 mm

Attic pitched roof

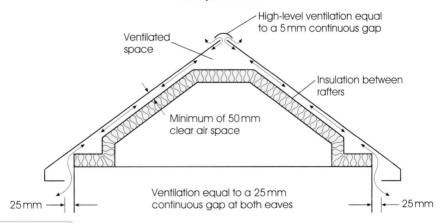

High-level ventilation equal to a 5 mm continuous gap

Ventilated space

Insulation between rafters

Minimum of 50 mm clear air space

Ventilation equal to a 25 mm continuous gap at both eaves

25 mm

25 mm

▲ **Figure 2.77** Pitched roof ventilation

Other roof types

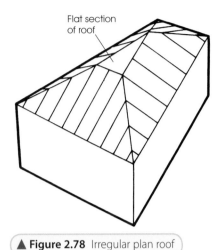

▲ Figure 2.78 Irregular plan roof

Flat section of roof

▶ Figure 2.79 Split-ridge details

Irregular plan roofs

In general most buildings are regular on plan and have equally pitched roofs (i.e. they have square corners and both sides are of equal slope). Sometimes, to create an architectural feature or to make use of a limited tapering site, buildings are constructed with an irregular plan. This, therefore, requires the use of irregular roof shapes and unequal pitches.

An equal-pitched hip-end roof on a building whose long sides are not parallel is shown in Figure 2.78. All of the rafters are kept square on plan to their respective wall plates, to avoid compound plumb and seat cuts to the common rafters. This necessitates a **split ridge** and a flat section to the top of the roof. The alternatives to this, which are impractical, would result in a sloping ridge and cut courses of roof tiles, or sloping eaves. Figure 2.79 shows a method of forming the split ridge, where the flat section is a maximum of about 300 mm. Two ridge boards can be used with bearers positioned between them at each rafter. For wider flat sections, a tapering ridge framework should be used see Figure 2.80.

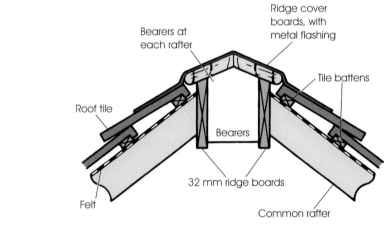

Ridge cover boards, with metal flashing

Bearers at each rafter

Tile battens

Roof tile

Bearers

32 mm ridge boards

Felt

Common rafter

▼ Figure 2.80 Split-ridge detail for wider flat section

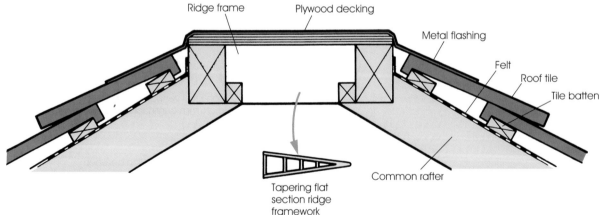

Ridge frame

Plywood decking

Metal flashing

Felt

Roof tile

Tile batten

Common rafter

Tapering flat section ridge framework

Mansard roofs

These are constructed so that accommodation can be provided in the roof space. Traditionally, the roof was constructed from framed Mansard roof trusses, which were a combination of a king post truss and a queen post truss, see Figure 2.81. Today, they will normally be formed using Mansard-trussed rafters.

A method of forming a Mansard roof using on-site cut timbers is shown in Figure 2.82. This detail is suitable for roofs up to 6 m in span. Also shown are two alternative means of forming the overhanging eaves.

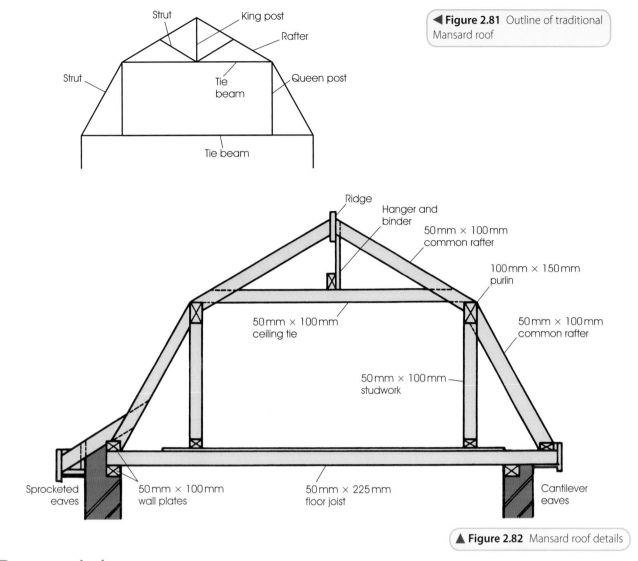

◀ **Figure 2.81** Outline of traditional Mansard roof

▲ **Figure 2.82** Mansard roof details

Dormer windows

These are vertical windows that project from a sloping roof surface to provide daylight and ventilation to a room in the roof space. The triangular sides of the dormer (cheeks) above the main roof are framed out with studding and covered with tiles, sheet metal or timber cladding. The actual roof over the dormer window, which extends from the main roof, can be constructed in various ways, see Figure 2.83, although flat or pitched roofs are most commonly used. Figure 2.84 is an isometric view of the basic framework construction for a flat roof dormer. Twin rafters are used on either side of the opening to carry the increased load. Figure 2.85 illustrates a section through a dormer window showing typical construction details.

Related information

Also see Book 1, Chapter 9, Applied geometry, pages 376–7, which refers to dormer geometry.

▼ **Figure 2.83** Dormer roofs

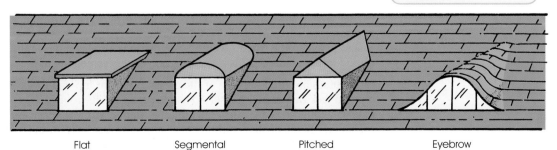

Flat   Segmental   Pitched   Eyebrow

▶ **Figure 2.84** Flat roof dormer window construction

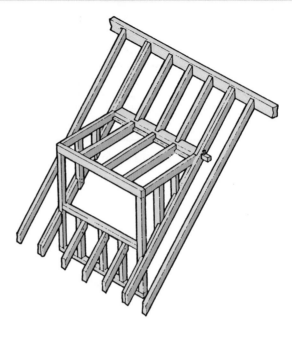

▶ **Figure 2.85** Dormer details

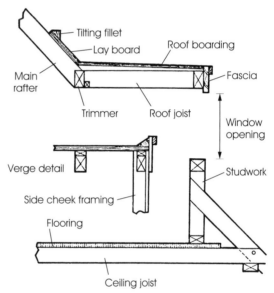

Bolted roof trusses

Traditionally, the component parts of bolted trusses had overlapped joints, which were connected up using bolts and toothed plates. They were normally prefabricated in a shop and arrived on site in two easily managed halves, which had to be bolted together once the truss had been positioned. Figure 2.86 shows a typical bolted truss with enlarged joint details. The trusses were spaced along the roof at about 1.8 m centres to support the roof components. The purlins and ceiling binders were then fixed to these. The spaces between the trusses were in-filled with normal common rafters and ceiling joists. Figure 2.87 is a line diagram of this arrangement.

Half-trusses are used in place of the crown rafter for hip-end roofs, normal hip and jack rafters being inserted as required. Timber roof trusses for industrial buildings are rarely encountered in new work as these are almost exclusively made of steel, although the type of trusses shown in Figure 2.88 may be met in maintenance and rehabilitation work.

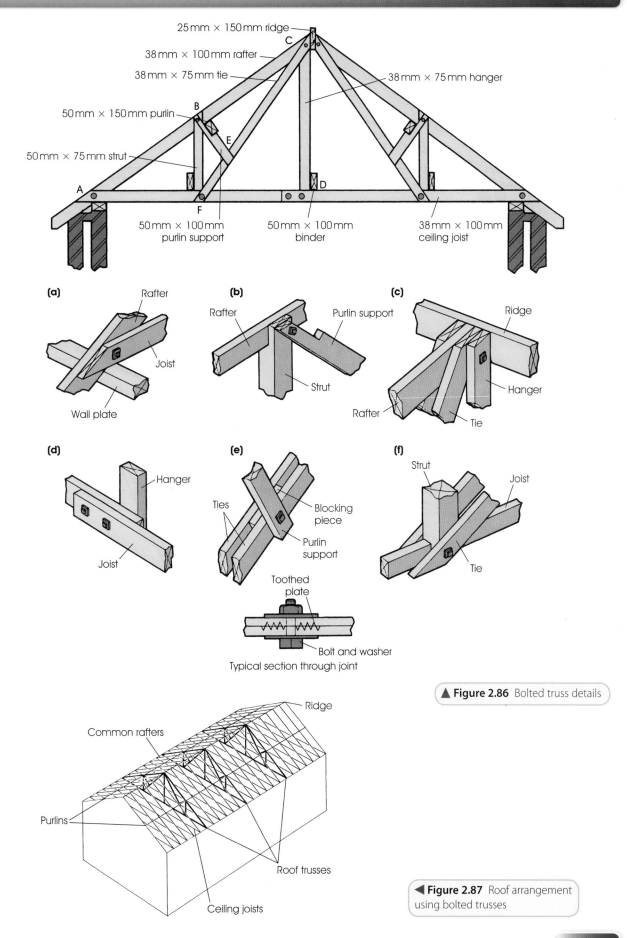

25 mm × 150 mm ridge

38 mm × 100 mm rafter

38 mm × 75 mm tie

38 mm × 75 mm hanger

50 mm × 150 mm purlin

50 mm × 75 mm strut

50 mm × 100 mm
purlin support

50 mm × 100 mm
binder

38 mm × 100 mm
ceiling joist

(a)
Rafter
Joist
Wall plate

(b)
Rafter
Purlin support
Strut

(c)
Ridge
Hanger
Rafter
Tie

(d)
Hanger
Joist

(e)
Ties
Blocking piece
Purlin support

Toothed plate
Bolt and washer
Typical section through joint

(f)
Strut
Joist
Tie

▲ **Figure 2.86** Bolted truss details

Ridge
Common rafters
Purlins
Roof trusses
Ceiling joists

◀ **Figure 2.87** Roof arrangement
using bolted trusses

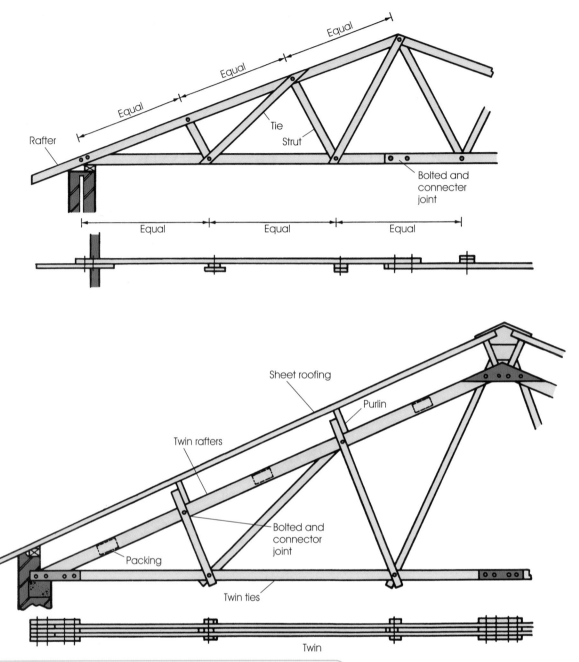

▼ Figure 2.88 Half elevation and plan of two alternative industrial roof trusses

Ornamental roofs

This classification includes turrets, spires and domes. These are all forms of fairly small decorative roofs, normally weatherproofed in sheet lead, copper or zinc and used to cover structures such as bell towers and church steeples. The construction of each roof is basically the same irrespective of its plan shape, which may be square, polygonal or circular.

Figure 2.89 shows the plan and section of a hexagonal turret roof, which may be mounted on a brick or timber structure. It consists of a wall plate, halved together to form the plan shape. Hip rafters are birdsmouthed over the wall plate at their lower end and are tenoned into a final post at the top. The hips should be backed to enable the roof to be boarded. Jack rafters are used between the hips.

The main differences in the construction of turrets, spires and domes may be stated briefly. Spires tend to be taller and more steeply pitched than turrets. These, therefore, require special consideration in the method of tying down to the main structure because of the considerable wind pressures they are subjected to. Domes have exactly the same construction details as turrets, except that the top edges of the hips and jacks will have been pre-cut to the required curve before erection, see Figure 2.90.

Heritage Link

Ornamental roofs were popular as garden features as well as for roofing over towers etc in the main building.

The one shown below is a Victorian style summer house that has a lead covered octagonal dome roof with bell cast eaves.

Finial post

Ply boarding

Hip rafter

Louvred ventilator

Wall plate

Fascia

Brick or timber structure

Section

Plan

▲ **Figure 2.89** Hexagonal turret roof details

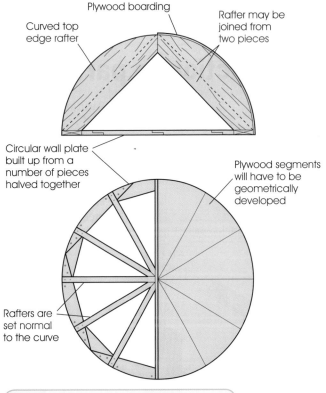

Plywood boarding

Curved top edge rafter

Rafter may be joined from two pieces

Circular wall plate built up from a number of pieces halved together

Plywood segments will have to be geometrically developed

Rafters are set normal to the curve

▲ **Figure 2.90** Simple dome roof construction

3

Chapter Three

Partition walls

This chapter covers the work of the site carpenter and joiner. It is concerned with principles of partition wall construction. It includes the following:

- Introduction to partitions.
- Timber stud partition construction.
- Metal stud partition construction.
- Proprietary panel partition construction.

Partitions are internal walls used to divide space into a number of individual areas or rooms. The cutting and erection of timber partitions is carried out after the building has been made watertight and is classified as a 'first-fixing' operation.

Introduction to partitions

Partition walls in modern construction are normally of a non-structural, non-load-bearing nature; they are not intended to be used as support for structural members above, such as floor or ceiling joists. However, suitable provision must be made at floor level to carry and transfer the partitions' own weight. In older buildings, partitions were often of a load-bearing nature.

Partitions are commonly formed from studwork framing, proprietary prefabricated systems or lightweight blockwork, see Figure 3.1.

▼ **Figure 3.1** Partition walls

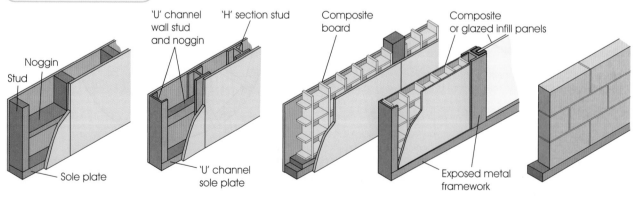

Timber studwork framing **Metal studwork framing** **Proprietary systems** **Lightweight blockwork**

Studwork framing partitions are commonly known as stud partitions or stud partitioning. These are walls built using timber or metal studs, fixed between a sole and head plates, often incorporating noggins for stiffening and fixing, see Figure 3.2. Plasterboard is mainly used as the covering material fixed by nailing to timber partitions or screwed to metal ones. Boards can be dry finished where the joints are filled and taped or wet finished using a skim coat of plaster.

■ *Stud.* A vertical timber or metal member of a partition wall fixed between the sole plate and head plate. The main member of a partition, it provides a fixing for the covering material.

■ *Sole plate/head plate.* A horizontal timber fixed above or below studs to provide a fixing point for the studs and ensure an even distribution of loads.

■ *Noggin.* A short horizontal piece fixed between vertical studs of a partition. Its use stiffens the studs, provides an intermediate fixing point for the covering material and, in addition, a fixing point for heavy items that may be hung on the partition (WC cistern, hand basin and so on).

Proprietary partitions of panel construction are available from many manufacturers and can be divided into two groups.

■ Demountable partitions, which are capable of being taken down and reused in different locations, are often used to divide up the large spaces in commercial and industrial buildings. They consist of an exposed metal framework to house typically composite plasterboard panels and glazing.

■ Non-demountable partitions are of a more permanent nature and are used in industrial, commercial and domestic buildings. They consist of composite plasterboard panels with a concealed framing to provide a fixing at floor, wall and ceiling intersections and at panel joints. Panels can be dry or wet finished.

Lightweight block work partitions consist of bonded blocks bedded in cement mortar and can be built directly off concrete floors. However, when built off a timber suspended floor, a sole plate should be introduced to spread the wall's load evenly. Where blockwork partitions are built off a suspended timber floor and run parallel with the joists, double joists should be incorporated under the partition to support its increased load. Block partitions in industrial and other heavily trafficked buildings may be left unplastered, (termed fair faced) to provide a hardwearing finish. Extra care must be taken with the joints as these will be exposed. Paint grade blocks are available for use in fair-faced partitions where a painted finish is required. In other locations they are more often finished using a two-coat plaster system or have plasterboard bonded to the surface (dry lining). 'Dot and dab' is the term given to bonding plasterboard to a wall surface, it involves applying dots of plasterboard adhesive to the wall surface or the back of the plasterboard and pushing the board directly onto the wall.

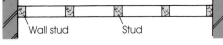

Figure 3.2 Stud partition components

Head plate · Noggin · Sole plate · Wall stud · Stud

Related information

Also see Book 1, Chapter 11, which gives further details of finishing walls.

Timber stud partitions

Timber grade: Sawn timber was traditionally used for making studwork framing. Today, prepared timber is mostly specified as it is easier and safer to handle and also provides consistently sized sections that are easier to joint, aids the plumbing of the partition and also provides a flat fixing surface for the partition-covering materials. Prepared timbers used for partitioning are: PAR (planed all round); machined timber, also termed regularised (timber machined to a consistent width by re-sawing or planing one or both edges); or ALS/CLS (American/Canadian Lumber Stock) processed timber with pencil-rounded corners may also be specified (see Figure 3.3).

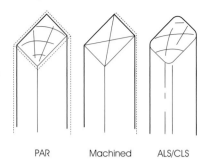

PAR   Machined   ALS/CLS

▲ **Figure 3.3** Consistently sized timber sections

Covering materials: The standard covering material for partition walls is plasterboard. Typical for stud partition use are thicknesses of 9.5 mm or 12.5 mm and sheet sizes of either 900 mm × 1800 mm or 1200 mm × 2400 mm. Tapered edge boards are best for a dry finish. These should be filled and taped to provide a smooth seamless joint that will not show through the subsequent paint or wallpaper finish. Square edge boards can also be used for a filled and taped dry finish, although they are primarily used for surfaces that are skimmed with a thin coat of wet board finish plaster (see Figure 3.4).

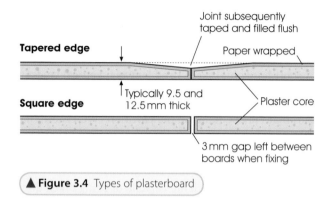

▲ **Figure 3.4** Types of plasterboard

Jointing timber partition members: Traditionally, timber partitions were framed using basic joints to locate members and provide strength, see Figure 3.5. Studs were often housed into the head plate, slotted over a batten at the sole plate and mortised or tenoned at openings. However, present day techniques calling for speed of erection and economy result in the majority of partitions being simply butt jointed and skew nailed (see Figure 3.6). An alternative to both methods would be the use of butt joints reinforced with metal framing anchors, see Figure 3.7. These provide a quick yet strong fixing, although they are rarely specified.

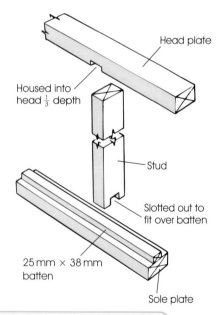

▲ **Figure 3.5** Traditional joint details

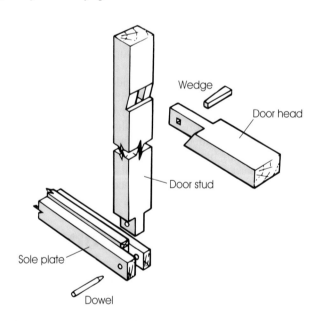

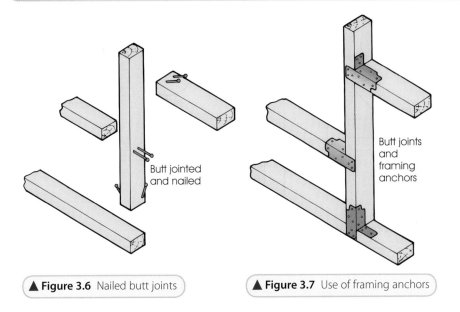

▲ **Figure 3.6** Nailed butt joints

▲ **Figure 3.7** Use of framing anchors

Constructing *in situ* timber stud partitions

Timber stud partitions may be constructed *in situ* or be pre-made and later erected.

- ■ *In situ* **partition.** Components are cut, assembled and fixed in situation on site.
- ■ **Pre-made partition.** A ready-assembled partition (either on site or in a factory) for later site erection.

1. Mark the intended position of the partition on the ceiling. Ideally this should either be at right angles to the joists or positioned under a joist or double joist. Where the joists run in the same direction as the partition and it is not directly under a joist, noggins will have to be fixed between joists, to provide a fixing point (see Figure 3.8).

2. Fix the head plate to ceiling, using 100 mm wire nails or oval nails at each joist position or 400 mm centres as appropriate. In older buildings, where ceilings might be in a poor position or easily damaged by nailing, 100 mm screws can be used as an alternative.

3. Plumb down from the head plate on one side at each end, to the floor, using either a straightedge and level or a plumb bob and line, as shown in Figure 3.9. This establishes the position of the sole plate.

4. Fix the sole plate in position with 100 mm wire nails, oval nails or screws, as appropriate. Ideally, as with head plates, this should be either at right angles to the joists or over a joist or double joist. Where the joists run in the same direction as the partition, but not directly under it, noggins will have to be fixed between joists to provide a fixing point, see Figure 3.10. When the sole plate is fixed to a concrete floor, it should be plugged and screwed. Alternatively, it may be possible to 'shot fix' both head and sole plates to concrete surfaces. Refer to your supervisor for permission/instruction.

5. Cut, position and fix the end wall studs, see Figure 3.11. These may be either:
   - plugged and screwed to brickwork and blockwork using 100 mm screws into proprietary plugs or twisted timber pallets
   - nailed to blockwork or into brickwork mortar joints using 100 mm cut nails
   - nailed directly to brickwork using 75 mm hardened steel masonry nails. (It is essential that eye protection is used when driving masonry nails as they are liable to shatter.)

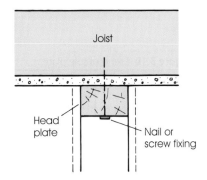

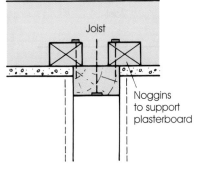

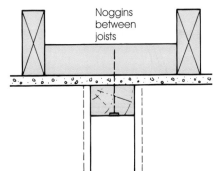

▲ **Figure 3.8** Fixing of head plate

▶ **Figure 3.9** Plumbing partition

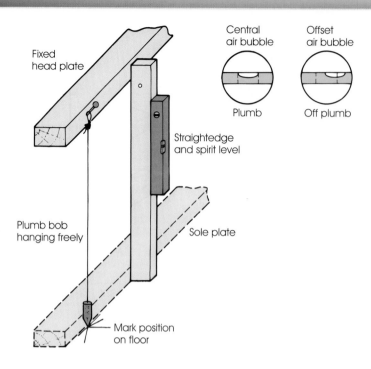

Fixed
head plate

Central
air bubble

Offset
air bubble

Plumb

Off plumb

Straightedge
and spirit level

Plumb bob
hanging freely

Sole plate

Mark position
on floor

▼ **Figure 3.10** Fixing of sole plate

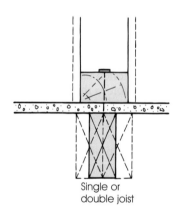

Nail or
screw fixing

Sole plate

Joist

Single or
double joist

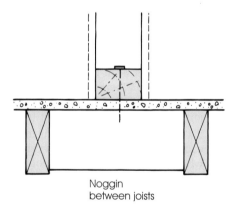

Noggin
between joists

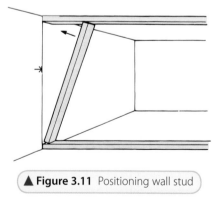

▲ **Figure 3.11** Positioning wall stud

6. Mark the positions of vertical studs on the sole and head plates. Studs will be required at each joint in the covering material and using 1200 mm wide sheets at 400 mm centres for 9.5 mm plasterboard or at 600 mm centres for 12.5 mm plasterboard. Using 900 mm wide sheets, all studs should be at 450 mm centres (see Figure 3.12). Similar centres may be employed for other covering material/cladding. Assuming 12.5 mm × 1200 mm covering, the second stud is fixed with its centre 600 mm from the wall, the third with its centre 1200 mm from the wall and the remaining studs at 600 mm centres thereafter.

7. Measure, cut and fix each stud to head and sole plates, using 100 mm wire or oval nails, driven at an angle (skew nailed, see Figure 3.13). Each stud should be measured and cut individually as the distance between the plates may vary along their length. Studs should be a tight fit: position one end, angle the stud and drive the other end until plumb. The length of each stud may be measured using either a tape or pinch rods, see Figure 3.14.

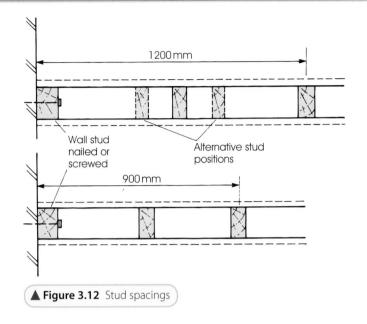

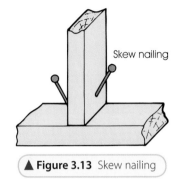

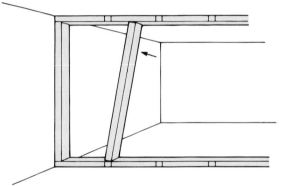

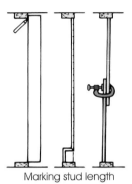

Marking stud length

▲ **Figure 3.12** Stud spacings

▲ **Figure 3.13** Skew nailing

◄ **Figure 3.14** Marking and positioning studs

8. Mark noggin centre line positions. These will vary depending on the specification. Typically they are fixed at vulnerable positions where extra strength is required: at knee height 600 mm up from floor; at waist height 1200 mm up from floor; and at shoulder height 1800 mm up from floor, as shown in Figure 3.15. Where deep section skirting is to be fixed, an extra noggin may be specified near the top edge of the skirting for fixing purposes. Additional noggins may also be required where heavy items are to be hung on the wall, such as where a fixing is required for electrical mounting boxes and where the covering sheet material is jointed in the height of the partition (see Figure 3.16).

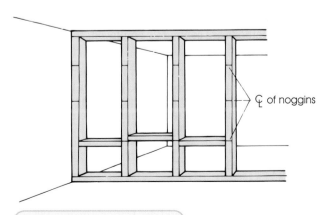

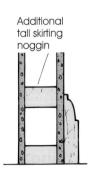

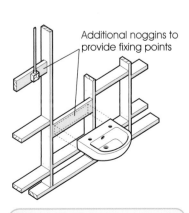

▲ **Figure 3.15** Noggin positions

▲ **Figure 3.16** Additional noggins

9. Cut and fix noggins (see Figure 3.17), either by skew nailing using 100 mm wire or oval nails or staggering either side of the centre line and through nailing again using 100 mm wire or oval nails.

► **Figure 3.17** Fixing noggins

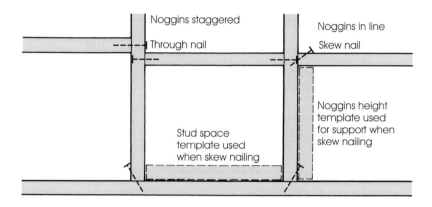

Openings in partitions

Where door, serving hatch or borrowed light (internal glazing) openings are required in a studwork framing, studs and noggins should be positioned on each side to form the opening (see Figures 3.18 and 3.19). The sole plate will require cutting out between the door studs. Linings are fixed around the framed opening to provide a finish and provision for hanging doors or glazing.

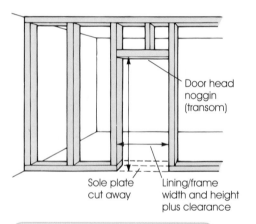

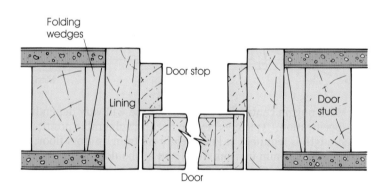

▲ **Figure 3.18** Door opening in a stud partition and horizontal cross-section

Returns in partitions

Should a return partition forming internal or external angles be required, stud positions must be arranged to suit. Consideration must be given to providing a support and fixing for the covering material around the return intersection. Two alternative methods are commonly employed, see Figures 3.20 and 3.21.

The particular method adopted will depend on whether the carpenter is fixing the covering material as the work proceeds or the plasterer is fixing it after the carpentry work is complete. Sole and head plates in long partitions and those containing return intersections will require joining, preferably using halving joints secured with screws (see Figure 3.22).

Cutting and fixing plasterboard

Plasterboard is normally marked by its manufacturer to indicate which face is suitable for plastering or dry decoration. Typically the following is marked on the rear face: 'Use other face for plastering and decoration'. In general, the ivory face is fixed to the outside.

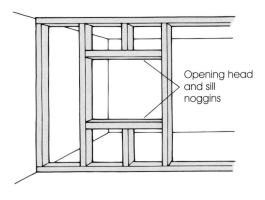

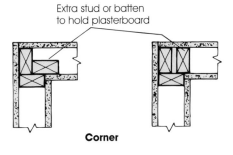

▲ **Figure 3.19** Borrowed light opening in a stud partition and vertical cross-section

Head noggin

Lining

Glazing bead pinned or screwed

Glass

Glazing compound

Sill noggin

Opening head and sill noggins

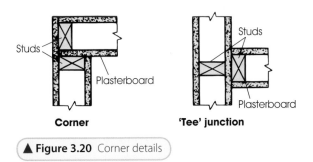

Studs

Plasterboard

Studs

Plasterboard

Corner

'Tee' junction

▲ **Figure 3.20** Corner details

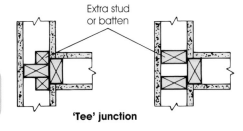

Extra stud or batten to hold plasterboard

Corner

Extra stud or batten

'Tee' junction

▶ **Figure 3.21** Stud treatment to provide support for covering at returns

▼ **Figure 3.22** Sole and head plate joints

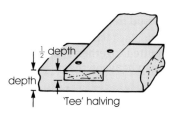

$\frac{1}{2}$ depth

depth

'Tee' halving

$\frac{1}{2}$ depth

Depth

Corner halving

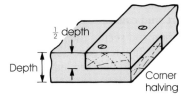

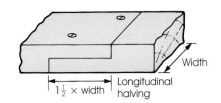

Width

$1\frac{1}{2}$ × width

Longitudinal halving

Plasterboard can be cut with a saw or with a replaceable blade craft knife (see Figure 3.23):

■ When sawing use the saw at a low angle to the sheet.
■ When using a knife, score a fairly deep line into the face along a straightedge. Apply downward pressure to snap the sheet along the scored line. Run the knife along the snapped line on the underside to cut the paper and separate the two pieces.
■ Holes for electrical fittings may be cut using a pad saw or with a craft knife.

Plasterboard sheets are secured to the studs using plasterboard nails or plasterboard screws at approximately 150 mm centres around the sheet edges and intermediate studs. To avoid distortion, nailing should commence from the centre of the sheet working outwards. Nails should be driven just below the surface, but taking care not to break or damage the paper face of the sheet. See Figure 3.24 for the fixing procedure.

Use hand saw at a low angle

Score line on face using knife against a straight edge

Snap board on scored line, use knife to separate

Cut holes using a pad saw

▲ **Figure 3.23** Cutting plasterboard

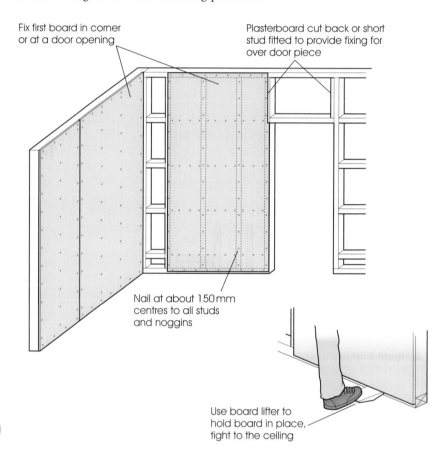

Fix first board in corner or at a door opening

Plasterboard cut back or short stud fitted to provide fixing for over door piece

Nail at about 150 mm centres to all studs and noggins

Use board lifter to hold board in place, tight to the ceiling

▶ **Figure 3.24** Fixing plasterboard

1. Pre-cut the sheets to length, making them 10 mm to 15 mm shorter than the floor to ceiling height.
2. Starting either from the corner or a doorway opening, use a board lifter to position and hold the board in place up against the ceiling; secure to all studwork members.
3. A narrow strip must be cut from the edge of the board over a doorway, to enable two boards to butt on the centre of the stud. Alternatively, a short stud can be fixed to the side of the door stud to provide a fixing.
4. Position and secure the remaining full boards in place working across the position. Tapered edge boards can be lightly butted together, square edge boards should be fixed leaving about a 3 mm gap between them.
5. Cut final board to width, position and secure.
6. Repeat the procedure to cover the other side of the partition.

Where services are to be accommodated in the partition, they should be installed either before any plaster boarding is done or before the second face is boarded.

Pre-made partitions

These may be made either by the joiner in a workshop or by the carpenter on site. The method of jointing will normally be either: studs housed into plates with noggins butt-jointed and nailed, or all joints butted and nailed. In addition, occasionally pre-made partitions may be jointed using framing anchors.

This type of partition must be made undersize in both height and width, in order for it to be placed in position. Once in position, folding wedges are used to take up the positioning tolerance prior to fixing (see Figures 3.25 and 3.26). Plates and wall studs are fixed using the same methods and centres as are applicable to *in situ* partitions.

▼ **Figure 3.25** Pre-made partitions must be constructed undersized to allow for positioning tolerances

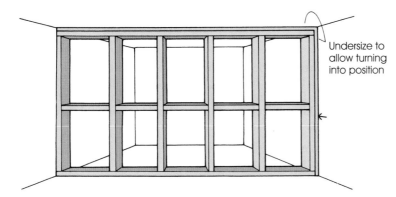

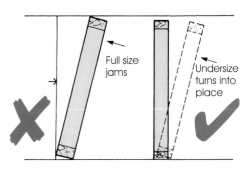

Drilling and notching partitions

Service cables and pipes for water, gas and electricity are often concealed within stud partitions. In common with floor joists, the positioning of holes and notches in studs has an effect on strength. It is recommended that holes and notches in studs should be kept to a minimum and conform to the following, see Figure 3.27.

- **Holes.** Up to one quarter (0.25) of the stud's width, drilled on the centre line (neutral stress line) and located between 25 per cent and 40 per cent (0.25 and 0.4) of the stud's height from either end are permissible. Adjacent holes should be separated by at least three times their diameter measured centre to centre.

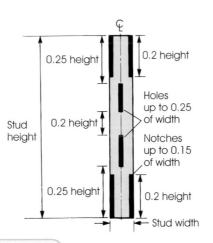

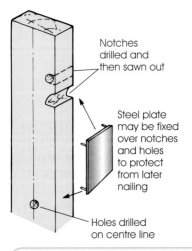

▲ **Figure 3.27** Hole and notch details and positioning

▲ **Figure 3.26** Use of folding wedges to take up positioning tolerances

■ **Notches.** Permitted on either edge of the stud up to 15 per cent of the stud's width and located up to 20 per cent (0.2) of the stud's height from either end.

━━━ *Example*
The position and sizes for holes and notches in 100 mm width studs, 2400 mm in length are:
- *Holes.* Between 600 mm and 960 mm in from either end of the stud and up to 25 mm in diameter.
- *Notches.* Up to 480 mm in from each end and up to 15 mm deep.

Excessive drilling and notching outside these permissible limits will weaken the stud and may lead to failure.

In addition, holes and notches should be kept clear of areas where the services routed through them are likely to get punctured by nails and screws, e.g. behind skirtings, dado rails and kitchen units and so on, or metal plates can be fitted for protection.

Fire resistance, thermal insulation and sound insulation

Depending on the partition's location it may be required to form a fire-resisting and/or a thermal-insulating and/or a sound-insulating function. See Figure 3.28. Specific requirements and methods of achieving them are controlled by the Building Regulations. These should be referred to for further information as may be required.

▶ **Figure 3.28** Fire, heat and sound insulation of stud partition

Related information

Also see Book 1, Chapter 11, page 471, which gives further details on insulation materials.

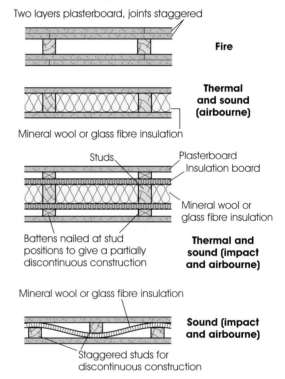

However, the general principles for achieving these functions are as follows.

■ **Fire resistance** can be increased by a double covering of plaster board, the second outer layer being fixed so that the joints overlap those in the lower layer.

■ **Thermal insulation** can be increased by filling the space between studs with either mineral wool or glass fibre quilt.

■ **Sound insulation** can be increased by a discontinuous construction to avoid impact and vibration and the use of lightweight infilling material, such as mineral or glass fibre quilt, to absorb sound energy.

Metal stud partitions

Various types of metal stud partition systems are available from a number of manufacturers. A typical system for domestic use consists of:

- galvanised 'U' section channels for fixing to the floor and ceiling
- 'U' or 'H' sections for intermediate studs
- 'U' section channels for wall studs and use at openings or junctions
- 'U' section channels for noggins.

Typical sections and sizes are shown in Figure 3.29.

All members are pre-slotted to accommodate electrical cables and other service runs. Rubber grommets must be used where electrical cables pass through members in order to prevent chafing. Services are normally installed after one side of the partition has been covered.

A range of typical fixing details for various junctions is shown in Figure 3.30. Timber sole plates may be specified under the floor channel especially where uneven or out-of-level floors are encountered. 12.5 mm or 15 mm plasterboard is recommended for covering metal stud partitions. Fitting two layers of plasterboard with the joints staggered and filling the spaces between the studs with insulating material will improve the partition's thermal-, sound- and fire-insulation properties.

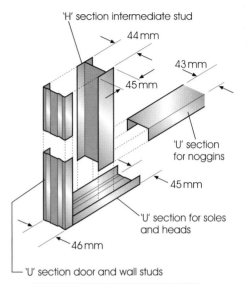

▶ **Figure 3.29** Metal stud partition members

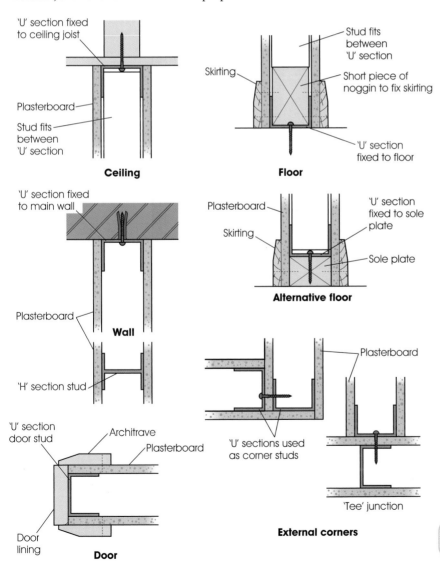

◀ **Figure 3.30** Metal stud junction details

Erection sequence

Figure 3.31 shows the erection sequence of metal stud partitions:

1. Mark the position of the partition on the floor, ceiling and walls.
2. Screw fix floor and ceiling channels and end wall studs at 600 mm centres. A continuous 6 mm bead of acoustic mastic/sealant applied to the back of all members before positioning will enhance the partition's sound-insulating properties.
3. Cut intermediate studs to length and insert studs between floor and ceiling channels at 900 mm centres.
4. Insert skirting blocks centrally between each pair of studs. These may be a standard item or cut from a noggin section.
5. Insert noggins at a height of 1200 mm.

▶ **Figure 3.31** Metal stud erection

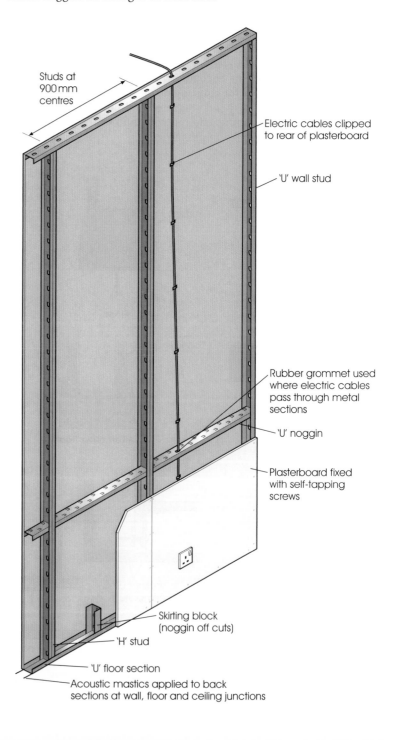

Studs at 900 mm centres

Electric cables clipped to rear of plasterboard

'U' wall stud

Rubber grommet used where electric cables pass through metal sections

'U' noggin

Plasterboard fixed with self-tapping screws

Skirting block (noggin off cuts)

'H' stud

'U' floor section

Acoustic mastics applied to back sections at wall, floor and ceiling junctions

Depending on the actual make of partition components being used, members may friction fit together, have interlocking lugs or may require fixing together with self-tapping screws:

1. Cut, position and screw fix plasterboard to one side of the partition. Use self-tapping bugle-head plasterboard screws at up to 400 mm centres.
2. Cut and fix plasterboard on the other side of the partition.

Door openings

At the head of door and other openings, channels will have to be fabricated to fit between the studs, see Figure 3.32:

1. Cut a length of floor or ceiling channel to span the opening plus about 150 mm extra at each end.
2. Cut through the side flanges at each end and bend at right angles.
3. Use self-tapping screws to fix the door head channel.

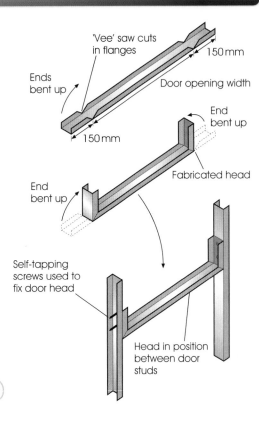

▶ **Figure 3.32** Fabricated door head

Proprietary panel partitions

Various proprietary partition systems are available from a number of manufacturers, but probably the most popular for domestic work is the cellular core dry partition system. This is constructed using prefabricated composite panels, which consist of two sheets of plasterboard separated by and bonded to a polystyrene, polyurethane or cardboard cellular core, see Figure 3.33.

These are available in panel widths of 900 mm and 1200 mm, thicknesses of 57 mm and 63 mm, and heights of 2350 mm, 2400 mm and 2700 mm. The 57 mm thick panels use 9.5 mm plasterboard and the 63 mm boards use 12.5 mm.

The panels rest on timber sole plates fixed to the floor and fit over timber battens, which are fixed to the wall and ceiling. Battens are also fixed between adjacent panels (see Figure 3.34). All battens that fit into the panels should be a press fit and, therefore, may require some easing. A bevel may be applied to the leading edges of battens to facilitate the easy entry into the panel edge.

A range of typical fixing details for various junctions is shown in Figure 3.35.

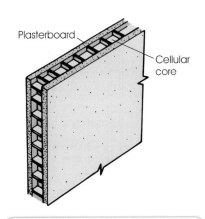

▲ **Figure 3.33** Proprietary dry partition board

Erection sequence

1. Mark the position of the partition on the floor, ceiling and walls.
2. Fix the panel width sole plate to floor along the total length of the partition.
3. Fix the 19 mm × 37 mm bevelled edge battens to the ceiling and end walls.
4. Fix 19 mm × 37 mm × 150 mm location blocks to the sole plate at the wall ends.

▶ **Figure 3.34** Proprietary panel partition

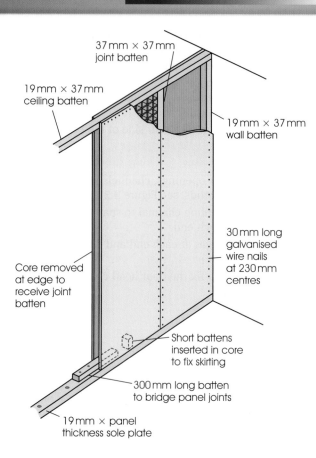

37 mm × 37 mm joint batten

19 mm × 37 mm ceiling batten

19 mm × 37 mm wall batten

30 mm long galvanised wire nails at 230 mm centres

Core removed at edge to receive joint batten

Short battens inserted in core to fix skirting

300 mm long batten to bridge panel joints

19 mm × panel thickness sole plate

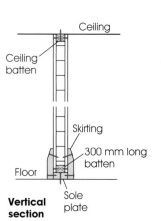

Ceiling

Ceiling batten

Skirting

300 mm long batten

Floor

Sole plate

Vertical section

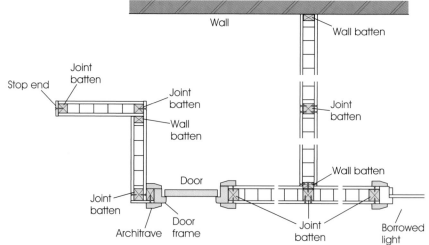

Wall

Wall batten

Joint batten

Stop end

Joint batten

Wall batten

Joint batten

Wall batten

Door

Joint batten

Architrave

Door frame

Joint batten

Wall batten

Borrowed light

▲ **Figure 3.35** Proprietary partition details

5. Use a handsaw to cut first panel to length; this should be 3 mm shorter than the distance between the top of the sole plate and ceiling.

6. Insert skirting fixing battens into the bottom edge of the panel, one for 900 mm panels and two for 1200 mm panels.

7. Locate the top of the panel over the ceiling batten about 200 mm from the wall; swing the panel to the vertical position, resting on the sole plate; slide the panel along the sole plate to locate over the location block and wall batten.

8. Insert 300 mm long location block half way into bottom corner of panel and fix to sole plate.

9. Cut to length and insert a 37 mm × 37 mm joint batten. Tap halfway into the edge of the panel.

10. Nail or screw fix the panel to all battens, using either 38 mm plasterboard nails at up to 150 mm centres or 38 mm bugle head plasterboard screws at up to 300 mm centres.
11. Cut and fix remaining panels using the previous procedure.

It is not normally necessary to rake out the panel core before inserting battens. Gently tapping/pushing the batten into the panel edge/end should force back sufficient core to accommodate the batten. Alternatively, the core on the top and two long edges may be removed prior to positioning by raking out with the claw of a hammer.

Services

Electric cables and other similar services up to about 30 mm in diameter can be accommodated in the cellular core. Use a spiked rod to hollow out the core to the position of the electrical outlet before positioning the panel. The wires should be fed down the panel to the outlet box hole as the panel is offered up to the ceiling batten and slid into place.

Timber plugs (37 mm × 37 mm) can be inserted into the cellular core in various positions during erection to provide later fixing points for cisterns, radiators, washbasins and so on.

Door openings

At door and other openings the joint batten should be tapped into the panel until it is flush with the edge. Fix short lengths of 19 mm batten to the upper end of the joint battens at either side of the door head. Cut the door head panel from an offcut, slide into position and fix with nails or screws, see Figure 3.36.

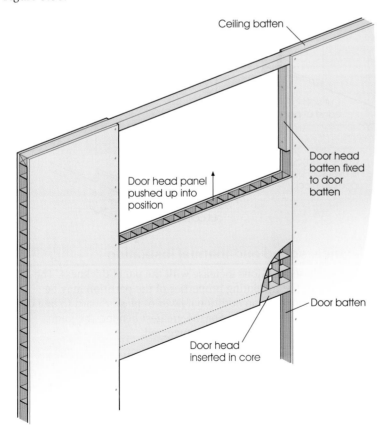

Ceiling batten

◄ **Figure 3.36** Door head panel

Door head panel pushed up into position

Door head batten fixed to door batten

Door batten

Door head inserted in core

Fixing of final panel

Where a partition is required between two walls and it does not contain a door opening or offset, a different procedure is required to position and fix the final panel, see Figure 3.37:

1. Use the previous erection sequence, but work from both end walls, making the final panel somewhere in mid run between two fixed panels.
2. Rip out the core on both long edges of the final panel to allow the joint batten to be inserted flush with the edges of the panel. Also rip out the core on the edges of both fixed panels to a depth of 19 mm.
3. Cut the two bottom corners off the final panel to allow it to fit over the sole plate location blocks. Also cut two 20 mm slots in both edges of the panel, about 450 mm from the top and bottom.
4. Place joint battens halfway into the edge of the final panel and partly insert a nail or screw into the battens at each slot position. Push the battens fully into the panel edges.
5. Position final panel over ceiling batten and swing into place. Use the protruding nails or screws to draw the joint battens into the edges of the two fixed panels.
6. Secure the panels to all battens and finally refit and fix the two cut-out pieces of panel at the sole plate.

▶ **Figure 3.37** Fixing of final panel

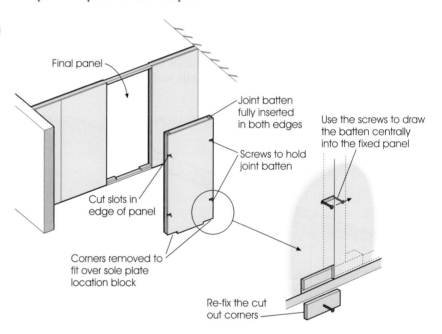

Fire resistance, sound and thermal insulation

The properties of these partitions increase with the panel thickness. The sound-, thermal- and fire-insulating properties of the partition may be further improved by bonding an additional layer of plasterboard to one or both sides with a suitable adhesive. For the greatest benefit, the joints between the panels and sheets should be staggered.

4

Chapter Four

Timber-framed construction and timber-engineered components

This chapter covers the work of the site carpenter and bench joiner. It is concerned with principles of timber-frame construction and timber-engineered components. It includes the following:

- Development of timber framing.
- Construction and erection of modern timber-framed buildings.
- Details of curtain walling.
- Details of glulam components, stressed skin panels and built-up beams.
- Construction and erection of shell roofs.

Timber framing has been used for hundreds of years as a construction method; the details that follow provide an overview of its development and an in-depth understanding of modern forms.

Development of timber framing

Cruck frame buildings (see Figure 4.1) were constructed by the Saxons. They were a sea-faring nation and so applied their boat-building skills to construction. Their early timber-frame dwellings resembled inverted boat hulls and used curved tree trunks split down the middle to make the pointed arch end-frames known as **crucks**. These end-frames were joined at their apex by a ridge pole, which often needed a centre post for additional support. This framework of the early cruck houses was covered in with wattle and daub

▼ **Figure 4.1** Cruck frame buildings

Ridge
Centre post
Crucks

King post
Rafter
Tie beam
Wall post
Wall plate

Thatch roof
Wattle and daub pan infill

(basketwork of timber lathing covered in a mixture of clay, lime, animal fat and cow dung) at the lower levels and thatch or turf for the upper roof.

Later, upright side walls were developed by the use of tie beams, wall plates and vertical wall posts. The spaces between the posts, which were filled with wattle and daub, are known as pans, so the method is known as **post and pan work**. The central post supporting the ridge restricted the internal space and was replaced by a short **king post** resting on a central tie beam. Rafters were laid from the wall plates to the ridge and weathered with thatch.

Wind holes, later to become known as windows, were made in the pans on the prevailing wind side of the house to give a good draught to the central fire and so gave rise to the term **window pane**.

Early entrance doorways were covered by simply hanging a length of animal skin across the opening. Later solid doors were made from vertical planks battened together with cross-pieces and pivot-hung using metal or wooden pins.

The chiefs' house or 'hall' was the largest in a village. It often included upper storeys that were formed by laying large planks spanning across the wall plates. Access to the upper storey was provided by a simple internal ladder or an external staircase. These were developed further by the Normans into larger halls by adding further crucks to extend the length and extending the walls and roof past the crucks forming an aisle or outshut.

Post and beam buildings (see Figure 4.2) were built on a box-frame principle using heavy section oak horizontal beams and vertical posts, secured with pinned mortise and tenon framing joints and cogged bearing joints. Upper floors often overhung the lower walls for stability, forming jetties. The pans between the posts were filled with wattle and daub. Sometimes **pargetting** (decorative patterned plasterwork) was used to finish pans or later they could be in-filled with brickwork. Windows were glazed with small leaded lights. Roofs were built with their gables facing the street and finished with decoratively carved bargeboards. Roof coverings were mainly thatch, although stone or slate tiles were also used for some buildings.

There are a number of specialist companies still using this method today. The main difference between the original is that the panes will now be in-filled with highly insulated panels, utilising rendered blockwork, face brickwork or timber cladding as the surface finish.

▶ **Figure 4.2** Jettied post and beam building

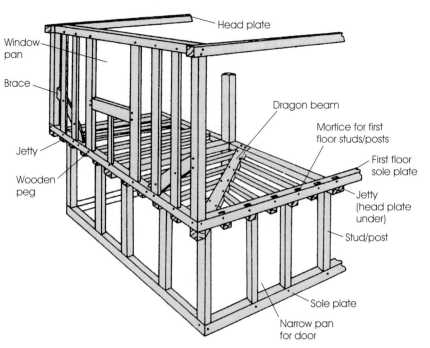

Platform and balloon timber-framed buildings (see Figure 4.3): Both types are constructed using timber studwork wall panels, with plywood or oriented strand board (OSB) sheathing on the outside for strength. Studwork panels are filled with rockwool or fibreglass thermal insulation normally after erection, the internal surfaces are lined in plasterboard and either receive a skim coat of plaster or are taped and filled. Externally, the building can be weathered using a variety of claddings such as an outer brick skin, timber boarding and tile hanging or cement rendering. Advantages of timber frame include: factory manufactured in controlled environment; rapid site erection; and high levels of thermal insulation. The main difference between platform and balloon timber-framed buildings is the size and arrangement of the panels. Platform panels are single-storey height; upper floors extend over the wall panels to create a platform on which the next set of wall panels are erected. Balloon panels are the full height of the building from the ground level up to the eaves, with intermediate floors being supported by ribbons housed in the studs.

Structural insulated panel construction (see Figure 4.4): This is a fairly recent development in panel construction, which is based on the platform frame method. Timber-based structural insulated panels, known as SIPs, are manufactured using a rigid polyurethane insulation panel sandwiched between two layers of OSB. The panels, which simply interlock together, are used to form the walls, floors and roof, giving a building shell with good strength properties and superior insulation values. One of the main advantages of this method over standard platform frames is the absence of timber studwork in the panel, which can act as a thermal bridge. Externally, SIP buildings can be completed using the same weathering as standard platform or balloon frames.

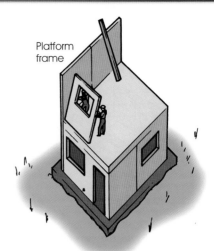

Platform frame

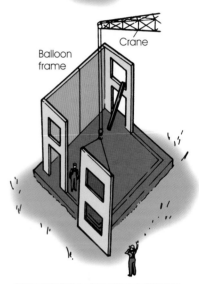

Balloon frame

Crane

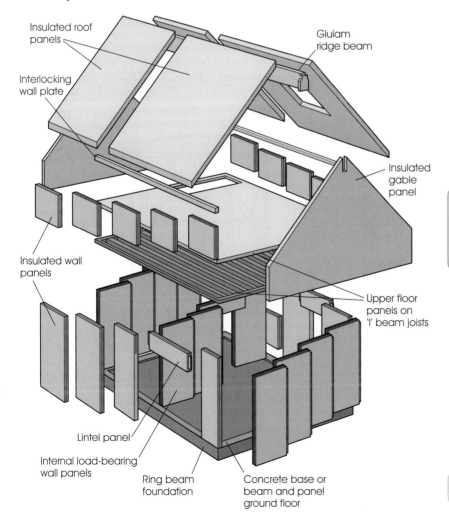

Insulated roof panels

Glulam ridge beam

Interlocking wall plate

Insulated gable panel

Insulated wall panels

Upper floor panels on 'I' beam joists

Lintel panel

Internal load-bearing wall panels

Ring beam foundation

Concrete base or beam and panel ground floor

▲ **Figure 4.3** Platform and balloon timber-framed buildings

Related information

Also see Book 1, Chapter 11, page 483, which compares modern methods of construction (MMC).

◀ **Figure 4.4** Structural insulated panel construction

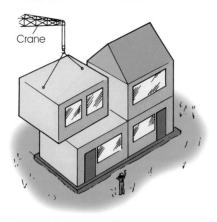

▲ **Figure 4.5** Volumetric construction

Volumetric construction (see Figure 4.5). This is a variation of the platform frame method and can be constructed using either studwork panels or SIPs. They utilise off-site factory construction to a much larger extent as complete box or room-size units, known as pods, are delivered to the site and stacked together to form the completed structure. These are normally fully decorated and have all services installed, so reducing the on-site construction and finishing time to an absolute minimum. They have the disadvantage of requiring the use of a large, powerful crane. Again, externally, these buildings can be completed using the same weathering as standard platform or balloon frames.

Modern timber-frame buildings

These can be divided into two main groups:

■ Temporary or relatively short lifespan buildings for use as garden sheds, stores, summerhouses, garden playhouses or home offices.
■ Permanent buildings used for dwellings, office and industrial accommodation.

▲ **Figure 4.5** Volumetric construction

Temporary or relatively short lifespan buildings

▼ **Figure 4.6** Temporary or relatively short lifespan buildings

The use of each building will determine its size and the standard of facilities required. Garden sheds and stores need to be well ventilated and waterproof, buildings for other uses may, in addition, require thermal insulation, lining and the provision of heating and lighting.

The carcass of a typical shed is shown in Figure 4.6. The framing sections for the walls, roof and floors are made up using a number of studwork panels that bolt together.

The external wall cladding may be either timber boarding or weather and boil-proof (WBP) plywood. Where timber boarding is to be used for the cladding, a moisture barrier can be fixed between the cladding and the studs to prevent the penetration of wind-driven rain. Where plywood is used for the

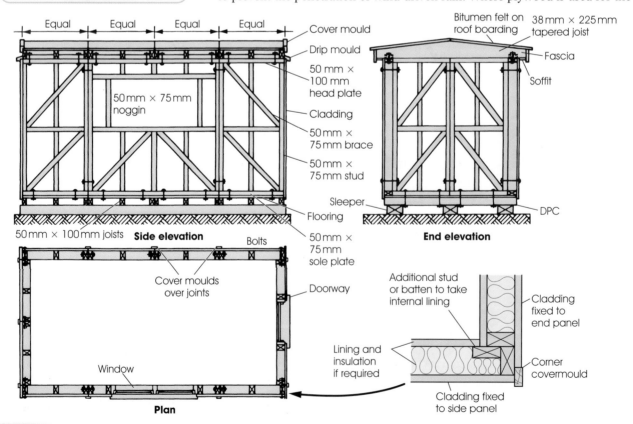

cladding, the diagonal brace can be omitted, as the sheets themselves will provide sufficient rigidity. A fairly flat roof is shown, which has been formed using double tapered joists. However, it could equally be formed using parallel joists with firring pieces or as a pitched roof using rafters. Plywood or tongued and grooved (T&G) boarding can be used for the floor and roof coverings, with the roof normally being weathered using bitumen felt. Standard details for a casement window, door frame and a ledged and braced door can be used to complete the building.

The carcass should be erected on a level base of large section preservative-treated sleepers or a concrete sub-base. A damp-proof course (DPC) can be inserted between the sub-base and the floor panels/joists to prevent dampness and decay.

Buildings for uses other than as a garden shed or store should be thermally insulated with rockwool or fibreglass and lined using WBP plywood or MRMDF (moisture-resistant medium-density fibreboard).

Permanent timber-framed buildings

In the UK, permanent buildings are normally prefabricated under factory conditions and transported to the site for erection on prepared foundations. In other countries, the 'stick built' method is more commonly utilised. This is where the timber is delivered to the site as separate loose timbers and framed up adjacent to where it will be erected subsequently.

In timber-framed buildings, all of the structural parts above the DPC, that is the walls, floor and roof, are constructed from timber and plywood. The external cladding merely provides a weatherproof finish and may include bricks, stone, tiles, cement rendering and various types of timber boarding.

Timber-framed construction methods have a number of advantages over other forms of construction. These include:

- Reduction in on-site erection time.
- Improved standards of thermal insulation.
- A wide variety of external finishes are available.
- Dry construction methods allow quicker occupation as no drying out time is required.
- No subsequent remedial work as a result of shrinkage is necessary.
- Adverse weather conditions do not greatly affect the construction progress.
- Prefabricated components are manufactured under ideal factory conditions.
- Timber is an environmentally friendly, naturally renewable building material that can be easily reused and or recycled at the end of its life, along with any waste wood from manufacturing operations.

Platform frame

This is by far the most used method of timber frame construction. The walls are built in single-storey sections. Once the first wall sections have been erected on the ground floor, the first-floor joists and floor coverings are fixed on top of them. This creates a platform on which the second-storey walls can be erected. Further floors are formed by continuing this process. Finally, the roof is erected on top of the upper wall panels. A variation of the platform frame method, which utilises factory construction to a much larger extent, is volumetric housing units. In this, complete box or room-size units are delivered to the site and stacked together to form the completed structure.

▲ **Figure 4.7** Platform frame house under erection

Balloon frame

This method uses timber frames that are the full height of the building. The studs are continuous from DPC level to the eaves. The wall panels are, therefore, erected to their full height in one operation, with intermediate floors being supported on ribbons notched into the continuous studs.

Methods of construction

The platform frame method is normally preferred because of its smaller, more easily managed components and the fact that a working platform is erected at each stage. The panels, which comprise head and sole plates, studs and sheathing, are normally related to the maximum size and weight that can be handled by two people. This gives panel sizes of single-storey height by approximately 3.6 m. Unsheathed panels for internal partitions may be considerably larger. In situations where a crane is available on site to erect the superstructure, the maximum panel size restriction is removed. This enables the use of balloon frames and larger platform frames. This method is often known as the large-panel erection system. It allows full house width and depth panels to be used. In addition, these panels are often more highly factory finished, including windows, glazing, insulation, internal lining and, where appropriate, external cladding. This has the advantage of reducing the on-site time and enables more rapid completion.

External walls

Typical examples of the external wall construction, using both platform and balloon frame methods, are shown in Figure 4.8. The majority of timber-framed buildings are constructed with solid concrete ground floors. Where a suspended timber ground floor is required, the detail shown in Figure 4.8(c) may be used.

Upper floors

The upper floors of timber-framed buildings are normally simply constructed on site using loose joists and boarded over with either T&G boarding or sheet material. However, where a contractor is using a crane for large panel erection, the floor panels may be prefabricated in large sections known as cassettes, using either standard joists and sheet material or specially designed stressed skin panels.

Partitions

The internal partitions of a timber-framed building are normally erected along with the main structure. They are constructed in exactly the same way as the external wall panels, including lintels over door openings and so on. The only difference is that they will have no structural sheathing prefixed. This tends to make them unstable when handled. Therefore, it is normal to fix a temporary diagonal brace across the face of the studs to stiffen and keep the panels square during erection.

Roofs

The roofs of timber-framed buildings are the same as those used for other methods of construction and so the roof details covered in the previous chapter apply.

Erection sequence

The sequence of operations for the erection of platform and balloon frames is similar. Once the foundations are complete, the major components may be delivered from the factory. These can then be erected in one or two days to provide a fully weatherproof structure.

Finishing work can be carried out inside at the same time as the external cladding is being undertaken.

Related information

Also see Chapter 1 in this book, which covers ground and upper floors, and Chapter 2 in this book, which covers flat and pitched roofs.

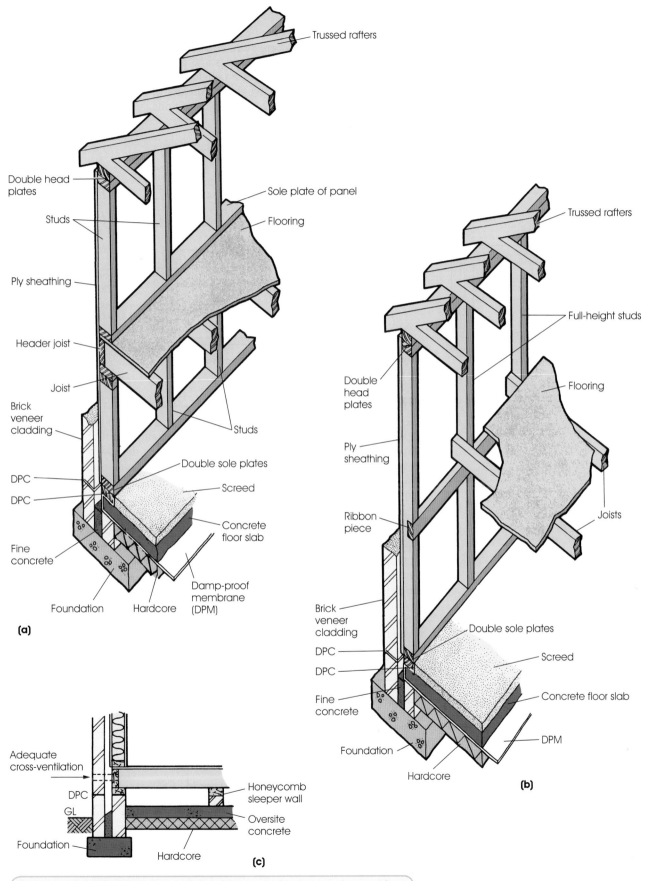

▲ Figure 4.8 (a) Platform frame, (b) balloon frame and (c) suspended ground floor

Figure 4.9 shows the major components and erection sequences for both methods of construction.

Care must be taken during erection, as until the shell is fully completed, including floor decking, roof coverings and glazing, the timber frame may be structurally unstable. This creates an obvious risk during the erection; to a great extent this risk may be avoided by the use of temporary diagonal braces and struts, particularly in windy conditions.

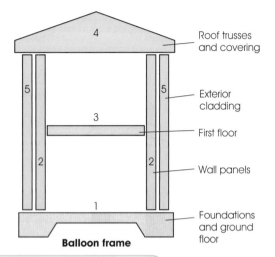

Balloon frame

Roof trusses and covering

Exterior cladding

First floor

Wall panels

Foundations and ground floor

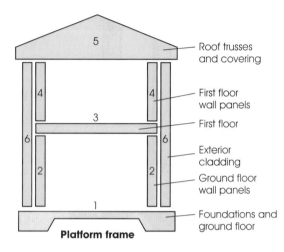

Platform frame

Roof trusses and covering

First floor wall panels

First floor

Exterior cladding

Ground floor wall panels

Foundations and ground floor

▲ **Figure 4.9** Components and erection sequence for timber-framed buildings

Fixings to foundations

The first item to be fixed is the lower sole. Its purpose is to form a level base on which subsequent components can be erected and also to transmit vertical and wind loads down to the foundations. As the lower sole provides an important link between the substructure and the superstructure, care taken in accurately positioning and fixing it will prevent any inaccuracies being repeated throughout the structure.

There are two alternative methods of fixing the sole plate, either by nailing with a cartridge-operated fixing tool or by using some form of rag bolt or expanding bolt, see Figure 4.10. It can be clearly seen that the use of either method will hole the DPC. With nails this problem is not considered significant as the DPC tends to self-seal around the tight perforation made by the nail and, as the sole plates are pressure impregnated with preservative, any minute amount of moisture will have no harmful effect. Where bolts are used, any gap between the DPC and bolt can be sealed with mastic or a fibre washer.

The fixing down of the sole plate, the nailing of the panels to it and the nailing of adjacent panels to each other are a critical part of the structural design. Each timber frame designer will, therefore, produce a comprehensive **nailing schedule**, see Figure 4.11, which must be fully complied with both during manufacture and on site.

▼ **Figure 4.10** Sole plate fixing

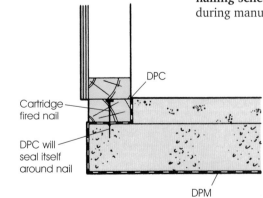

DPC

Cartridge fired nail

DPC will seal itself around nail

DPM

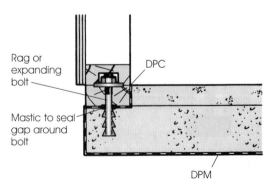

Rag or expanding bolt

DPC

Mastic to seal gap around bolt

DPM

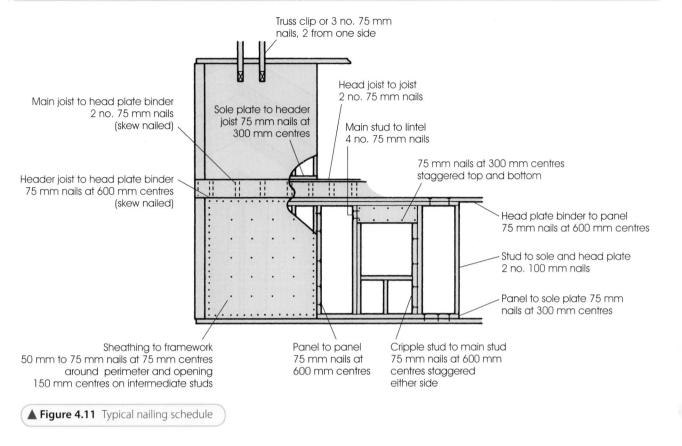

▲ Figure 4.11 Typical nailing schedule

External wall panels

These normally consist of prepared 50 mm × 100 mm strength-graded softwood studwork. The studs, which are spaced at 400 mm or 600 mm centres to suit the sheet sizes, are simply butt jointed and nailed to the head and sole plates. No noggins or diagonal braces are required as the plywood sheathing, which is nailed to the studs, provides the required rigidity. An additional head plate is used to bind the panels together, line them and provide additional strength, so joist and trussed rafter spacing are not dependent on the stud positions. Where wall panels intersect, additional studs should be used to provide a fixing for the internal lining (see Figure 4.12). It should be noted that the head plate joints are staggered at wall panel intersections to tie them together.

Lintels must be provided over openings where these occur in external walls. This is normally two pieces of 50 mm × 150 mm timber fixed together and stood on edge. This is supported on either side of the opening by double studs, one stud being cut shorter (cripple stud) to provide a bearing for the lintel (see Figure 4.13). Where wall panels contain a door opening it is advisable to leave the sole plate in position during the erection process to avoid panel distortion and damage.

On-site construction (stick built)

Although in general the majority of timber-framed buildings are produced in factory conditions often using computer-controlled cutting and assembly machines, timber frame is equally suited to the on-site stick-built method. This is especially true of the one-off design constructed by a small builder. All of the previous details will still apply to the on-site method and, if care is taken in the making of the panels, the same accuracy in sizes as achieved by the factory method is possible (e.g. +0 mm and –3 mm in both directions).

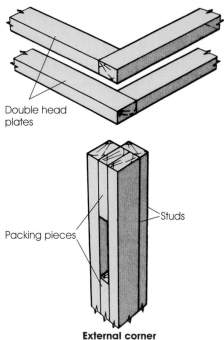

Double head plates

Packing pieces

Studs

External corner

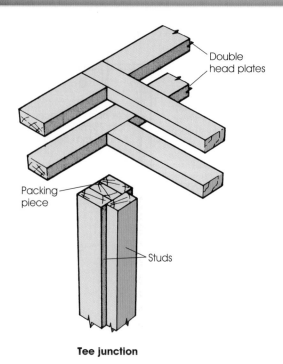

Double head plates

Packing piece

Studs

Tee junction

▲ **Figure 4.12** Corner details

A jig suitable for the on-site manufacture of timber frame panels consists of a plywood deck on timber bearers, supported by trestles, see Figure 4.14. Fixed stops and wedges are used to locate the member and cramp up the panel.

Preservative treatment

All external wall timbers, sole plates, battens, plywood, cavity barriers and timber cladding should be preservative treated to protect them against decay and insect attack, preferably using a vacuum or pressure treatment. All timbers cut after treatment, whether in the factory or onsite, must be given two flood brush applications of the preservative on the cut surfaces.

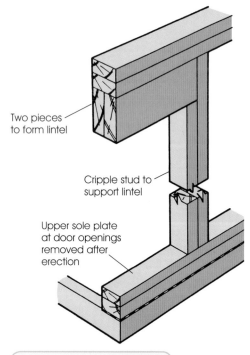

Two pieces to form lintel

Cripple stud to support lintel

Upper sole plate at door openings removed after erection

▲ **Figure 4.13** Opening details

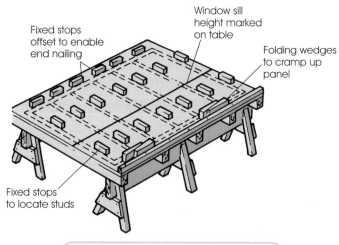

Window sill height marked on table

Fixed stops offset to enable end nailing

Folding wedges to cramp up panel

Fixed stops to locate studs

▲ **Figure 4.14** Jig for on-site panel making

Barriers

Two essential barriers must be incorporated in the construction. These are a vapour barrier and a moisture barrier.

Vapour barrier

It is most important that a vapour barrier is used on the warm side of the insulation. Its purpose is to minimise the amount of water vapour passing from the moist interior of the building into the structure and condensing (interstitial condensation). If this were allowed to take place, the wall materials would become saturated, making the insulation ineffective and creating the ideal conditions for the development of timber decay.

The vapour barrier, which is normally of 250–500 gauge polythene sheet, should be situated between the insulation and plasterboard. It should be fixed to the head and sole plates, around openings and overlaps with staples at approximately 250 mm. All joints should be overlapped by at least 100 mm and occur over a stud.

Care should be taken in fixing the polythene to ensure that it covers the entire wall surface, including the sole plate, head plate, studs, lintels and reveals. Any holes around services or tears in the barrier must be repaired with a suitable adhesive tape to seal any gaps before the plasterboard is fixed.

It is most important that the moisture content of the framework is checked prior to the fixing of the vapour barrier so as not to seal moisture within the wall structure. This should always be less than 20 per cent.

It is impossible for a vapour barrier to be 100 per cent effective because of the joints in it and fixings through it. Therefore, a breather-type moisture barrier is installed on the outside of the timber frame under the cladding so that any water vapour that has penetrated through the vapour barrier is allowed to escape or 'breathe' to the outside air.

Moisture barrier

A moisture barrier must be waterproof but at the same time must allow water vapour to pass through it. Fixed on top of the sheathing, it 'weathers' the timber frame shell immediately after erection. In addition, it also provides a second line of protection against the penetration of wind-driven moisture that might penetrate the cladding.

Moisture barriers also seal any gaps in the panel joints and so act as draughtproofers.

Moisture barriers are usually special lightweight building paper or felt, designed to be both waterproof and vapour permeable (able to breathe). They are fixed to the sheathing, starting at the bottom of the wall, in horizontal widths around the building. They should be fixed to the head and sole plates, studs, around openings and overlaps with staples, at approximately 250 mm centres. Horizontal laps should be at least 100 mm, with the upper layer overlapping the lower one. Vertical laps should be at least 150 mm and staggered in adjacent layers (see Figure 4.15).

Staples may be driven through narrow plastic strips. This avoids the barrier pulling over the staples, particularly when exposed to windy conditions. In addition, the strips also serve to indicate the stud positions for subsequent wall tie fixing.

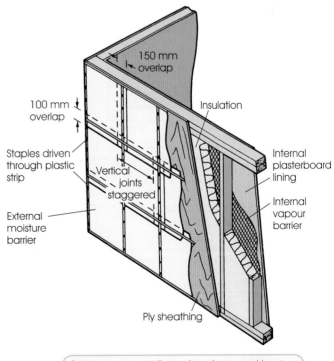

▲ **Figure 4.15** Wall panel insulation and barriers

Related information

Also see Book 1, Chapter 11, page 471, which gives further details about thermal insulation materials.

Insulation

High levels of thermal insulation are easily achieved in timber-framed buildings by installing fibreglass or mineral wool insulation in the spaces between the studs of the external walls and between the ceiling joists in the roof space. Suspended ground floors may be insulated between the joists with fibreglass or mineral wool insulation supported on galvanised wire or plastic mesh fixed to the underside of the joists. Solid ground floors can be insulated by placing a suitable material, such as rigid polyurethane or polystyrene sheets, between the floor slab and screed.

Fire stops and cavity barriers

Fire stops and cavity barriers must be fitted in the timber frame construction to control the spread of flame or smoke through cavities.

A fire stop is a non-combustible material used to divide the separating wall cavities of semi-detached or terraced buildings from the external wall or roof cavity (see Figure 4.16). In addition, fire stops are also required where services pass through the timber structure.

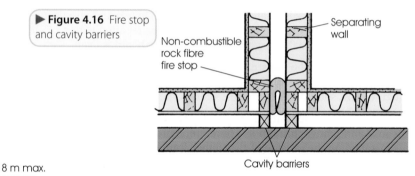

► **Figure 4.16** Fire stop and cavity barriers

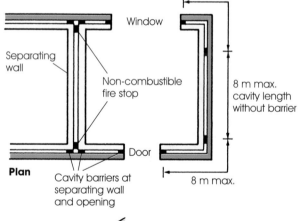

▼ **Figure 4.17** Cavity barriers

Timbers of at least 38 mm section are often used as cavity barriers. They must be fixed to seal the cavity between the timber frame and cladding, horizontally at each floor and roof level, around all door and window openings and vertically at intervals of not more than 8 m. In addition, vertical cavity barriers are required to separate the external wall cavities of semi-detached or terraced buildings (see Figure 4.17).

Cavity barriers should be protected from the external wall by a DPC, see Figure 4.18

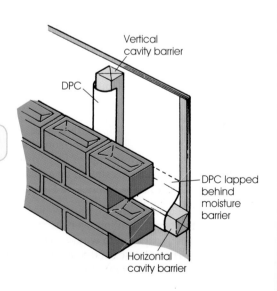

► **Figure 4.18** Protecting cavity barriers

Cladding

Various types and arrangements of material can be used for cladding the exterior of timber-framed buildings, see Figure 4.19. The cladding provides a visually acceptable and sufficiently durable weatherproof finish.

The most popular cladding is brick veneer, which gives the building a traditional appearance, although this is often used in conjunction with large areas of timber cladding, tile hanging or cement rendering.

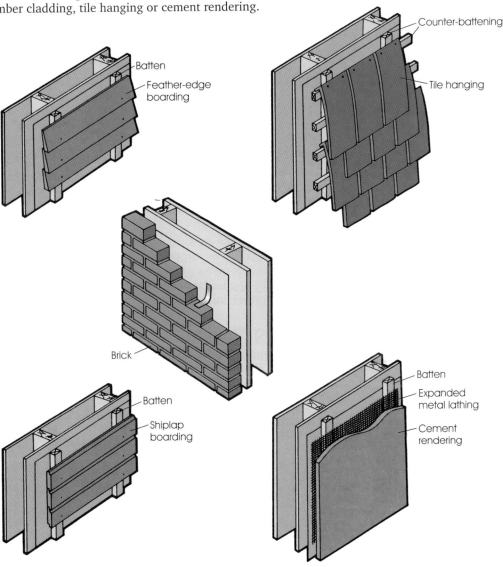

▲ **Figure 4.19** Main types of cladding

Brick cladding

This must be effectively connected to the timber frame for stability. Since there will be differing amounts of movement between the timber frame and the brick cladding, flexible wall ties must be used.

In addition, gaps should be left between the brickwork and the timber to allow for movement at the eaves and around window and door frames that are fixed to the timber frame. These gaps around the window and door frames should be sealed with compressible mastic to prevent rain penetration.

The cavity between the brick cladding and the timber frame should be ventilated at the eaves and below the DPC level, by raking out the vertical brick joints (perpends) at 1500 mm intervals. The lower open joints also serve as weep-holes to drain the cavity.

Other lightweight cladding

Timber, tile hanging and cement rendering are preferably fixed to vertical battens nailed to the studs or horizontal and vertical counter-battening. This allows for free movement of air so that the cladding is ventilated on both sides. When timber boards are used, care must be taken when nailing to ensure that the boarding is nailed near one edge only and that the nails do not penetrate two boards. This is to ensure that each board is able to move more freely, without danger of splitting, when subsequent moisture movement takes place. Horizontal timber boarding is nailed near the bottom edge of the board. The top edge is held by the overlapping edge of the board above.

The length of nails used to fix the cladding must be 2½ times the cladding thickness and should be treated to resist corrosion or be made of a non-corrosive metal.

Naturally durable timber cladding, such as Western red cedar, may be used without preservative treatment or any subsequent finish. Other timbers that are not naturally durable must be preservative treated. They may also have a further finish applied for decorative reasons or to prevent natural fading.

Paints and varnishes are the commonest finish. However, they are not particularly suitable as cracks in the film soon appear along the large number of joints in the cladding. This allows moisture to penetrate, causing the paint or varnish to peel.

Exterior wood stains are available in a wide range of colours. These are more suitable than paints or varnishes as they give the cladding a water-repellent surface without forming a film and, therefore, cannot crack and peel off. Stains may be applied by brush, or by pressure impregnation at the same time as the preservative treatment.

Cladding details

Careful detailing to allow for differing movement and to prevent rain penetration is required where claddings finish around door and window openings or where different claddings join. Figure 4.20 illustrates a number of typical cladding details.

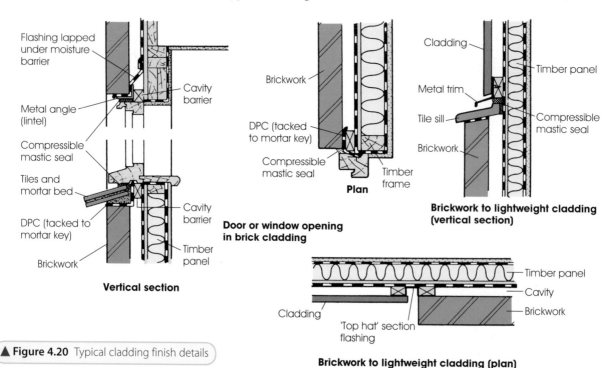

▲ **Figure 4.20** Typical cladding finish details

Curtain walling

Timber-framed panels can be used as non-load-bearing in-fill wall units (curtain walling) for skeleton frame structures of either steel or reinforced concrete and for brick or concrete cross-wall construction.

These panels may be constructed as large window frames using a combination of glazed openings and insulated in-fill sections as shown in Figure 4.21.

▼ **Figure 4.21** Curtain walling details

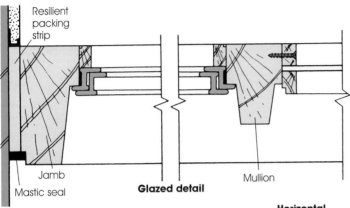
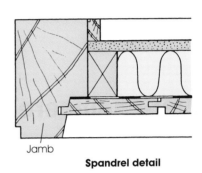

Resilient packing strip

Jamb

Mastic seal

Glazed detail

Mullion

Horizontal

Jamb

Spandrel detail

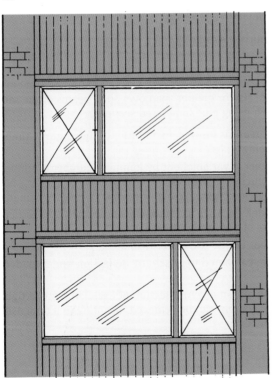

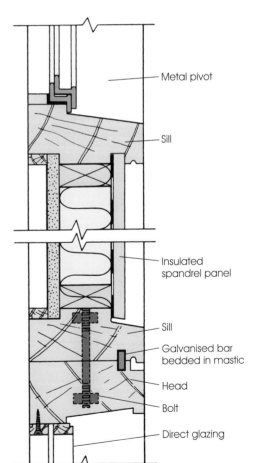

Metal pivot

Sill

Insulated spandrel panel

Sill

Galvanised bar bedded in mastic

Head

Bolt

Direct glazing

Vertical

The curtain walling frames are broken down into units bolted together at the head and sill. These horizontal joints incorporate a galvanised water bar bedded in mastic to produce an effective seal. A similar detail can be used to join frames side by side in order to form a long run of curtain walling.

Standard in-fill panels extend from head height of one frame to the normal windowsill height of the frame above. An alternative to the insulated timber clad panel shown, is the insulated, enamelled steel or glass-reinforced plastic-faced (GRP-faced) panels used in some proprietary curtain walling systems.

In this example, metal pivot-hung windows have been used, these are bedded in mastic and screwed to the main frame. The direct fixed glazing shows the use of both internal and external beads. The external ones should be permanently fixed with a suitable synthetic resin adhesive and pins. The internal beads are recess-cupped and screwed in place, so permitting the glazing to be carried out from the inside of the building.

The panels must be securely fixed to the structural frame or cross-walls, normally using metal brackets, straps or expanding bolts.

Careful detailing is required in order to allow for the differential movement between the curtain walling and the main structure while at the same time maintaining a weatherproof joint. This can be achieved by filling the joint with a resilient strip and sealing with compressible mastic.

Timber-engineered components

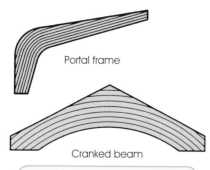

▲ **Figure 4.22** Examples of glulam

Glue-laminated timber (glulam)

Structural timber members, which are of a large cross-section, long in length, or shaped, can be fabricated by the process of glue lamination (glulam). Examples of this are shown in Figure 4.22.

A glulam member comprises boards (termed laminates) of small cross-section, glued and layered up either horizontally or vertically. The boards are layered with a parallel grain direction and cramped up in special jigs, which are set up to provide the straight or curved profile of the finished member. A reduction in sectional size or greater span over solid beams of the same section is possible when using glue-laminated sections.

There are a number of reasons for this, in addition to the grade of timber used, the main ones being:

- The strength of timber is variable, but by joining a number of timbers together to act as one beam, the average strength of each layer (laminate) will be greater than the strength of the weakest laminate.
- Strength-reducing defects, such as large knots or shakes, are restricted in size to the thickness of one laminate.
- Higher-quality timber can be used in critical areas.
- Members can be pre-cambered to offset any deflection under loaded conditions.
- Because of the relatively small cross-section of each laminate, a tightly controlled moisture content can be achieved, resulting in a dimensionally stable finished member.

Preparing laminates

In straight glulam members the thickness of each laminate can be up to 50 mm. When forming curved members, the laminate must be flexible enough to enable it to bend around the curve without breaking, so the maximum thickness is dependent on the radius of the curve. As a guide, the radius can be divided by 100 to give the maximum laminate thickness (MLT) for bending softwood and by 150 for hardwood.

> **Example**
> The MLT for a softwood arched beam having a radius of 2 m
> MTL = radius ÷ 100
>   = 2000 ÷ 100
>   = 20 mm

Laminates can be jointed to produce full length and full width laminations, see Figure 4.23. Lengthening joints are either scarf or finger joints.

For maximum strength, scarf joints should have slope of 1:6 for laminates in the compression area and 1:12 for laminates in the tension area. Edge jointing should only be done on the inner laminates and must be separated in adjacent laminates by about 25 mm or the laminate thickness if this is greater.

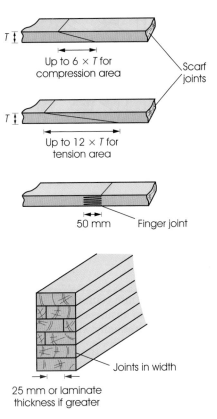

◀ **Figure 4.23** Jointing laminates

Laminates should be planed and jointed just before they are to be assembled, to achieve maximum glue penetration.

Before gluing and cramping up a member, the laminates should be dry assembled to correctly position the laminates and stagger joints. Wet assembly and cramping can then follow. The type of jig used will depend on the size of member and the type of firm, which can range from the occasional assembly of small glulam members in a joiners shop to large specialist timber plants using complex machinery.

A small cramping jig suitable for a limited number of members is shown in Figure 4.24. Where larger numbers are to be produced, cramping frames bolted to the floor can be used, see Figure 4.25.

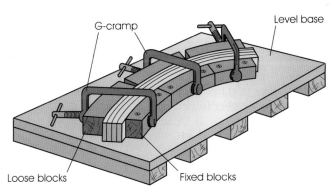

◀ **Figure 4.24** Small cramping jig

▶ **Figure 4.25** Use of cramping frames

Related information

Also see Book 1, Chapter 1, page 77, which covers the properties and uses of adhesives.

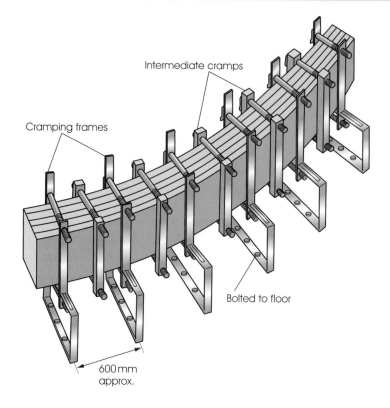

Intermediate cramps

Cramping frames

Bolted to floor

600 mm approx.

Cramps can be spaced at up to 400 mm centres for straight work, but this must be reduced for curved work to ensure a thin uniform glue line. Centre cramps should be tightened first, gradually working out towards the ends.

The type of adhesive used will depend on the intended use; casein or urea formaldehyde is suitable for use in low-heat, low-humidity internal conditions; phenol or resorcinol formaldehyde can be used in conditions where humidity or temperature levels are not controlled.

After the adhesive has cured, the member can be removed from the cramps and, after a period of conditioning, it can be finished by planing, sanding and jointing if required.

Site erection

Glulam members are often delivered to site shrink-wrapped in a plastic or waterproof covering. This should be left in place after erection to protect the member from the effects of weather and subsequent wet and finishing trades. Where members are to be jointed or fitted into metal shoes and similar, the protective covering can be temporarily pulled back.

The site work for glulam members normally involves jointing, erection, plumbing and fixing. In order to prevent accidental collapse, temporary ties and braces must be used until the structure is stable. Care must be taken to avoid distortions when lifting the members. For large structures, manufacturers will provide or specify lifting points in certain positions to ensure that any stresses in erection do not exceed the permissible stresses in any member.

A range of typical joint and fixing details are shown in Figure 4.26. The main method of securing a member is with bolts and metal plates. The feet of arched or portal frames and posts are fixed to the base by the use of a metal shoe. This method has the added advantage of preventing moisture rising up the timber.

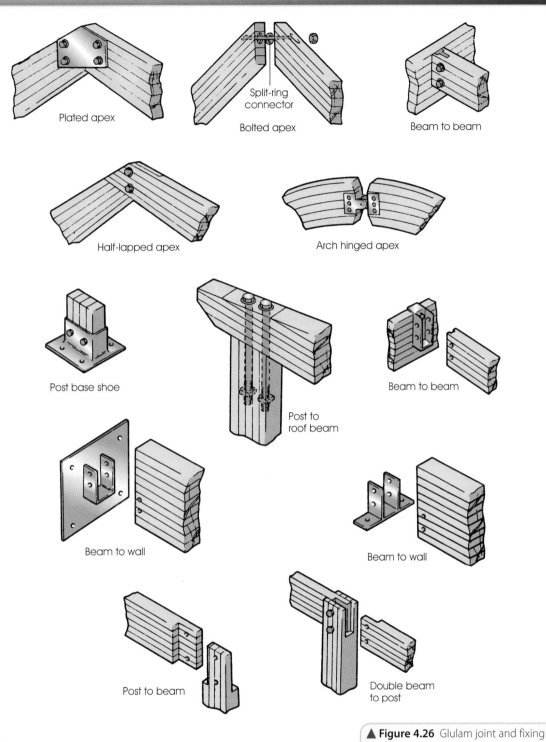

Plated apex

Split-ring connector

Bolted apex

Beam to beam

Half-lapped apex

Arch hinged apex

Post base shoe

Post to roof beam

Beam to beam

Beam to wall

Beam to wall

Post to beam

Double beam to post

▲ **Figure 4.26** Glulam joint and fixing details

Built-up beams

Built-up nailed and glued beams have a distinct advantage over solid timber sections or glulam. Figure 4.27 shows typical details of four different methods of forming built-up beams. Each consists of top and bottom flanges, spaced by either ties and struts; or of plywood webs and stiffeners; or vee-shaped galvanised steel webs. It is possible to use built-up beams for far greater spans than solid timber, and they have a much better weight to strength and stiffness ratio than either solid timber or glulam sections.

They may be used as purlins or support beams in roof construction, as binders or main support beams in double or framed floors, and as joists in suspended floors and flat roofs.

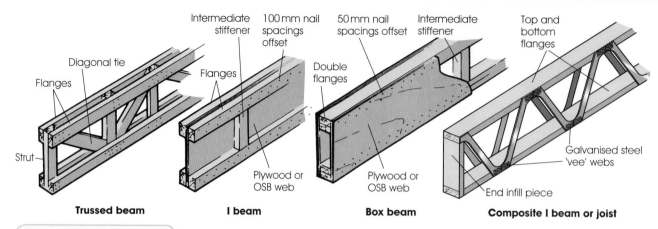

Figure 4.27 Built-up beams

Stressed-skin panels

A stressed-skin plywood panel consists of timber-framing members called webs, onto which a skin of plywood is fixed. The panels can be single skin for most situations or double skin for use in high-load conditions. A double-skin panel is shown in Figure 4.28. The panels are used as prefabricated floor or roof panels (cassettes), especially where long uninterrupted spans are required.

Stressed-skin panels differ from normal joists and boarding in that the plywood, in addition to supporting the load between the framing, also contributes to the strength and stiffness of the framing itself. This is because the plywood is glued and nailed to the framing. So the panels function as an efficient structural unit transmitting the stresses between the plywood and framing.

It is possible to manufacture portal frames using the stressed skin method of construction, as shown in Figure 4.29.

▼ **Figure 4.28** Stressed-skin panel

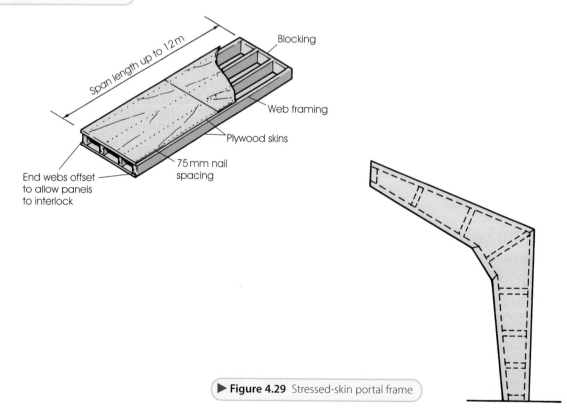

▶ **Figure 4.29** Stressed-skin portal frame

Shell roof construction

A timber shell roof consists of a thin curved lightweight membrane acting as a stressed skin that obtains its strength as a result of its curvature. Laminated edge beams are incorporated around the perimeter to carry the membrane's forces and strengthen the edge of the shell.

Apart from the aesthetic appeal of this exciting geometric form of roof, shells also have the advantage of being able to form large clear spans without the need for intermediate support. In addition, their lightweight construction enables the enclosing walls to be of a mainly non-load-bearing nature, which in turn allows smaller more simple foundations.

Timber shell roofs have been used in various forms to cover many different types of building, including schools, churches, factories, libraries, sports complexes and leisure centres.

The form of shell roof most often used is the hyperbolic paraboloid. It is created by raising the two diagonally opposite corners of a square to a higher level than the other two corners (see Figure 4.30). The roof derives its name from the curves produced between its diagonally opposite corners. Between the low corners a convex curve is produced and between the high corners a concave curve, whereas all cross sections parallel to the edges of the shell will be straight lines.

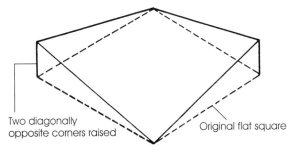

Two diagonally opposite corners raised

Original flat square

▶ **Figure 4.30** Hyperbolic paraboloid roof formation

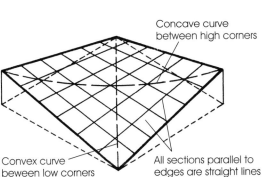

Concave curve between high corners

Convex curve beween low corners

All sections parallel to edges are straight lines

In general, shells with a greater curvature (i.e. those with a greater rise between high and low corners) are stiffer than shells of the same plan size but with a lesser curvature (i.e. a smaller rise between high and low corners).

Three layers of T&G boards glued with a synthetic resin adhesive and nailed together with square twisted-shank or annular nails form the hyperbolic paraboloid timber shell roof, see Figure 4.31. Scaffolding is erected before construction to allow access to the underside of the roof, and, more importantly, to provide support for the lower layer of boarding. The middle layer is nailed to the lower layer, while the nails in the upper layer should penetrate right through the middle layer and into the lower layer. It can be seen that these boards are laid diagonally across the roof with the grain direction at right angles to its adjacent layers.

▶ **Figure 4.31** Hyperbolic paraboloid shell roof

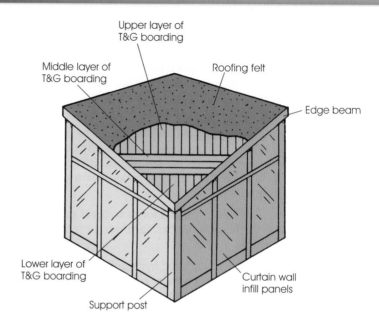

Upper layer of T&G boarding

Middle layer of T&G boarding

Roofing felt

Edge beam

Lower layer of T&G boarding

Curtain wall infill panels

Support post

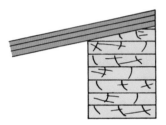

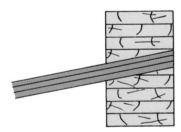

▲ **Figure 4.32** Shell roof edge beams

Any end jointing of the boards in the upper layers may be simple butts, although for strength it is normally recommended that the bottom layer is finger jointed. It is also advisable to use a better-quality timber on the bottom layer as this will be varnished to provide the finished ceiling.

Two layers of boarding may be used for the membrane of limited span shell roofs.

The edge beams, normally glulam, can be either made in one piece and positioned below the boarding or made in two pieces and positioned half above and half below the boarding, see Figure 4.32.

In order to stop the shell flattening out, the low corners are tied together with the aid of a mild steel tie bar that has been threaded at each end. Figure 4.33 shows this tie bar and also the method of fixing the edge beams in position.

Alternatively, some form of buttress can be used to withstand the outward diagonal thrust of the shell at its low corners.

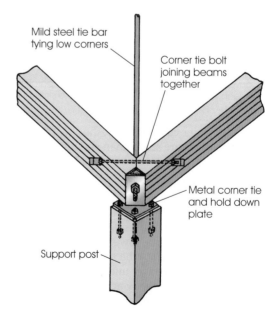

Mild steel tie bar tying low corners

Corner tie bolt joining beams together

Metal corner tie and hold down plate

Support post

▶ **Figure 4.33** Shell roof tie bar

Wherever possible, the edge beams should be given support by posts, walls or curtain walling mullions. In any case, the low corners must be well supported as these take the main load of the roof. Also, the high corners must be tied down to prevent lifting by wind forces.

Erection sequence

The sequence of operations for the erection of a hyperbolic paraboloid shell roof is as follows:

1. Erect corner support posts. Assemble scaffolding to support the first layer of boards.
2. Position lower half of beams and bolt to corner support posts.
3. Glue and nail each layer of boarding in position at right angles to each other, taking care to ensure that a close contact exists between all meeting surfaces.
4. Position upper half of beams. These are glued and fixed using either long square twisted-shank nails or coach screws.
5. Fix angle fillets to facilitate the dressing of the roofing felt over the edge beam. See Figure 4.34 for a typical edge detail.
6. Fix fascia.
7. Dismantle scaffolding.
8. Felt over the roof to make it watertight.

Alternative forms of shell

Multiple hyperbolic paraboloid shell roofs can be formed by joining a number of shells side by side (see Figure 4.35). Many other forms of shell roof are possible, some of which are also shown in the figure. Although vastly different in shape, these shells have a similar formation in that they all consist of a curved layered membrane.

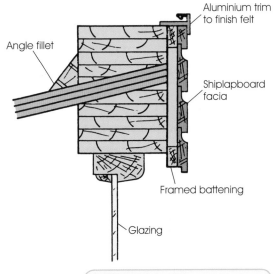

▲ **Figure 4.34** Shell roof edge detail

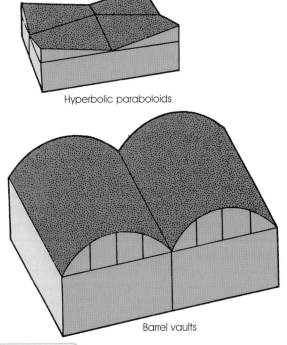

Hyperbolic paraboloids

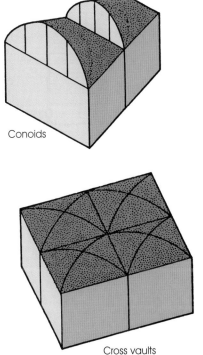

Conoids

Barrel vaults

Cross vaults

▲ **Figure 4.35** Shell roof forms

5

Chapter Five

Stairs

This chapter covers the work of the site carpenter and bench joiner. It is concerned with principles of stair construction and includes the following:

- An introduction to types of stairs.
- Stairway construction and assembly.
- Installation of stairs.
- Construction of landings and railings.
- Protection of completed work.

Stairway identification

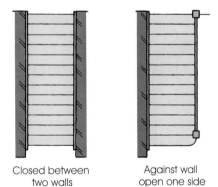

Closed between two walls

Against wall open one side

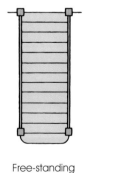

Free-standing open both sides

▲ **Figure 5.1** Straight-flight stairs

Timber stairs are normally made and partially assembled offsite by the joiner. The final on-site assembly and fixing of timber stairs is carried out by the carpenter, after the building has been made watertight and is classified as a **first fixing** operation. A stairway can be defined as a series of steps (combination of tread and riser) giving floor-to-floor access. Each continuous set of steps is called a flight. Landings are introduced between floor levels either to break up a long flight, so giving a rest point, or to change the direction of the stair where there is a restricted access.

Stairs can be classified according to their plan shape and can be further classified by their method of construction.

Straight-flight stairs

These run in one direction for the entire length. Figure 5.1 shows three different variations.

- The flight, which is closed between two walls, also known as a cottage stair, is the simplest and most economical to make. The handrail is usually a simple section fixed either directly on to the wall or on brackets.
- The flight fixed against one wall is said to be open one side. This open or outer string is normally terminated and supported at either end by a newel post. A balustrade must be fixed to this side to provide protection. The infilling of this can be either open or closed and is usually capped by a handrail. Where the width of the flight exceeds 1 m, a wall handrail will also be required.
- Where the flight is free standing, neither side being against a wall, it is said to be open both sides. The open sides are treated in the same way as for the flight open one side.

Quarter-turn stairs

As its name suggests, this type of stair changes direction 90° to the left or right by means of either a quarter-space landing or tapered steps (see Figure 5.2).

Tapered steps, also termed winders, although economising on space because of the introduction of the extra steps in place of the landing, are potentially dangerous owing to the narrowness of the treads on the inside of the turn. They should, therefore, be avoided where at all possible, especially in situations where they are likely to be used by young children or elderly persons, or at least be located at the bottom rather than at the top of the flight.

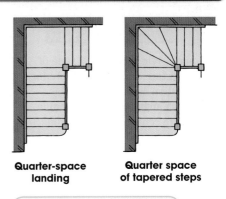

Quarter-space landing **Quarter space of tapered steps**

▲ **Figure 5.2** Quarter-turn stairs

Half-turn stairs

This stair reverses its direction through 180° normally by a half-space landing. There are two main types, see Figure 5.3.

- The dog-leg stair, where the outer strings of the upper and lower flights are joined into a common newel immediately above each other. This stair takes its name from the appearance of its sectional elevation.

- The open-newel stair. This is also known as an open-well stair. Two newels are used at landing level. This separates the string of the upper and lower flights and so creating a central space or well.

Half-space landings

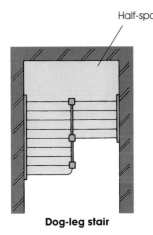

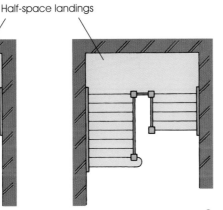

Quarter-space landings

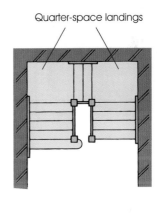

Dog-leg stair   **Open-newel stairs**

▲ **Figure 5.3** Half-turn stairs

A short flight is often introduced between the two newels at landing level, thereby creating two quarter-space landings and at the same time economising on space.

In addition, tapered steps could also be used instead of landings to negotiate turns in both dog-leg and open-newel stairs.

Geometrical stairs

The previous more robust types of stair utilise newels to change direction and also to terminate and support the outer strings, giving them an image of strength and rigidity. Geometrical stairs present a graceful, more aesthetic appearance, see Figure 5.4.

- Wreathed string stairs have an outer string and a handrail that are continuous from one flight to another throughout the entire stairway. As both the string and handrail rise to suit the stairs throughout the curve they are said to be 'wreathed'. Although the use of newels is not essential, they are sometimes used at the top and bottom of a stair to support the scrolled handrail above. This type of stair is the most expensive and is so rarely encountered, except in 'one-off' very high-quality construction and refurbishment work.

Wreathed string stair   **Helical stair (spiral)**

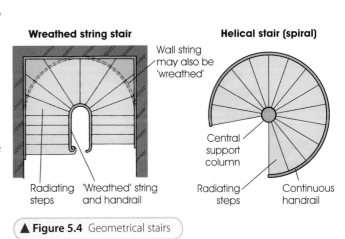

▲ **Figure 5.4** Geometrical stairs

■ Helical stairs, often termed as spiral stairs, are normally 'string less' and use a central column to joint and support the radiating steps.

Stair construction

In addition to their shape on plan, stairs may also be further classified by their method of construction, see Figure 5.5.

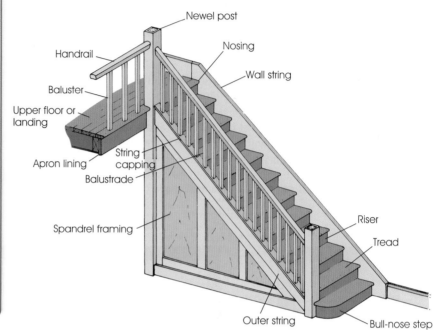

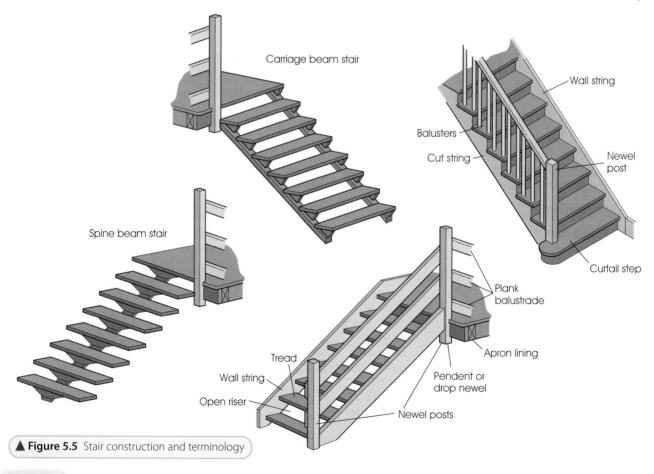

▲ **Figure 5.5** Stair construction and terminology

- **Close string:** having parallel strings with the treads and risers being housed into their faces and secured by gluing and wedging.
- **Cut string:** having one or more strings that have been cut to conform to the tread and riser profile. Treads sit on the strings' horizontal cut portion and the risers are mitred to the vertical portion.
- **Open riser:** stairs with no risers, also termed open-plan stairs. Treads can be housed or mortised into strings, or supported on carriages or spine beams.

Stair terminology

- **Apron lining.** The boards used to finish the edge of a trimmed opening in the floor.
- **Balustrade.** The handrail and the infilling between it and the string, landing or floor. This can be called either an open or closed balustrade, depending on the infilling.
- **Baluster.** The short vertical infilling members of an open balustrade. Also termed spindles, where they have been turned to a circular shape on a lathe.
- **Bull-nose step.** The quarter-rounded end step at the bottom of a flight of stairs.
- **Carriage.** This is a raking timber fixed under wide stairs to support the centre of the treads and risers. Brackets are fixed to the side of the carriage to provide further support across the width of the treads. Also, the raking members used to support the treads of stringless open plan stairs.
- **Commode step.** A step with a curved tread and riser normally occurring at the bottom of a flight.
- **Curtail step.** The half-round or scroll-end step at the bottom of a flight.
- **Newel.** The large sectioned vertical member at each end of the string. Where an upper newel does not continue down to the floor level below, it is known as a *pendant* or drop newel.
- **Nosing.** The front edge of a tread or the finish to the floorboards around a stairwell opening.
- **Riser.** The vertical member of a step.
- **Spandrel.** The triangular area formed under the stairs. This can be left open or closed in with spandrel framing to form a cupboard.
- **String.** The board into which the treads and risers are housed or cut. They are also named according to their type, for example, wall string, outer string, close string, cut string and wreathed string.
- **Tread.** The horizontal member of a step. It can be called a parallel tread, an alternating tread or a tapered tread, depending on its shape.

Stair regulations

Both the design and construction of stairs are very closely controlled by the Building Regulations. These lay down different requirements for stairs, depending on the use of the building. These requirements are summarised and illustrated in Table 5.1.

Table 5.1 Stair requirements (Building Regulations)

Requirement	Type of building where stairs are located		
	Private: Stairs for domestic use in a dwelling occupied by a single household.	**Institutional and assembly:** Stairs serving places where a substantial number of people gather.	**Other:** Stairs serving all other buildings.
Pitch	Maximum of 42°.	Governed by tread and riser requirements.	
Number of risers	Good practice not to exceed 16 without an intermediate landing. Consecutive flights of more than 36 risers must change direction between flights by at least 30°.	Maximum of 16 for assembly buildings. Consecutive flights of more than 36 risers must change direction between flights by at least 30°.	Maximum of 16 for shops. Consecutive flights of more than 36 risers must change direction between flights by at least 30°.
Rise of step	Maximum of 220 mm.	Maximum of 180 mm.	Maximum of 190 mm.
Going of step	Minimum of 220 mm (Note: maximum rise cannot be used with the minimum going as it will be over the pitch requirement).	Minimum of 280 mm. May be reduced to 250 mm if the floor area served by the stair is less than 100 m².	Minimum of 250 mm.
Combined rise and going	Twice the rise plus the going 2R + G to fall between 550 mm and 700 mm.		
Width of stair	No minimum requirement. Except when used as a means of escape, then the minimum width is determined by the number of people assessed as using the stair in an emergency: up to 50 people minimum width of 800 mm; 51 to 150 people minimum width of 1000 mm; 151 to 220 people minimum width of 1100 mm. Minimum widths for over 220 people depend on whether there is to be a phased or simultaneous evacuation (see Building Regulations for calculation formula). In all cases, stairs for the disabled where there is not a lift must be a minimum of 1000 mm between handrails.		
Headroom stair	Minimum of 2 m. May be reduced for loft conversions in private dwellings.		

Pitch diagram: Pitch

Number of risers diagram: Landing — Landing — At least 30 — Up — Max 16 risers per flight — Max 36 risers before change of direction

Rise of step diagram: Rise — Top of tread to top of tread

Going of step diagram: Going — Nosing to nosing — Going — Face of riser to face of riser

Combined rise and going diagram: G — R — R = 200 mm — G = 275 mm — 2R+G = 675 mm OK for private use

Width of stair diagram: Wall to handrail — Handrail to handrail — Wall to handrail on balustrade — Minor projections ignored

Headroom stair diagram: Measured vertically from nosing line — Stair headroom — Nosing line

continued

Headroom landing

Minimum of 2 m. May be reduced for loft conversions in private dwellings.

Landings

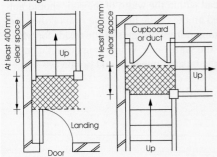

Must be provided at the top and bottom of every flight. The length and width of landings must be at least the width of the stair; part of the floor may count as a landing. Landing must be free from permanent obstructions. A door may swing over the landing at the bottom of a flight, but must leave a clear space of 400 mm across the full width of the flight. Doors to cupboards and ducts may open in a similar way over landings at the top of a flight.

Guarding

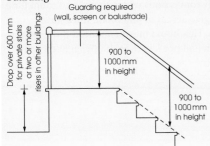

Guarding in the form of a wall screen or balustrade is required to protect the sides of flights and landings in all buildings. Guarding need not be provided in private dwellings where the drop is 600 mm or less, or in other buildings where there are fewer than two risers. In buildings likely to be used by children under 5 years old, this guarding should not be easily climbed or permit the passage of a 100 mm sphere. Heights for all buildings are the same as for handrails, between 900 mm and 1000 mm measured vertically from the pitch line or floor to the top.

Handrail height and provision

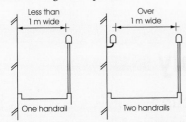

Handrail required on one side of stair if under 1 m wide, and on both sides if over 1 m. Height for all handrails should be between 900 mm and 1000 mm measured vertically above the pitch line or landing to the top of the rail. Stairways in public buildings over 1800 mm wide should be subdivided by a handrail so that sub divisions do not exceed 1800 mm. Handrails are not required over the bottom two steps in private stairs, except when intended for disabled use.

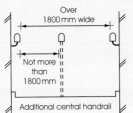
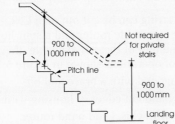

Balustrading

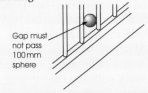

Balustrading can be used to form the required guarding to the sides of flights and landings. Again in buildings likely to be used by children under 5 years old, the balustrade should not be easily climbed or permit the passage of a 100 mm sphere.

continued

Open riser stairs

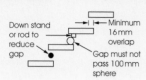

Open risers permitted, but for safety the treads should overlap by at least 16 mm. In all stairways that are likely to be used by children under 5 years old the gap between treads must not permit the passage of a 100 mm sphere.

Tapered tread stairs

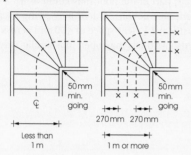

The minimum going at any part of the tread within the width of the stair should not be less than 50 mm. The going is measured on the centre line of the stairs if they are not more than 1 m wide or at points 270 mm in from the ends of the treads for wider stairs. All consecutive treads should have the same taper. Where stairs contain both tapered and parallel treads, the going of the tapered treads should not be less than the parallel ones.

Stairs to loft conversions

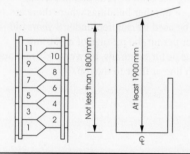

Alternating treads may be permitted for loft conversions where there is no room for a proper staircase. They can only be used to give access to one room, must have handrails on both sides and non-slip treads. Headroom may be reduced if the height at the centre of the stair is at least 1900 mm and not less than 1800 mm at the side.

Stairway construction and assembly

Close string stair construction

The construction of any staircase with close strings follows the same basic procedure, with slight variations depending on the particular type of stair.

The housings in the string can be cut out on a CNC machine, spindle moulder or by using a portable router and a stair housing template (see Figure 5.6). Where neither of these is available, they can be cut by hand using the sequence of operations shown in Figure 5.7.

1. Bore out at nosing end with a brace and bit.
2. Clean out nosing and cut edges of housing with a tenon saw.
3. Clean out waste with chisel and hand router.

The riser of bull-nosed and curtail steps can be formed using built up layers of thin plywood or from solid timber. Figure 5.8 shows a method of forming a bull-nose step using solid timber. The curved section of the riser is reduced to 2 mm thickness and bent around a laminated block. The wedges tighten the riser around the block and hold it there until the glue has set. The reduced section of the riser should be steamed before bending. It can then be bent around the block fairly easily without the risk of breaking. Splayed end steps are sometimes used in cheaper quality work with the riser mitred and tongued at the joints (see Figure 5.9).

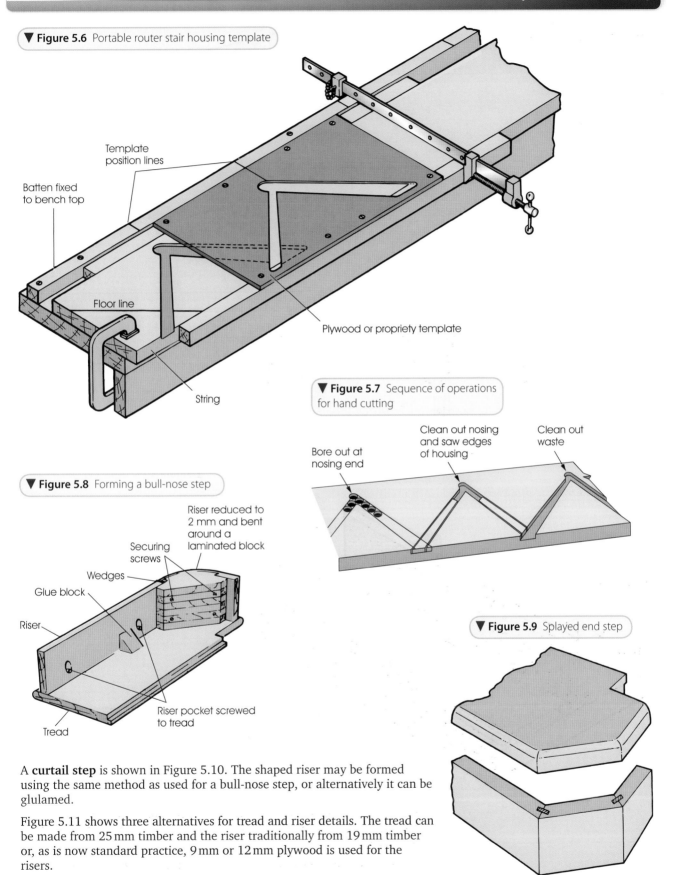

▼ **Figure 5.6** Portable router stair housing template

Template position lines

Batten fixed to bench top

Floor line

Plywood or propriety template

String

▼ **Figure 5.7** Sequence of operations for hand cutting

Bore out at nosing end

Clean out nosing and saw edges of housing

Clean out waste

▼ **Figure 5.8** Forming a bull-nose step

Riser reduced to 2 mm and bent around a laminated block

Securing screws

Wedges

Glue block

Riser

Tread

Riser pocket screwed to tread

▼ **Figure 5.9** Splayed end step

A **curtail step** is shown in Figure 5.10. The shaped riser may be formed using the same method as used for a bull-nose step, or alternatively it can be glulamed.

Figure 5.11 shows three alternatives for tread and riser details. The tread can be made from 25 mm timber and the riser traditionally from 19 mm timber or, as is now standard practice, 9 mm or 12 mm plywood is used for the risers.

Each step (tread and riser) is made up in a jig before being fixed to the strings, see Figure 5.12. Glue blocks strengthen the joint between the tread and the riser. The absence or loosening of these often results in squeaky stairs.

▶ **Figure 5.10** Round end step (curtail)

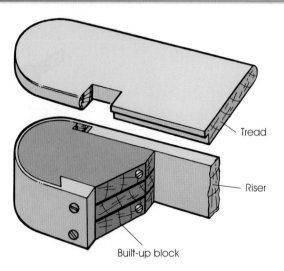

Tread

Riser

Built-up block

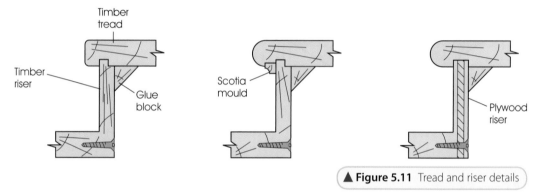

Timber tread

Timber riser

Glue block

Scotia mould

Plywood riser

▲ **Figure 5.11** Tread and riser details

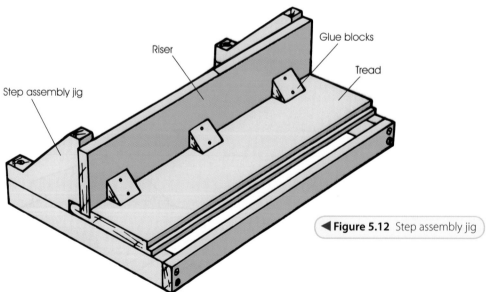

Riser

Glue blocks

Tread

Step assembly jig

◀ **Figure 5.12** Step assembly jig

A part view of the steps fixed into a string is shown in Figure 5.13. The treads and risers are glued and securely wedged into their positions in the string housing.

The outer string and handrail are normally mortised into the newels at either end, see Figure 5.14.

The newels will also require housing out to receive the treads and risers (see Figure 5.15).

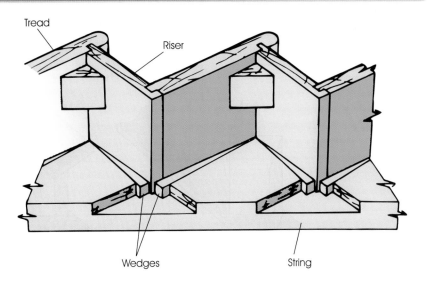

▶ **Figure 5.13** Fixing of steps into string

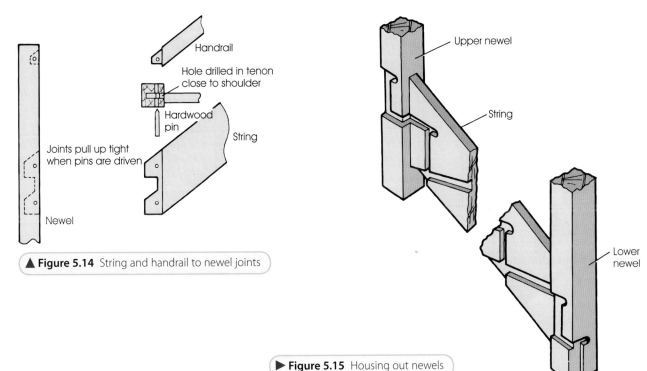

▲ **Figure 5.14** String and handrail to newel joints

▶ **Figure 5.15** Housing out newels

Assembly of close string stairs

This on-edge assembly employs a cramping jig to cramp up the strings and hold the staircase square.

The following procedure is shown in Figure 5.16:

1. Clean up the inner-housed faces of the string, the upper face and edge of tread and exposed face of the risers.
2. Pre-assemble the treads and risers including glue blocks.
3. Position one string on the base of the cramping up frame.
4. Apply glue to the string housings and tread riser intersections of the pre-assembled steps. Place the steps in position, see (a).
5. Apply glue to the housings of the other string and place in position over the steps, see (b).

Related information

Also see Chapter 7 in Book 1, which deals with the marking and setting out of stairs.

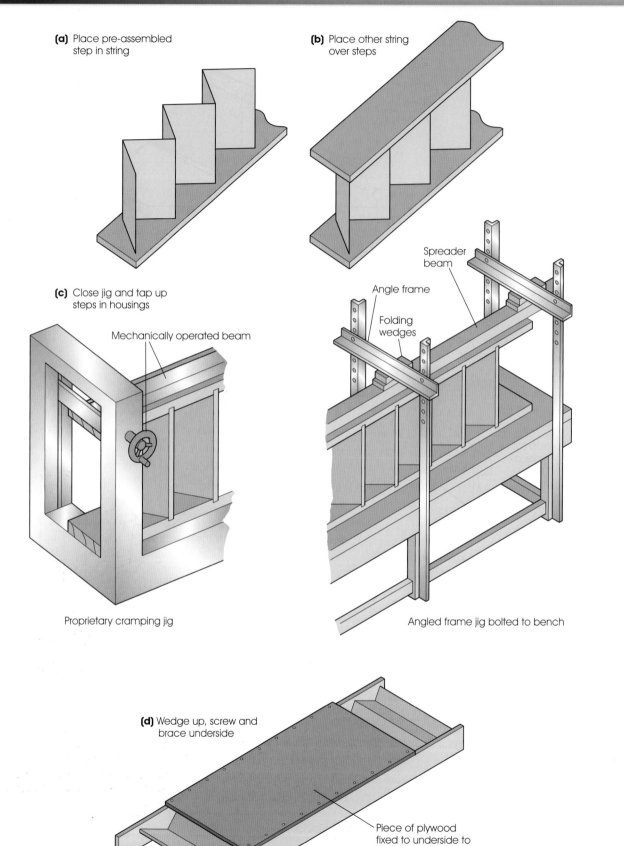

(a) Place pre-assembled step in string

(b) Place other string over steps

(c) Close jig and tap up steps in housings

Mechanically operated beam

Proprietary cramping jig

Spreader beam

Angle frame

Folding wedges

Angled frame jig bolted to bench

(d) Wedge up, screw and brace underside

Piece of plywood fixed to underside to prevent raking

▲ **Figure 5.16** On-edge stair assembly

6. Close the cramping frame and tap up all the treads and risers, making sure they are all fully home in their housings, see (c).

7. Glue and drive the wedges for the treads. Trim off any surplus length with a chisel, so that they clear the riser housings.

8. Glue and drive the wedges for the risers.

9. Screw bottom edge of risers to back edge of treads at about 225 mm centres.

10. Temporarily brace underside of stair, with diagonal braces or a piece of sheet material to prevent it racking out of square before installation, see (d).

11. Finally, remove any squeezed out surplus glue from seen faces with a damp cloth. Clean up outer face and top edge of string. Remove any sharp arrises.

It is normal practice to assemble the flight of stairs to this stage only, for ease of handling and installation. Each flight will be separate, with the bottom bull-nose step, top nosing, newels, handrail and balustrade supplied loose, ready for on-site completion.

Cut string stairs

Figure 5.17 shows the construction details of a cut and bracketed string. This type is used as the outer string of more decorative flights. The string is cut to the shape of the step and the treads are pocket screwed to them. The risers extend past the face of the string and are mitred with thin plywood brackets, which are glued and screwed in place.

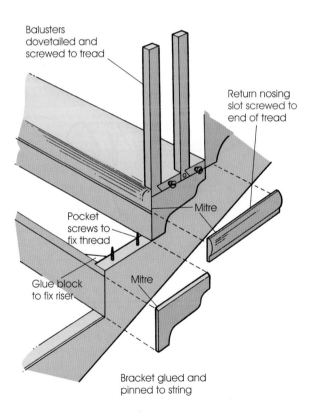

◀ **Figure 5.17** Cut-and-bracketed stair

Balusters dovetailed and screwed to tread

Return nosing slot screwed to end of tread

Mitre

Pocket screws to fix thread

Glue block to fix riser

Mitre

Bracket glued and pinned to string

Most of the assembly will be completed in the workshop with the exception of the following which will be completed on site: the balusters, which are dovetailed and screwed to the tread, and the return nosings, which are slot screwed to the end of the tread.

Open-riser stairs

Figure 5.18 shows a method used to join the treads to the strings in a riser-less flight of stairs (open plan). The through tenons should be wedged on the outside of the string. Alternatively, the treads could be simply housed into the string, and metal ties used under every third or fourth tread to tie the flight together.

Where carriages instead of strings are used in an open-plan flight, the treads are supported by two timber brackets, which are either doweled or tenoned into the carriage, see Figure 5.19. Alternatively, these may be replaced by purpose-made metal brackets.

The balustrade of open-plan flights is often of the **straightforward ranch style**. This is a number of planks either screwed and pelleted directly to the inside stair face of the newels or mortised and tenoned between them.

On certain flights where the gap between the treads is restricted, a partial riser tongued to either the top or underside of the tread can be used. These partial risers also provide additional support to thinner treads. Alternatively, the gap may be reduced by the introduction of metal or timber rods positioned midway between each tread, see Figure 5.20.

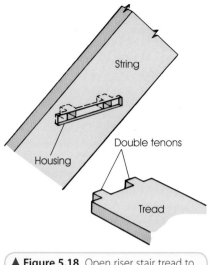

▲ **Figure 5.18** Open riser stair tread to string joint

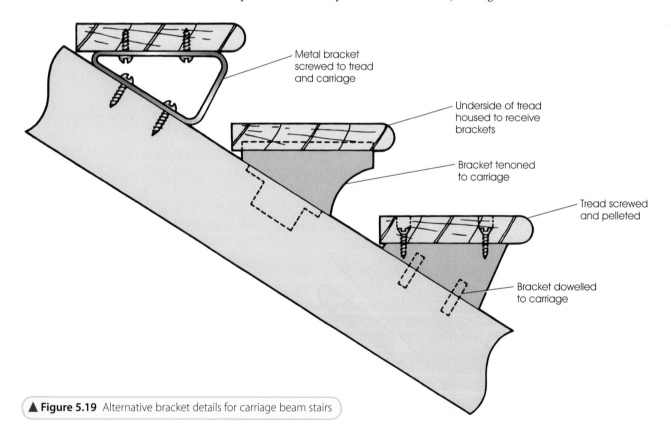

▲ **Figure 5.19** Alternative bracket details for carriage beam stairs

Alternating tread stairs

A form of open-riser stair having paddle-shaped treads, one part of the tread is cut away, with the wide portion alternating from one side to the other on each consecutive tread. They should only be used to give access to one room of a domestic **loft conversion**, where there is insufficient space to accommodate a proper staircase, see Figure 5.21. In addition, a handrail should be fitted to both sides and the treads should have a non-slip surface.

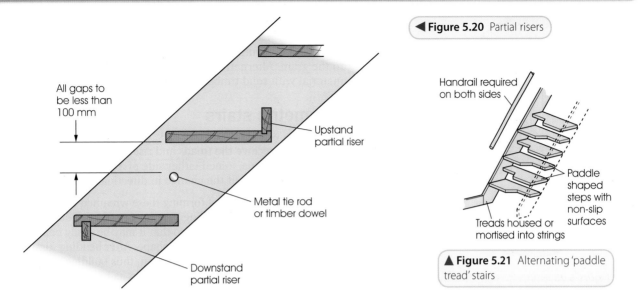

◀ **Figure 5.20** Partial risers

All gaps to be less than 100 mm

Upstand partial riser

Metal tie rod or timber dowel

Downstand partial riser

Handrail required on both sides

Paddle shaped steps with non-slip surfaces

Treads housed or mortised into strings

▲ **Figure 5.21** Alternating 'paddle tread' stairs

Tapered-step stairs

Before marking out tapered steps, they should be set out full size on a sheet of ply, as shown in Figure 5.22. This enables the shapes of the strings and tapered treads to be determined. The risers must normally radiate from a point outside the stair width in order to achieve the 50 mm minimum going. Shaped easing pieces are glued and tongued onto the string where the extra width is required. The two wall strings should be tongued and grooved together where they join.

▶ **Figure 5.22** Setting out tapered steps

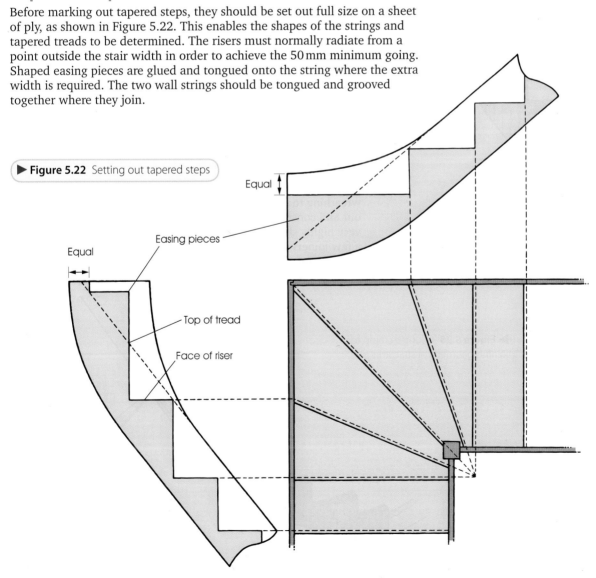

Equal

Easing pieces

Top of tread

Face of riser

Equal

Equal

133

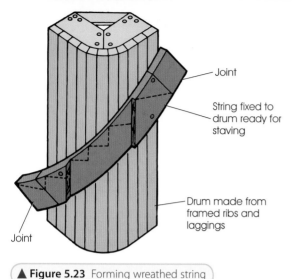

▲ **Figure 5.23** Forming wreathed string

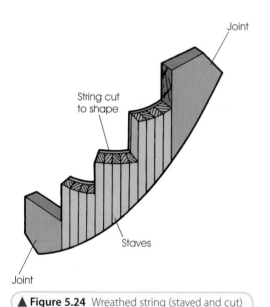

▲ **Figure 5.24** Wreathed string (staved and cut)

The increased width of tapered steps necessitates them being made up from a number of pieces tongued together. Care should be taken when cutting to ensure that the nosing is kept parallel to the grain. Alternatively, the treads may be cut from sheet material with solid timber nosing glued to their front edge.

Geometric stairs

These have wreathed outer strings, which are normally cut and bracketed to receive the treads and risers. These strings consist of straight sections, joined either side of a curved section, which rises and turns to suit the change in direction.

The two main methods of forming these wreathed strings are by glulam construction or staving. Staving involves reducing the curved section to about 2 mm so that it may be easily bent around a former or drum to the required shape. After bending, timber staves are glued to the reduced section, thus building up the string to a uniform thickness and retaining it in its bent position.

Figure 5.23 shows a quarter-turn **wreathed string** fixed around a drum ready for staving. After the staves have cured sufficiently the string may be removed from the drum, cut to the shape of the steps and prepared for jointing to the straight strings (see Figure 5.24).

These are jointed one rise past the turn with a cross-grained or loose tongue. A counter cramp can be used to pull the two ends tightly together, see Figure 5.25. This also has the effect of stiffening the stair and preventing movement at the joint. The cramp consists of three short pieces mortised to take the wedges. They are screwed to the strings initially only at one end. The two outside pieces are screwed to the wreathed string and the middle piece with its mortise slightly off-centre to the straight string. After driving the wedges to pull up the joint, the other ends can be screwed.

The handrail associated with a geometric stair also requires wreathing to follow the line of the string below. The setting out and construction of wreathed strings and handrails is a very highly skilled operation, carried out in general by only a few joinery works who specialise in purpose-made staircase manufacture.

▶ **Figure 5.25** Counter-cramp for joining geometrical stair strings

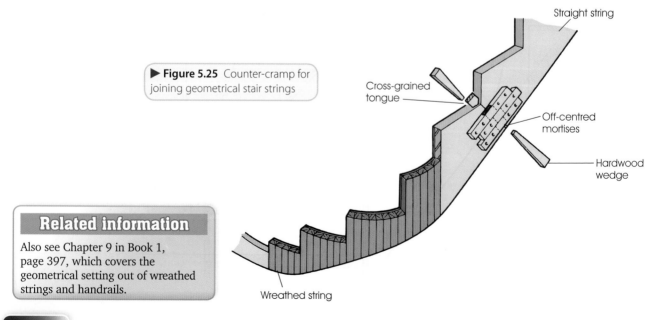

Related information

Also see Chapter 9 in Book 1, page 397, which covers the geometrical setting out of wreathed strings and handrails.

Installation of stairs

The installation of any staircase on site follows the same basic procedure, although there will be differences depending on the type of stair and the nature of the site.

Stairs are normally delivered to site assembled as far as possible. However, for ease of handling, each flight will be separate, its newels, handrail, balustrade and any tapered treads being supplied loose ready for on-site completion.

For maximum strength and rigidity, the stairs should be fixed using the following details:

1. The top newel is notched over the landing or floor trimmer and either bolted or coach screwed to it (see Figure 5.26).

2. The lower newel should be carried through the landing or floor and bolted to the joists.

3. Inserting a steel dowel partly into the newel and grouting this into the concrete can fix the lower newel on a solid ground floor (see Figure 5.27).

4. The outer string and handrail are mortised into the newels at either end. With the flight in position these joints are glued and then closed up and fixed using hardwood draw pins.

5. The wall string is cut over the trimmer at the top and cut nailed or screwed to the wall from the underside (see Figure 5.28).

6. The wall string is also cut at either end to provide an abutment for and to suit the height of the skirting. All cut surfaces should be cleaned up with a block or smoothing plane, as there will be exposed end grain.

7. The bottom of a flight may be secured by screwing it to a batten fixed to the floor.

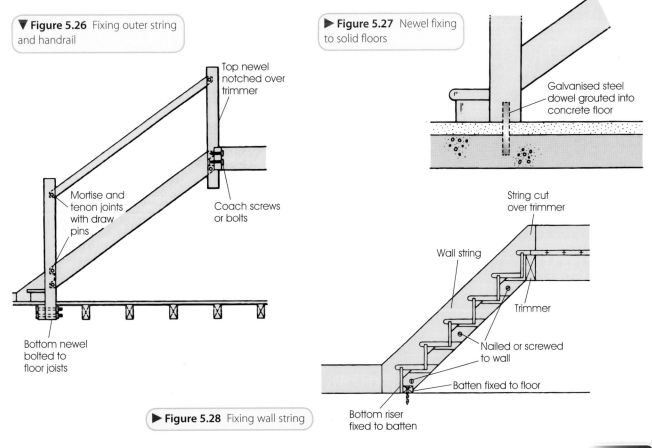

▼ **Figure 5.26** Fixing outer string and handrail

Top newel notched over trimmer

Mortise and tenon joints with draw pins

Coach screws or bolts

Bottom newel bolted to floor joists

▶ **Figure 5.27** Newel fixing to solid floors

Galvanised steel dowel grouted into concrete floor

String cut over trimmer

Wall string

Trimmer

Nailed or screwed to wall

Batten fixed to floor

Bottom riser fixed to batten

▶ **Figure 5.28** Fixing wall string

Finishing

Figure 5.29 shows a section across a flight fixed up against one wall showing typical finishing details. The trimmer around the stairwell opening is finished with an apron lining and nosing, see Figure 5.30.

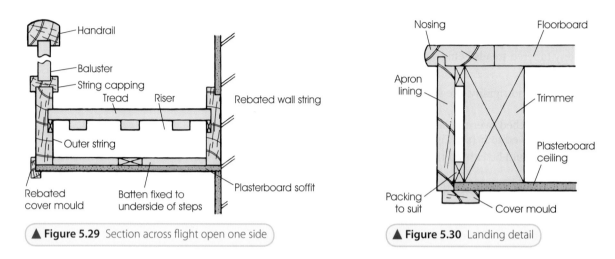

▲ **Figure 5.29** Section across flight open one side

▲ **Figure 5.30** Landing detail

Where the width of the stair exceeds about 1 m, a carriage may be fixed under the flight to support the centre of the treads and risers. To securely fix the carriage, it is birdsmouthed at both ends, at the top over the trimmer and at the bottom over a plate fixed to the floor. Brackets are nailed to alternate sides of the carriage to provide further support across the width of the treads (see Figure 5.31).

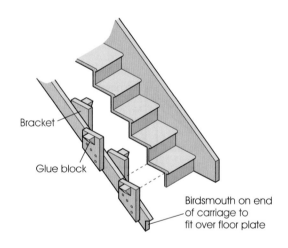

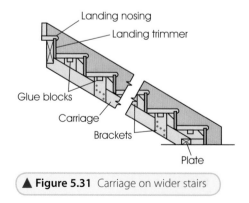

▲ **Figure 5.31** Carriage on wider stairs

Tapered treads

Any tapered treads are fixed after the main flights are in position. Figure 5.32 shows details of tapered steps at the bottom flight before and after the tapered treads have been completed. They should fit in place fairly easily as they will have been pre-fitted dry in the joiner's shop and disassembled for transportation. In practice, a certain amount of adjustment is often required to take account of on-site conditions, for example, slightly out-of-square or out-of-plumb brickwork and so on.

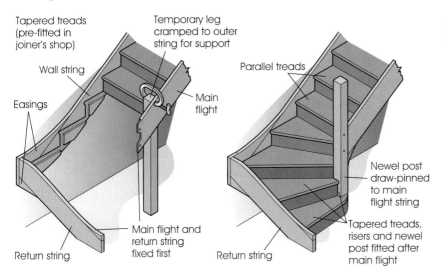

Tapered treads (pre-fitted in joiner's shop)

Temporary leg cramped to outer string for support

Wall string

Easings

Main flight

Parallel treads

Return string

Main flight and return string fixed first

Return string

Newel post draw-pinned to main flight string

Tapered treads, risers and newel post fitted after main flight

◀ Figure 5.32 Installation of tapered tread

After fitting the tapered treads and risers into the housings in the string, they should be wedged up and screwed. Because of their extra length some form of support is normally required under the tapered treads. This can be provided by bearers under the riser of each step, see Figure 5.33. These are fixed between the two strings or between the string and the newel. Also shown is how the carriages of the main flights are birdsmouthed over these bearers.

Landings

The construction of landings and the materials used are normally similar to the upper floor. Figure 5.34 shows details of a half-space landing. The trimmer joist spans between two walls and supports the end of the trimmed joists spanning the landing.

▼ Figure 5.33 Tapered treads

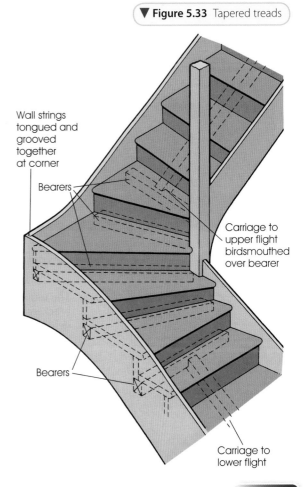

Wall strings tongued and grooved together at corner

Bearers

Carriage to upper flight birdsmouthed over bearer

Bearers

Carriage to lower flight

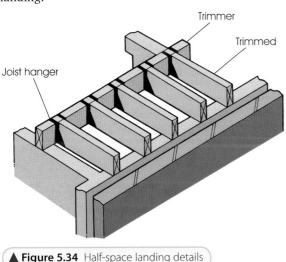

Trimmer

Trimmed

Joist hanger

▲ Figure 5.34 Half-space landing details

Extending the newel post down to the floor, see Figure 5.35, normally supports the free end of the trimmer for a quarter-space landing. Alternatively, either a landing frame or some form of cantilever landing could be used.

Figure 5.36 shows a typical landing support framework. This would be subsequently boxed in or made into a cupboard, creating extra storage space.

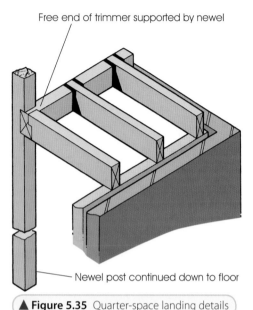

Free end of trimmer supported by newel

Newel post continued down to floor

▲ **Figure 5.35** Quarter-space landing details

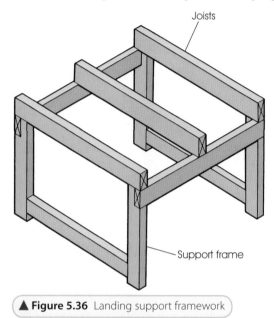

Joists

Support frame

▲ **Figure 5.36** Landing support framework

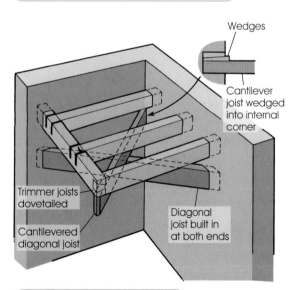

Wedges

Cantilever joist wedged into internal corner

Trimmer joists dovetailed

Cantilevered diagonal joist

Diagonal joist built in at both ends

▲ **Figure 5.37** Cantilever landing

The cantilever landing, see Figure 5.37 consists of a diagonal joist built into the wall at both ends. This provides a bearing midway along the cantilever joist, which is built into the internal angle of the wall.

The landing trimmer joists are dovetailed together and supported on their free end by the cantilevered joist. This method of support has the advantage of giving a clear space under the stair. It is essential that the timber used be well seasoned, as even slight shrinkage will cause the landing to move and creak in use.

Geometrical stair

These are supplied in a number of easily handled sections, the tapered treads being assembled on site as before. The main difference in the installation is the fixing of the 'wreathed' portion of the string. This will have been previously formed and permanently fixed at one end of the string. The other end of the wreathed string has to be fixed to its adjoining string on site.

The outer or wreathed string of a geometrical stair is usually of a cut-and-bracketed type, much of the assembly will have been completed prior to delivery with the exception of the balusters and return nosing. Where a section contains tapered treads they will normally be completely assembled on site.

Open-riser stair

The fixing of open-riser or open-plan stairs with close strings and newels is the same as the stairs previously mentioned. Where carriage or spine beam stairs are concerned, the method of installation differs. Both types can be completely prefabricated in the joiner's shop, requiring only fixing at the top and bottom on site. Alternatively, they can be delivered in knockdown form for on-site assembly.

Figure 5.38 shows a typical open-riser stair supported on carriages. The upper ends of the carriages are birdsmouthed around the trimmer, while the lower ends of the carriages are fixed to the floor with metal angle brackets. The treads are supported by and screwed and pelleted to timber brackets, which are in turn glued and dowelled to the carriages. Newel posts are fixed at either end of the flight, onto which the ranch-style plank balustrading is screwed and pelleted.

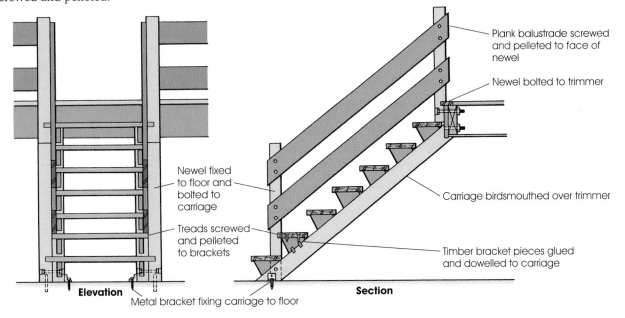

Plank balustrade screwed and pelleted to face of newel

Newel bolted to trimmer

Newel fixed to floor and bolted to carriage

Treads screwed and pelleted to brackets

Carriage birdsmouthed over trimmer

Timber bracket pieces glued and dowelled to carriage

Elevation

Metal bracket fixing carriage to floor

Section

Spine-beam stair

Figure 5.39 shows a spine beam stair, which is also known as a mono-carriage stair. The spine beam is a large glulam section that is often tapered or curved on its underside to reduce its somewhat bulky appearance. Both ends of the spine beam are fixed to metal brackets that have been cast in the concrete; because of the great stresses and likelihood of movement at the junction between the tread and beam, it is essential that these be securely fixed. The shaped bearers are bolted, screwed or fixed with metal brackets to the beam. The treads can then be screwed and pelleted onto the bearers.

▲ **Figure 5.38** Open-riser stair fixing details

◀ **Figure 5.39** Spine beam stair

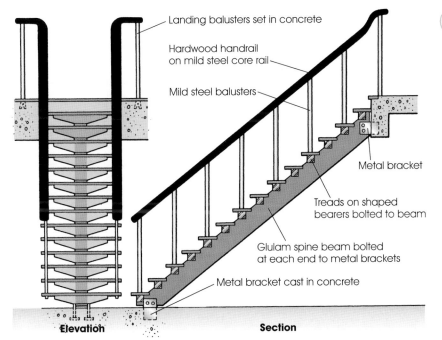

Landing balusters set in concrete

Hardwood handrail on mild steel core rail

Mild steel balusters

Metal bracket

Treads on shaped bearers bolted to beam

Glulam spine beam bolted at each end to metal brackets

Metal bracket cast in concrete

Elevation

Section

Metal rod balusters are often used for this type of stair. They are secured at one end to the tread. A mild steel core rail is used to secure the tops of the balusters. This is fixed using setscrews into the baluster after drilling and tapping. The hardwood handrail is grooved on the underside to conceal the core rail through which it is screwed. An enlarged detail of this is shown in Figure 5.40.

Treads of open-riser stairs are often found to be noisy and slippery in use. They can be made safer by fitting non-slip nosings to the front edges of the treads. Alternatively, to cut down on the noise and at the same time make them less slippery, carpet may be either wrapped around the treads between the carriages and tacked, or fitted into recesses that have been cut into the top of each tread (see Figure 5.41).

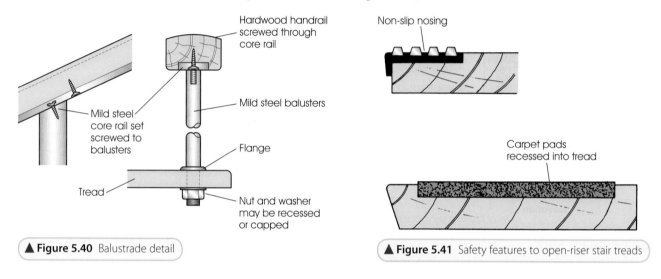

▲ **Figure 5.40** Balustrade detail

▲ **Figure 5.41** Safety features to open-riser stair treads

Balustrades and handrails

Balustrading

Stairs and landings can be fitted with balustrading in many ways other than the standard balusters, metal balusters and ranch-style plank balustrading covered in previous examples. Other methods of forming balustrades range from framed panelling with a variety of infill, including laminated or toughened glass, to decorative wrought ironwork in various designs.

Ranch-style plank balustrading is not recommended for buildings used by young people, as they can be easily climbed.

Standard balusters may be either stub tenoned into the string or fitted into a groove run into the string capping. At their upper end they are normally pinned into the groove run on the underside of the handrail (see Figure 5.42).

▼ **Figure 5.42** Fixing of balusters

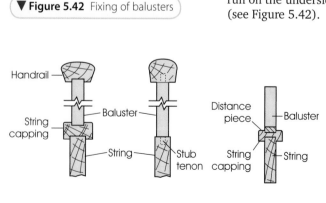

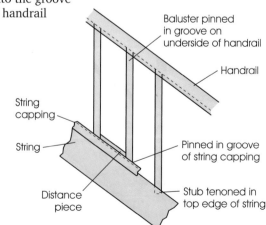

Handrails

Handrails to straight flights with newels, as shown in previous examples, are tenoned into the face of the newel posts. In better-quality work they will also be housed into the face of the newel by about 6 mm so that any shrinkage will not result in an open, unsightly joint. Off-the-shelf and replacement stair parts often use a bolted joint and cover cap in place of the mortise and tenon joint.

Wall handrails either may be fixed by plugging, screwing and pelleting direct to the wall, or may stand clear of the wall on metal brackets fixed at about 1 m centres. Figure 5.43 shows a traditional section and a modern built-up section fixed directly to the wall, as well as one on brackets.

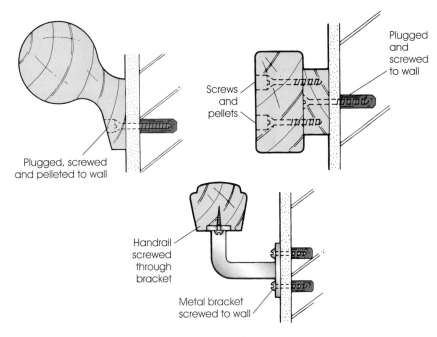

◀ Figure 5.43 Wall handrail sections

Where a handrail changes direction around a corner or from rake to level, the section fixed directly to the wall may simply be mitred. Handrails on brackets are not normally mitred but change direction with the aid of short tangential curved sections, jointed to the main straight lengths with handrail bolts and dowels. The nuts are set from the underside of the rail and the mortises are plugged with grain-matched inserts. Figure 5.44 shows a number of these curved sections.

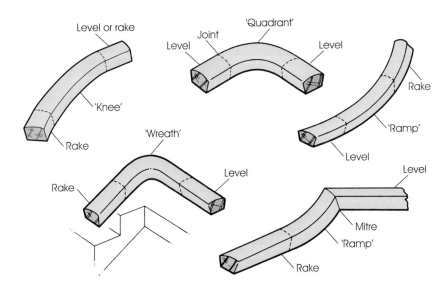

◀ Figure 5.44 Curved handrail views

Related information

Also see Chapter 9, Applied geometry, page 397 in Book 1, which covers the geometrical setting out for wreathed strings and handrails.

A quadrant is used to turn a level handrail around a 90° bend. Ramps are either concave or convex and are used to join raked to level handrails. Wreaths are double-curvature sections. They are used where a raking handrail turns a corner, for example, in stairs with tapered steps, or where a raking handrail turns a corner and changes to a level handrail, for example, at the junction of a stair and landing. Alternatively, a ramp and mitre may be used. A half-newel can be used to give support to the handrail and balustrade where it meets the wall or a landing (see Figure 5.45).

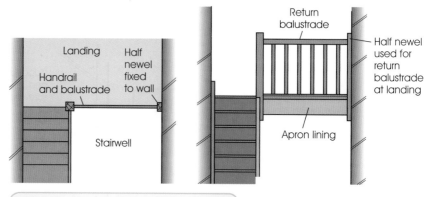

▲ **Figure 5.45** Return balustrades at landings

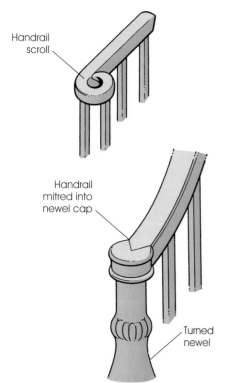

The end of a handrail in a geometrical stair is supported by the balusters and may either terminate in a scroll or be mitred into the top of a decorative turned newel post, see Figure 5.46. The setting out and making of geometrical handrails is the work of a highly specialised joiner.

Where the balustrading is of metal, as is often the case in present-day non-domestic construction, the handrail is fitted to a metal core rail. It is common practice for the handrail to be wreathed, grooved and jointed and then given to the metalworker to produce a suitable core rail.

Timber facings

Concrete stairs can be given a more pleasing appearance by the addition of hardwood treads, see Figure 5.47. These are screwed and pelleted to dovetail-shaped timber blocks that have been cast into the concrete.

▲ **Figure 5.46** Geometric handrail end details

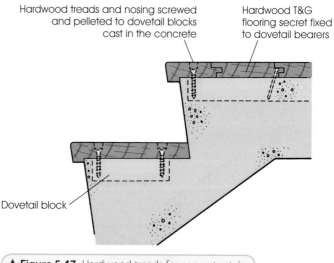

▲ **Figure 5.47** Hardwood treads for concrete stairs

Landings may be given the same finish by secret fixing hardwood T&G flooring to dovetail bearers that have again been cast into the concrete. In situations where the riser has to be faced, the detail shown in Figure 5.48 could also be used.

In certain circumstances a hardwood cut string is also used to finish the edge of a stair. This can be screwed and pelleted to cast-in fixing blocks.

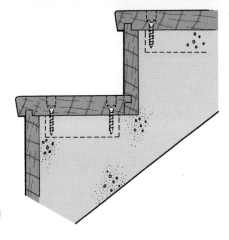

▶ **Figure 5.48** Hardwood risers for concrete stairs

Protection of completed work

After a new staircase has been installed, a short period spent taking measures to prevent damage during subsequent building work saves much more than it costs.

False treads made from strips of hardboard or plywood are pinned on to the top of each step, see Figure 5.49. The batten fixed to the strips ensures the nosing is well protected.

On flights to be clear finished the false treads should be held in position with a strong adhesive tape, as pin-holes would not be acceptable.

Newels and handrails should also be protected, see Figure 5.50. Strips of hardboard or plywood are also used to protect newel posts. These can be either pinned or taped in position depending on the finish.

Adequate protection of handrails and balustrades can be achieved by wrapping them in corrugated cardboard held in position with adhesive tape.

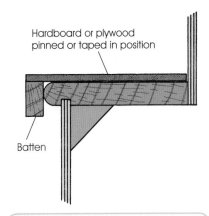

Hardboard or plywood pinned or taped in position

Batten

▲ **Figure 5.49** Temporary protection of treads

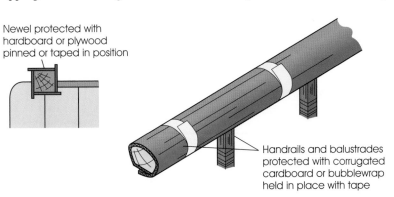

Newel protected with hardboard or plywood pinned or taped in position

Handrails and balustrades protected with corrugated cardboard or bubblewrap held in place with tape

▲ **Figure 5.50** Temporary protection of newels and handrails

6

Chapter Six

Doors, frames and linings

This chapter covers the work of both the site carpenter and bench joiner. It is concerned with principles of constructing and installing doors, frames and linings. It includes the following:

- An introduction to doors, frames and linings.
- The performance requirements of doors.
- Methods of door construction and manufacture.
- Hanging of doors.
- Types of door ironmongery.
- Installation of door ironmongery.
- Types of door frames and linings.
- Fixing of door frames and linings.

Introduction

Doors are moveable barriers used to cover an opening in a structure. Their main function is to allow access into and egress from a building, as well as the passage between the interior spaces. *Door frames and linings* surround the wall opening and provide a means of hanging the door. In general, external doors are hung in frames and internal doors in linings. See Figures 6.1 and 6.2.

▲ **Figure 6.1** External door in frame

▲ **Figure 6.2** Internal door in lining

Door performance requirements

In addition to fulfilling their main function of allowing access, egress and passage, doors also have a number of performance requirements to maintain, such as weather protection, fire resistance, sound and thermal insulation, security, privacy, ease of operation and durability. Care in both design and detailing is required so that any particular door is capable of fulfilling its function while at the same time maintaining the desired performance requirements.

A door will rarely be expected to fulfil all of the performance requirements listed above. In most situations, doors are only expected to maintain a limited number of them. The priority given to each requirement will differ depending on the situation in hand.

- **Weather protection.** This requirement applies to external doors. These may be exposed to wind, rain, snow, sunlight and extremes of temperature and must provide the same degree of weather protection as the remainder of the building. Openings are the weak point in a building as far as weather protection is concerned. Therefore, careful consideration of materials to be used, construction details, hanging, glazing and ironmongery is critical.

- **Fire resistance.** Building elements that have a separating compartment or other fire protection role must be able to contain a possible fire and provide protection to the side remote from the fire for a given period of time. Doors that permit passage through this element must provide the same degree of protection in order to ensure a safe means of escape for the building's occupants. Doors in industrial buildings may also be required to protect the contents of the area until the fire has been extinguished.

- **Sound and thermal insulation.** In general, as the size of the door is relatively small in relation to the surrounding wall area, doors are rarely designed specifically for these requirements. Although the use of draught stripping, sealed unit double-glazing and solid rather than hollow construction can all be used to an advantage.

- **Security.** Door security depends on the materials used, the soundness of construction and the selection and positioning of suitable ironmongery. Only external door security is normally required for domestic dwellings, although internal door security is often necessary for offices, shops, factories and so on.

- **Privacy.** Unglazed doors provide total privacy when closed and, if the layout has been carefully designed, the door can still provide this privacy when partly opened. For this reason, it is normal for doors in domestic dwellings to swing into the room rather than the wall. Glazed or partly glazed doors are used either to provide additional light through the doorway or to provide through vision for safety purposes. Partial privacy may be obtained by the use of obscure glass where through vision is not required.

- **Ease of operation.** A door's size, weight and position in a building will determine its method of operation, so correct selection of ironmongery is an important design consideration.

- **Durability.** A door, or any item of joinery for that matter, must be capable of giving satisfactory service, for a reasonable length of time, in the situation for which it was designed. In addition to the normal opening and closing, doors must also stand up to occasional slamming and other misuse. So there is a need for sound, stable materials, rigid construction, suitable protection and the selection and positioning of suitable ironmongery.

Door types

Doors may be classified by their method of construction: matchboarded, panelled, glazed, flush, fire resistant and so on, see Figure 6.3.

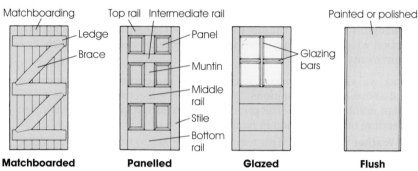

► **Figure 6.3** Door types by construction method

- **Matchboarded doors** are used mainly externally for gates, sheds and industrial buildings. They are simply constructed from matchboarding, ledges and braces clench nailed together. The bottom end of the braces must always point towards the hanging edge of the door to provide the required support. Framed matchboarded doors constructed with the addition of stiles and rails are used where extra strength is required.

- **Panelled doors** have a frame made from solid timber rails and stiles, which are jointed using either dowels or mortise and tenon joints. The frame is either grooved or rebated to receive one or more thin plywood or timber panels. Interior doors are thinner than exterior doors.

- **Glazed doors** are used where more light is required. They are made similar to panelled doors except glass replaces one or more of the plywood or timber panels. Glazing bead is used to secure the glass into its glazing rebates. Glazing bars may be used to divide large glazed areas.

- **Flush doors** are made with outer faces of plywood or hardboard. Internal doors are normally lightweight, having a hollow core, solid timber edges, and blocks, which are used to reinforce hinge and lock positions. New flush doors will have one edge marked 'LOCK' and the other 'HINGE'; these must be followed. External and fire-resistant flush doors are much heavier, as normally they have a solid core of either timber strips or chipboard. A variation on flush doors is to uses the same lightweight hollow core but covered with moulded or embossed facings to give the appearance of a traditional panel door. Internal doors use hardboard or plywood facings while plastic or metal facings are mainly for external use.

- **Fire-resisting doors** are mainly constructed as solid-core flush doors. The main function of this type of door is to act as a barrier to a possible fire by providing the same degree of protection as the element in which it is located. They should prevent the passage of smoke, hot gases and flames for a specified period of time. This period of time will vary depending on the relevant statutory regulations and the location of the door. Fire doors are not normally purpose-made, as they must have approved fire-resistance certification. It is advantageous to use proven proprietary products. Oversize fire door 'blanks' are available for cutting down to size, if required to suit specific situations.

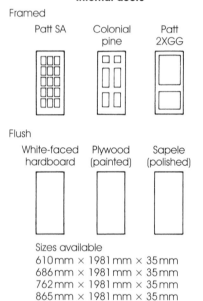

Internal doors

Framed

Patt SA   Colonial pine   Patt 2XGG

Flush

White-faced hardboard   Plywood (painted)   Sapele (polished)

Sizes available
610 mm × 1981 mm × 35 mm
686 mm × 1981 mm × 35 mm
762 mm × 1981 mm × 35 mm
865 mm × 1981 mm × 35 mm

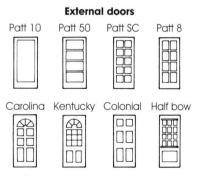

External doors

Patt 10   Patt 50   Patt SC   Patt 8

Carolina   Kentucky   Colonial   Half bow

Sizes available
762 mm × 1981 mm × 44 mm
835 mm × 1981 mm × 44 mm
813 mm × 2032 mm × 44 mm

▲ **Figure 6.4** Extract from a door manufacturer's door list showing the range of stock size doors

Door sizes and grading

Mass-produced doors may be purchased from a supplier in a range of standard sizes, see Figure 6.4. Special sizes or purpose-made designs are normally available to order from suppliers with joinery shop contacts.

The quality of a door will obviously affect its useful life (length of time that it is able to give satisfactory service). The majority of doors are, therefore, graded as being of either an **internal or external** quality.

In general, internal doors have a finished thickness of either 35 mm or 40 mm. The thickness of external doors and fire doors is normally increased to 44 mm, in order to withstand the extra stresses and strains that they are subjected to.

While the type of adhesive used in manufacture is a significant factor to be considered when making purpose-made doors, in practice it is of less significance when using standard doors, as most manufacturers use a synthetic adhesive for both grades. In addition, external doors should be preservative treated against fungal decay.

Related information

Also see Chapter 1 in Book 1, pages 77 and 40, which cover types of adhesive and the preservative treatment of timber.

Matchboarded doors

This group of doors involves the simplest form of construction. They are suitable for use both internally and externally, although they are mainly used externally for gates, sheds and industrial buildings.

Ledged and matchboarded or ledged, braced and matchboarded doors

The basic type of door shown in Figure 6.5 consists of matchboarding, which is held together by ledges. This type is little used because it has a tendency to sag and distort on the side opposite the hinges due to their own weight. In order to overcome this, braces are usually incorporated in the construction (see Figure 6.6). The use of braces greatly increases the rigidity of the door.

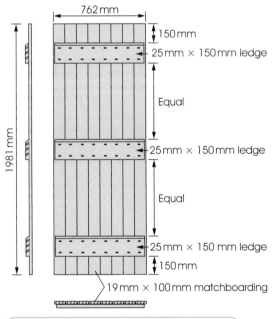

▲ **Figure 6.5** Ledged and matchboarded door

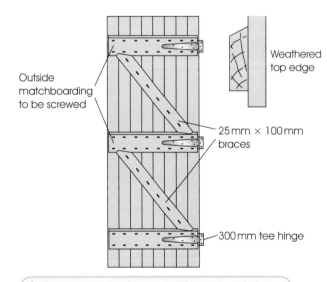

▲ **Figure 6.6** Ledged, braced and matchboarded door showing weathering if door is for external use

The bottom ends of the braces should always point towards the hinged edge of the door in order to provide the required support. Where these doors are used externally, the top edge of the ledges should be weathered to stop the accumulation of rainwater and moisture.

Three ledges are used to hold the matchboarding together. The outside pieces should be fixed with screws, while the remaining lengths of matchboarding are nailed to the ledges. Lost-head nails 6 mm longer than the thickness of the door are used for this purpose. The nails should be

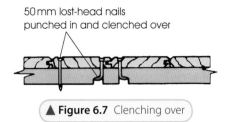

50 mm lost-head nails
punched in and clenched over

▲ **Figure 6.7** Clenching over

punched in and clenched over. Clenching over simply means bending the protruding part of the nails over and punching the ends below the surface (see Figure 6.7). The two braces, when used, are also fixed with lost-head nails, which are clenched over. Brace ends may be cut to the required angle and simply butt jointed to the ledge, but for greater support the let-in shouldered method is preferred, see Figure 6.8.

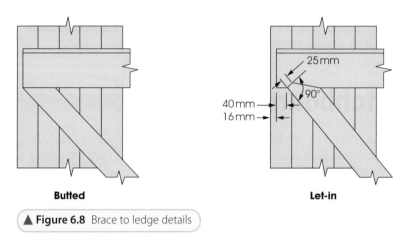

Butted

25 mm

40 mm
16 mm

90°

Let-in

▲ **Figure 6.8** Brace to ledge details

Framed, ledged, braced and matchboarded doors

These are an improvement on the ledged, braced and matchboarded door, as they include stiles, which are jointed to the top, bottom and middle rails with mortise and tenons.

The use of the framework increases the door's strength and resists any tendency to distort. Braces are optional when the door is framed, but their use further increases the door's strength.

Figure 6.9 shows the rear view of a typical framed, ledged, braced and matchboarded door. It can be seen from the section that the stiles and top rail are the same thickness, while the middle and bottom rails are thinner. This is so that the matchboarding can be tongued into the top rail over the face of the middle and bottom rails and run to the bottom of the door. As the middle and bottom rails are thinner than the stiles, barefaced tenons (tenons with only one shoulder) must be used. This joint is shown in Figure 6.10. These joints are normally wedged, although for extra strength draw pins can be used.

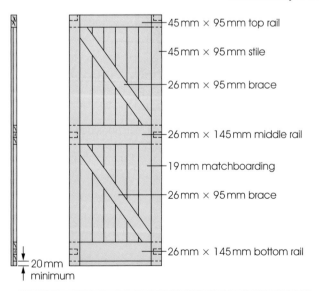

45 mm × 95 mm top rail

45 mm × 95 mm stile

26 mm × 95 mm brace

26 mm × 145 mm middle rail

19 mm matchboarding

26 mm × 95 mm brace

26 mm × 145 mm bottom rail

20 mm
minimum

▲ **Figure 6.9** Framed, ledged, braced and matchboarded door

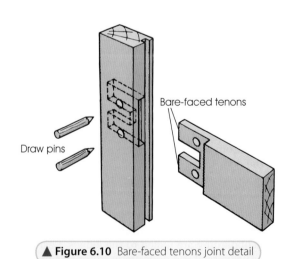

Bare-faced tenons

Draw pins

▲ **Figure 6.10** Bare-faced tenons joint detail

Purpose-made matchboarded doors

Large garage, warehouse and industrial doors were often traditionally made using the framed, ledged, braced and matchboarded principles. This is because this type of door is ideal in situations where strength is more important than appearance. A pair of doors suitable for a garage is shown in Figure 6.11.

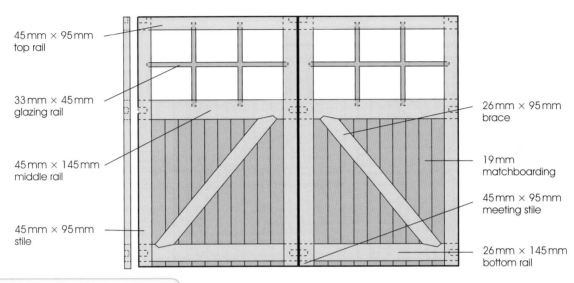

45mm × 95mm top rail

33mm × 45mm glazing rail

45mm × 145mm middle rail

45mm × 95mm stile

26mm × 95mm brace

19mm matchboarding

45mm × 95mm meeting stile

26mm × 145mm bottom rail

▲ **Figure 6.11** Part-glazed garage doors

The portion of the door above the middle rail can be glazed to admit a certain amount of light into the building. The top, middle and glazing rails must be rebated in order to receive the panes of glass that are held in the rebate by glazing sprigs and putty. When the upper portion of the door is glazed the middle rail will be the same thickness as the stiles and, therefore, bare-faced tenons are only used on the bottom rail.

The joint normally used at the intersections of the glazing rails is a scribed or mitred cross-halving, see Figure 6.12. As an alternative, a stub tenon can be used, but as the glazing rails are of a small section the joint is not as strong as the halving.

A section through the meeting stiles of a pair of garage doors is shown in Figure 6.13, along with an alternative means of joining the matchboarding to the stiles.

Very large industrial doors are normally made to slide rather than be side hung, because of their increased tendency to sag as weight and size are increased.

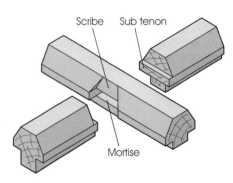
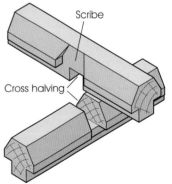

Scribe   Sub tenon

Mortise

Scribe

Cross halving

▲ **Figure 6.12** Glazing rail or bar alternative joint details

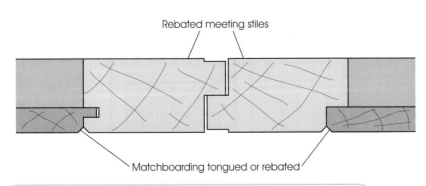

Rebated meeting stiles

Matchboarding tongued or rebated

▲ **Figure 6.13** Rebated meeting stiles and matchboarding to stile details

Details of a pair of large **industrial sliding doors** are shown in Figure 6.14. A small side-hung wicket door has been included to enable personal access without the need to open the main doors, although this is not common to all such doors. The main difference in the construction of these large doors is the increased sectional size of the members. In addition, since they are designed to slide, they are cross-braced to resist the tendency to distort when sliding in either direction.

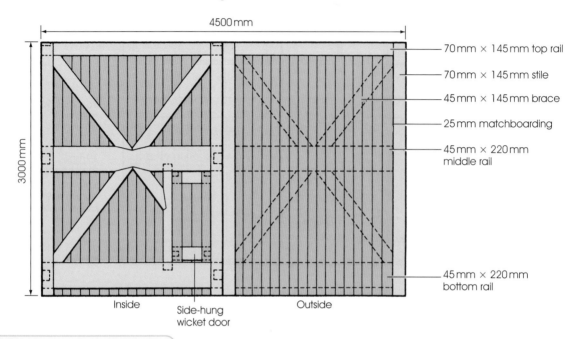

Inside

Side-hung wicket door

Outside

70 mm × 145 mm top rail

70 mm × 145 mm stile

45 mm × 145 mm brace

25 mm matchboarding

45 mm × 220 mm middle rail

45 mm × 220 mm bottom rail

4500 mm

3000 mm

▲ **Figure 6.14** Industrial sliding doors

A pair of external **Gothic design doors** made in oak and suitable for church use is shown in Figure 6.15. The internal and external elevations of each leaf are indicated in the one illustration. The construction of these doors differs only in the jointing of the shaped head from the standard framed, ledged and braced details.

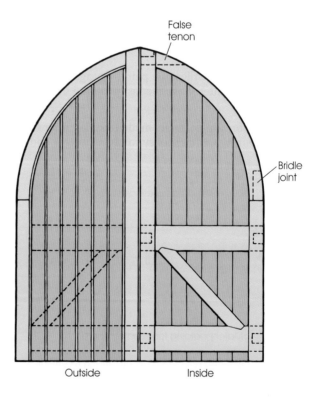

False tenon

Bridle joint

Outside

Inside

▶ **Figure 6.15** Gothic head doors

A pinned bridle joint is used between the outside stile and top rail (see Figure 6.16).

A false tenon is inserted to join the top rail to the meeting stile, as the rail's short grain at this point discounts the use of a normal tenon. In addition, the false tenon considerably strengthens the rail's short grain and minimises the possibility of it shearing.

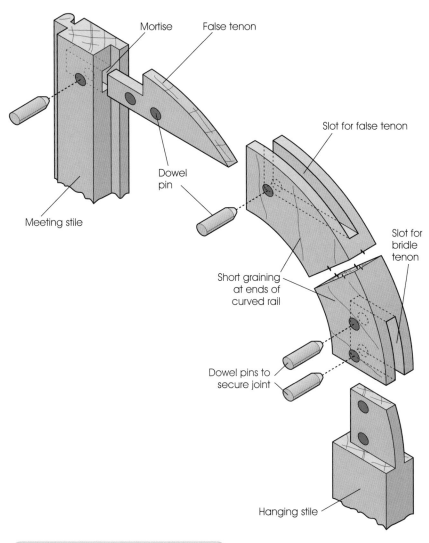

Mortise   False tenon

Slot for false tenon

Dowel pin

Slot for bridle tenon

Meeting stile

Short graining at ends of curved rail

Dowel pins to secure joint

Hanging stile

▲ **Figure 6.16** Jointing shaped rail to stiles

Assembly procedure for framed matchboarded doors

The main framework should first be assembled dry to check the fit of joints, sizes, square and winding.

Squaring up of a frame

This is checked with a squaring rod, which consists of a length of rectangular section timber with a panel pin in its end, see Figure 6.17. The end with a panel pin is placed in one corner of the frame, see Figure 6.18. The length of the diagonal should then be marked in pencil on the rod. The other diagonal should then be checked. If the pencil marks occur in the same place, the frame must be square. If the frame is not square, then sash cramps should be angled to pull the frame into square, see Figure 6.19.

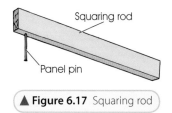

Squaring rod

Panel pin

▲ **Figure 6.17** Squaring rod

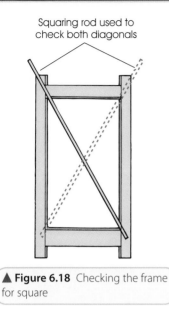

Squaring rod used to
check both diagonals

▲ **Figure 6.18** Checking the frame
for square

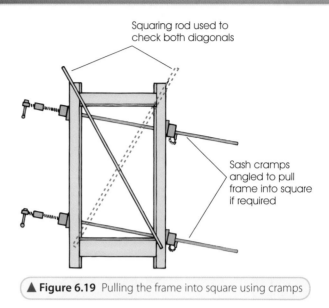

Squaring rod used to
check both diagonals

Sash cramps
angled to pull
frame into square
if required

▲ **Figure 6.19** Pulling the frame into square using cramps

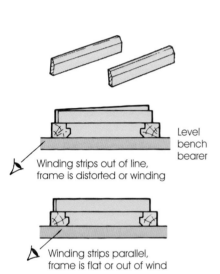

Level
bench
bearer

Winding strips out of line,
frame is distorted or winding

Winding strips parallel,
frame is flat or out of wind

▲ **Figure 6.20** Winding strips used
to check frame for winding

Winding of a frame

This is checked with winding strips, which are two parallel pieces of timber.
With the frame laying flat on a level bench, place a winding strip at either
end of the job. Close one eye and sight the tops of the two strips, see Figure
6.20. If they appear parallel the frame is flat or 'out of wind'. The frame is
said to be winding, in wind or distorted if the two strips do not line up.
Repositioning of the cramps or adjustment to the joints may be required.

Glue up, assemble and lightly drive wedges

A waterproof adhesive should be used for external joinery or where it is
likely to be used in a damp location. Ensure the overall sizes are within the
stated tolerances. Recheck for square and wind. Assuming all is correct,
finally drive in the wedges.

Figure 6.21 shows how the stiles and rails are assembled, glued and wedged
before the matchboarding is fixed.

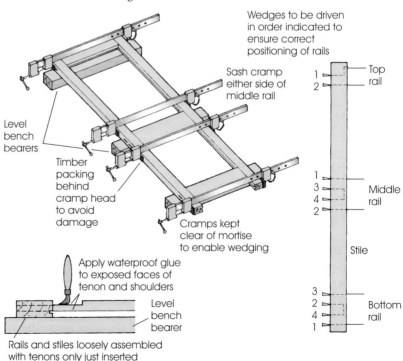

Wedges to be driven
in order indicated to
ensure correct
positioning of rails

Sash cramp
either side of
middle rail

Level
bench
bearers

Timber
packing
behind
cramp head
to avoid
damage

Cramps kept
clear of mortise
to enable wedging

Apply waterproof glue
to exposed faces of
tenon and shoulders

Level
bench
bearer

Rails and stiles loosely assembled
with tenons only just inserted

▶ **Figure 6.21** Assembling framed
matchboard door

Before the assembly of any matchboarded door, all concealed surfaces, such as the tongues and grooves of the matchboarding, back of rails, ledges and braces, should be coated either with a priming paint for painted work, or with a suitable sealer where a clear finish is required. Alternatively, preservative-treated timber should be used. This is in order to prevent moisture penetration and subsequent decay of the timber.

■ Arrange the boards so that the two outside ones are of equal width. They may either be tongued into the top rail and stiles or simply fit into a rebate (see Figure 6.22).

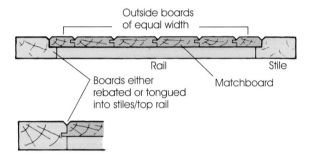

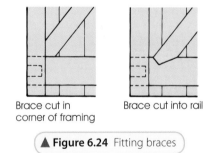

◀ Figure 6.22 Matchboarding details

■ Lay the boards on the assembled frame. Locate tongues and grooves so that the boards form an arc between the jambs. Place a short piece of timber across the door at either end. With assistance apply pressure at both ends of the door to fold the boards flat (see Figure 6.23).

■ Tap up the boards from the bottom to locate them correctly into the top rail and then clench nail or staple them (using a pneumatic nail gun) to all framing members.

■ Mark, cut and fix the braces. These may be either cut into the corners of the framework, or let into the rails (see Figure 6.24). The cut in the corner method is simpler, however, it has a tendency to push open the joints between the stiles and the rails.

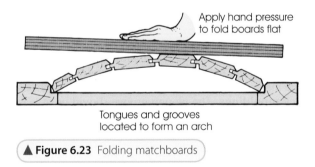

▲ Figure 6.23 Folding matchboards

Brace cut in corner of framing   Brace cut into rail

▲ Figure 6.24 Fitting braces

Panelled and glazed doors

The design and construction of panelled doors are very similar to those of glazed doors. They consist of a frame that has either a plough groove or rebate run around it to receive the panels or glazing. The framing members for these doors vary with the number and arrangements of the panels. They will consist of horizontal members and vertical members.

Rails, stiles and muntins

All horizontal members are called rails. They are also named according to their position in the door, such as top rail, middle rail, bottom rail, intermediate rail. The middle rail is also known as the lock rail and the upper intermediate rail is sometimes called a frieze rail. The two outside vertical members are called stiles, while all intermediate vertical members are known as muntins.

A typical panelled door, with all its component parts named is shown in Figure 6.25. The middle and bottom rails are of a deeper section as they serve to hold the door square and so prevent sagging. Muntins are introduced in order to reduce the panel width, therefore reducing the unsightly effect of moisture movement and the likelihood of panel damage.

▶ **Figure 6.25** A typical panelled door

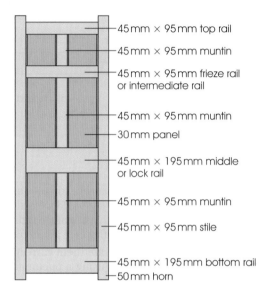

- 45 mm × 95 mm top rail
- 45 mm × 95 mm muntin
- 45 mm × 95 mm frieze rail or intermediate rail
- 45 mm × 95 mm muntin
- 30 mm panel
- 45 mm × 195 mm middle or lock rail
- 45 mm × 95 mm muntin
- 45 mm × 95 mm stile
- 45 mm × 195 mm bottom rail
- 50 mm horn

It is normal to leave at least a 50 mm horn on each end of the stiles. This serves two purposes:

■ It enables the joints to be securely wedged without fear of splitting out.

■ The horns protect the top and bottom edges of the door before it is hung.

Panels and mouldings

Figure 6.26 shows a ply panel that is held in a plough groove that is run around the inside edge of the framing. Two **ovolo mouldings** are also worked around the inside edges of the framing for decorative purposes. They are known as **stuck mouldings**. The plough groove should be at least 2 mm deeper than the panel. This is to allow for any moisture movement (shrinkage and expansion).

Figure 6.27 again shows a solid or plywood panel that is held in a plough groove that is run around the inside edge of the framing. Here a **planted mould** has been applied around the panel for decoration. This method avoids the need to scribe or mitre the shoulders of the rails, which applies with stuck mouldings. Planted moulds must not be allowed to restrict panel movement. Therefore they should be pinned to the framing and not the panel.

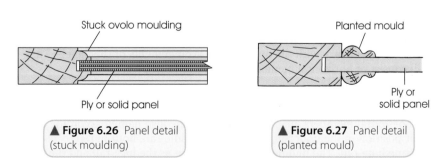

Stuck ovolo moulding

Ply or solid panel

Planted mould

Ply or solid panel

▲ **Figure 6.26** Panel detail (stuck moulding)

▲ **Figure 6.27** Panel detail (planted mould)

Figure 6.28 shows a timber panel that is tongued into a plough groove in the framing. This type of panel is known as a **bead butt panel** because on its vertical edges a bead moulding is worked, while the horizontal edges remain square and butt up to the rails.

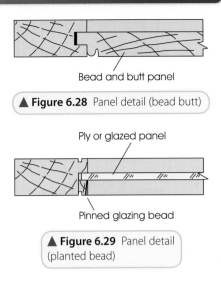

Bead and butt panel

▲ **Figure 6.28** Panel detail (bead butt)

Figure 6.29 shows a glazed panel that is located in the rebate. It is held in position by **planted glazing beads**, which are pinned into the framing. Where this type of door is used externally the planted beads should be replaced by face putty or placed on the inside of the door mainly for security reasons but also because when glazing beads are used externally water tends to get behind them. This makes both the beads and framing susceptible to decay.

Ply or glazed panel

Pinned glazing bead

▲ **Figure 6.29** Panel detail (planted bead)

Figure 6.30 shows a planted mould that is rebated over the framing in order to create an enhanced feature. This type is known as a **bolection mould**. In general, bolection moulds are fitted on the face and planted moulds on the reverse, although in the case of top-quality work bolection moulds could be used on both faces.

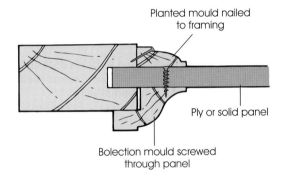

Planted mould nailed to framing

Ply or solid panel

Bolection mould screwed through panel

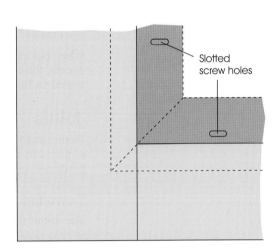

Slotted screw holes

▲ **Figure 6.30** Panel detail (bolection mould)

The bolection mould is fixed through the panel with screws. The holes for the screws should be slotted across the grain to permit panel movement without risk of splitting. The planted bead mould used on the other side to cover the screws should be skew nailed to the framing.

Decorative panels

In good-quality joinery, refurbishment or restoration work, the panels themselves may be decorated by working various mouldings on one or both of their faces. The portion around the edge of a panel is called the margin and the centre portion is known as the field. The small flat section around the edge of the panel is to enable its correct location in the framing.

Figure 6.31 shows the section and part elevation of the main types of decorative panels:

(a) **Raised or bevel raised panel**.

(b) **Raised and fielded panel**, also known as bevel raised and fielded, where the margin has been bevelled to raise the field.

(c) **Raised, sunk and fielded panel**, also known as bevelled, raised sunk and fielded. In this case the margin has been sunk below the field to emphasise it.

Related information

Also see Chapter 12, page 414, which covers the procedure for glazing using face putty and beads.

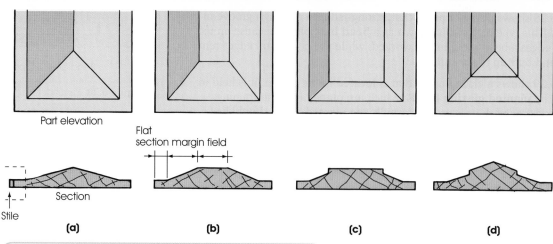

Part elevation

Flat
section margin field

Section

Stile

(a)   **(b)**   **(c)**   **(d)**

▲ **Figure 6.31** Decorative panels: (a) raised; (b) raised and fielded;
(c) raised, sunk and fielded; (d) raised, sunk and raised fielded

(d) **Raised, sunk and raised fielded panel,** also known as bevelled raised, sunk and bevelled raised fielded panel, where the field itself has been bevelled as a further enhanced detail.

When bolection moulds or planted moulds are used to finish decorated panels, the small flat section around the panel edge must be extended to provide a flat surface that will accommodate the mouldings.

Joints

Traditionally, the mortise and tenon joint was used exclusively in the jointing of panelled and glazed doors, but today the majority of doors are mass produced and, in order to reduce costs, the dowelled joint is used extensively.

Dowelled joints

The use of the dowelled joint reduces the cost of the door in three ways:

- The length of each rail is reduced by at least 200 mm.
- The jointing time is reduced, as holes only have to be drilled to accommodate the dowel.
- The assembly time is reduced, as no wedging has to be carried out.

Figure 6.32 shows a six-panel door, which has been jointed using dowels. These dowels should be 16 mm × 150 mm and spaced approximately 50 mm centre to centre. The following is the minimum recommended number of dowels to be used for each joint:

- Top rail to stile: two dowels.
- Middle rail to stile: three dowels.
- Bottom rail to stile: three dowels.
- Intermediate rail to stile: one dowel.
- Muntin to rail: two dowels.

Figure 6.33 shows an exploded view of a doweled joint between a top rail and stile. In addition to the dowel, a haunch is incorporated into the joint. This ensures that the two members finish flush. The use of a haunch also overcomes any tendency for the rail to twist.

Dowels should be cut to length, chamfered off at either end to aid location and finally a small groove should be formed along their length, to allow any excess glue and trapped air to escape when the joint is cramped up (see Figure 6.34). Alternatively, ready-made dowels may be used. These are available in a range of sizes; they have chamfered ends and multi-grooved sides.

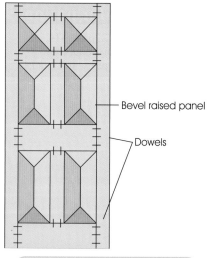

Bevel raised panel

Dowels

▲ **Figure 6.32** Dowel-jointed panel door

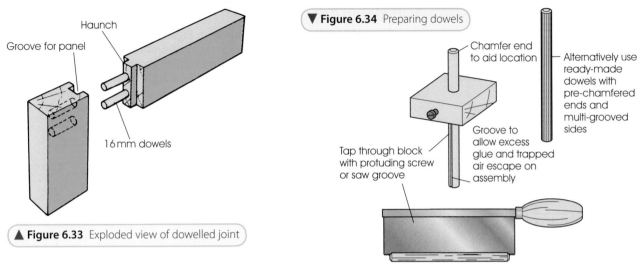

Haunch
Groove for panel
16mm dowels

▲ **Figure 6.33** Exploded view of dowelled joint

▼ **Figure 6.34** Preparing dowels

Chamfer end
to aid location

Alternatively use
ready-made
dowels with
pre-chamfered
ends and
multi-grooved
sides

Tap through block
with protuding screw
or saw groove

Groove to
allow excess
glue and trapped
air escape on
assembly

Mortise and tenon joints

Although the dowel joint is extensively used for mass-produced doors, the mortise and tenon joint is still used widely for purpose-made and high-quality door construction.

Figure 6.35 shows an exploded view of the framework for a typical six-panel door. Haunched mortise and tenons are used for the joints between the rails and stiles.

Related information

Also see Chapter 3 in Book 1, page 149, which covers the proportions and types of mortise and tenon joints.

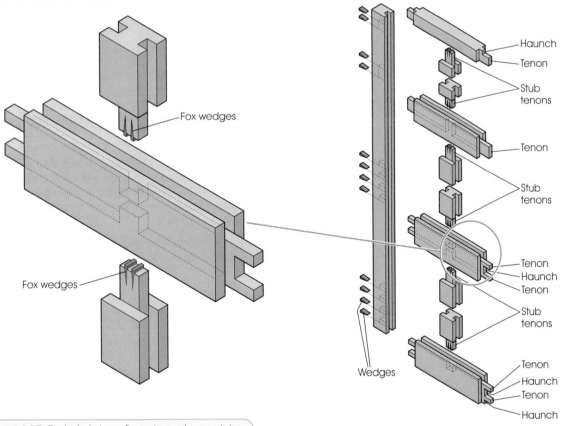

Fox wedges

Fox wedges

Wedges

Haunch
Tenon
Stub tenons
Tenon
Stub tenons
Tenon
Haunch
Tenon
Stub tenons
Tenon
Haunch
Tenon
Haunch

▲ **Figure 6.35** Exploded view of mortise and tenon joints

For joints between the muntins and rails, stub mortise and tenons are used. As these joints do not go right through the rails, they cannot be wedged in the normal way. Instead fox wedges are used. These are small wedges, which are inserted into the saw cuts in the tenon. When the joint is cramped up the wedge expands the tenon and causes it to grip securely in the mortise.

Half-glazed door

A traditional half-glazed door is shown in Figure 6.36. It is constructed with diminishing stiles, in order to provide the maximum area of glass and, therefore, admit into the building the maximum amount of daylight. This type of door is also known as gun stock stile door because its stiles are said to resemble the stock of a gun. The middle rail has splayed shoulders to overcome the change in width of the stiles, above and below the middle rail. An exploded view of this joint is shown in Figure 6.37.

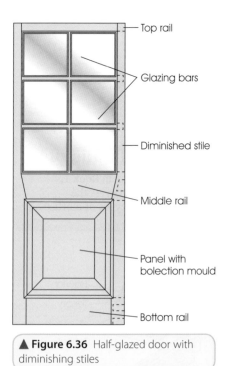

▲ **Figure 6.36** Half-glazed door with diminishing stiles

Labels: Top rail, Glazing bars, Diminished stile, Middle rail, Panel with bolection mould, Bottom rail

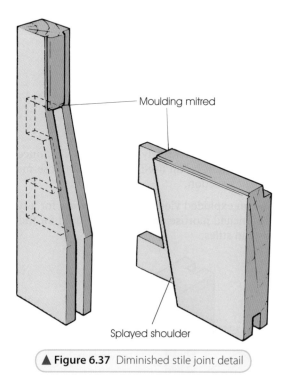

▲ **Figure 6.37** Diminished stile joint detail

Labels: Moulding mitred, Splayed shoulder

The top half of the door can either be fully glazed or subdivided with glazing bars as shown. When glazing bars are used they are normally stub tenoned into the stiles and rails. The joints between the glazing bars themselves are the same as used for the upper glazed portion of the matchboarded door, either stub tenoned or halved and scribed.

The bottom half of the door normally consists of a bevel raised sunk and fielded panel with planted bolection mouldings.

Other types of panelled door

Double margin door

This is a pair of narrow doors joined together at their meeting stiles to make a single door, with the appearance of a pair (see Figure 6.38). It is used for very wide openings where a single door would give an ill-proportioned, awkward appearance and where each half of a double door would not allow easy pedestrian passage.

The correct procedure for assembly is to glue and wedge the rails to the meeting stiles and then join together the two stiles with folding wedges. The panels are inserted from either side before assembling the outer stiles, cramping up and wedging. Figure 6.39 shows an exploded view of the framework for a typical double margin door.

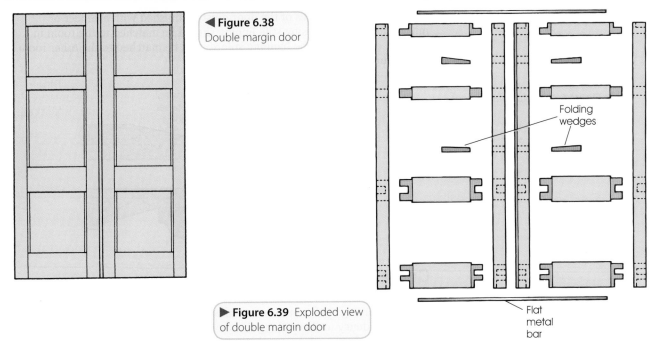

◀ **Figure 6.38**
Double margin door

▶ **Figure 6.39** Exploded view
of double margin door

Folding
wedges

Flat
metal
bar

For additional stiffness, a flat metal bar can be housed and screwed across the joint at both the top and bottom edges. Traditionally, the top and bottom rails would be continuous across the whole door width, with the meeting stiles bridle jointed to them.

Figure 6.40 shows a section through the meeting stiles. It illustrates the positioning of the folding wedges. In addition, it shows that the stiles are rebated together and a stuck bead is used to break the joint between the stiles. Alternatively, the two stiles can be tongued together with a loose tongue and the joint masked with either stuck or planted beads.

Line of folding wedges

◀ **Figure 6.40** Meeting stile detail

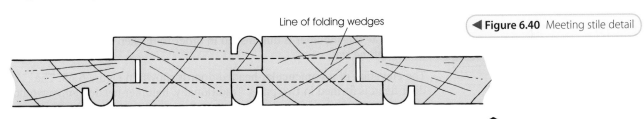

Double-faced doors

In certain situations, particularly in prestigious public buildings, stately homes and so on, adjacent rooms or corridors may be panelled in a different timber or to a different detail. In these cases it is necessary for the face of the door to match in timber and design the room or corridor in which it is situated. With flush veneered panelling all that is required is the application of different veneer. Where a framed panel door is required, the solution is to use two thin doors, each about 28 mm thick, which are fitted to each other with dovetail keys to form a door that matches its respective location.

The dovetail keys are screwed at intervals to the backs of the framing of one door and housed into the framing of the other (see Figure 6.41). The corresponding members (stiles, rails, etc.) of each door face are paired, glued and keyed together before the door itself is assembled.

Key

Housing

▶ **Figure 6.41** Dovetailed
key for double-faced doors

After assembly the edges of the door must be finished with a veneer or mitred lipping. The closing stile lipping should be matched to the room in which it opens, while the hanging stile should be matched to the other room (see Figure 6.42).

▶ **Figure 6.42** Double-faced door finishing detail

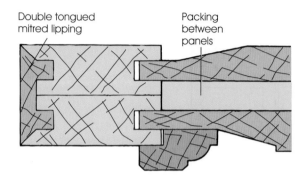

Curved rail doors

The main problem encountered with the construction of curved work is that of short graining. This causes a weakness in the member because of a marked tendency to sheer along the short grain. This risk can be reduced by taking care when marking out the member to ensure that there is as little short grain as possible (see Figure 6.43).

▶ **Figure 6.43** Marking out curved members

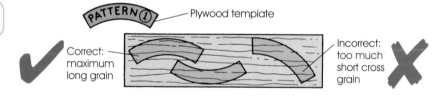

The design of the joint is also important, as this is often the area most affected. Figure 6.44 shows the joint between a shaped top rail and stile. It shows how the shoulder is diminished in order to thicken up the short grain section of the rail.

The elevation of a semicircular-headed door is shown in Figure 6.45. The door head is formed in two parts. Traditionally, hammer-headed joints would

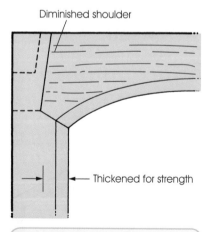

▲ **Figure 6.44** Shaped rail joint detail

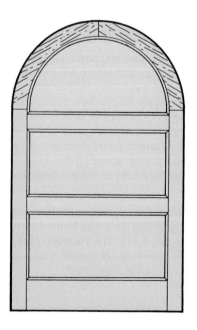

▶ **Figure 6.45** Semicircular-headed door

have been used, see Figure 6.46. A double-ended hardwood hammer-headed key is used at the crown. The wedges secure the joint and draw the two parts together. A hammer-headed tenon is used to join the stile and curved head. This is the same as one end of the key except that it is formed on one end of the stile.

An alternative method of jointing which reduces the amount of handwork is shown in Figure 6.47. This shows how the crown can be jointed using a false tenon insert and a bridle joint between the stile and rail. Both joints are secured using hardwood draw pins.

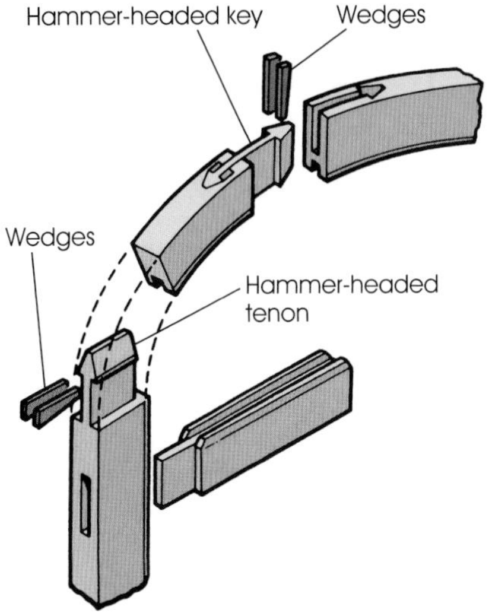

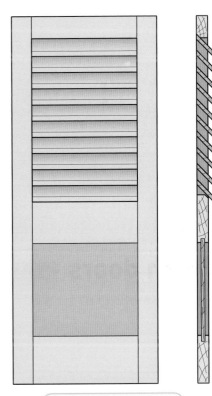

▲ **Figure 6.46** Traditional joint details, semicircular-headed door

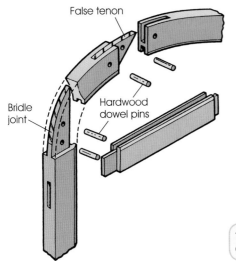

◀ **Figure 6.47** Alternative joint details, semicircular-headed door

Whichever method is used to construct the semicircular-headed door, the immediate rail should be kept about 50 mm below the head/stile joint. This is in order to ensure sufficient strength and minimise the possibility of shearing along the short grain between the mortise and shoulder.

Louvred doors

The frame of this type of door is similar in construction to that of a panel door, that is, mortise and tenons or dowels are used to join the rails to the stiles. Traditionally, louvres were only used as a means of ventilation, but they are also used as cupboard doors and decorative shutters.

External louvred doors. When used externally, the louvre slats or blades are normally set at an angle of 45°. They should also project beyond the face of the framing for weathering purposes. Figure 6.48 shows an elevation and section of an external half-louvred door.

There are two methods by which the projecting louvre slats can be jointed to the stiles for external use. Figure 6.49 shows the first method where the stiles have been mortised to receive the tenons, which have been formed on the ends

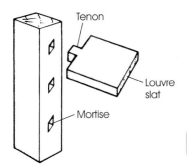

◀ **Figure 6.49** Louvre slat joint detail (mortise and tenon)

▲ **Figure 6.48** Louvre door

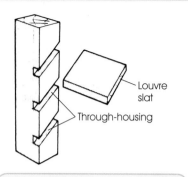

▲ **Figure 6.50** Louvre slat joint detail (though housing)

▶ **Figure 6.51** Louvre slat joint detail (stopped housing)

of the louvred slats. Figure 6.50 shows the second method. Here the stiles have been through-housed to accommodate the full thickness of the louvre slats.

In all cases the slats should have an overlap of 6 mm to prevent through vision.

Internal louvred doors. Where louvre doors are used internally it is not necessary for the louvre slats to project beyond the face of the framing. The internal slats are normally set at an angle of 60° for two main reasons: more of the face of the slat is visible, adding to the decorative appearance, and fewer slats are required, giving a more economical door.

Figure 6.51 shows the method normally used for jointing the louvre slats of an internal door to the stiles. The stiles are stop-housed on a high-speed router using a purpose-made jig. The housing must be cut to exactly the same size as the louvre slat, otherwise the slat will be a slack fit, or time will be wasted easing the housing to fit each individual slat.

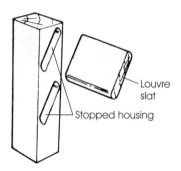

Assembly procedure for panelled and glazed doors

This can be carried out for all types of framed door, using the following assembly procedure:

1. Dry assemble to check fit of joints, overall sizes, square and winding.
2. Clean up inside edges of all framing components and both faces of panels.
3. Glue, assemble, cramp up and wedge. Re-check for square and winding.
4. Clean up remainder of item and prepare for finishing.

The assembly of a six-panel door is shown in Figure 6.52. The rails and muntins should be glued and assembled first. Panels can then be inserted dry, taking care to ensure that no glue has squeezed into the panel grooves. Next the stiles are positioned, glued, cramped and wedged, followed by final cleaning up.

Flush doors

Joinery works are rarely involved with the manufacture of flush doors, as mass produced ones are far more cost effective – except where special features, not available in standard mass-produced doors, are required.

Flush doors consist of either a hollow or solid core that is faced with sheets of hardboard, plywood or moulded plastic.

Hollow core doors

These are normally the cheapest type of flush door to produce. There are many ways in which the hollow core can be made. The three main methods are the skeleton core, the lattice core and the honeycomb core.

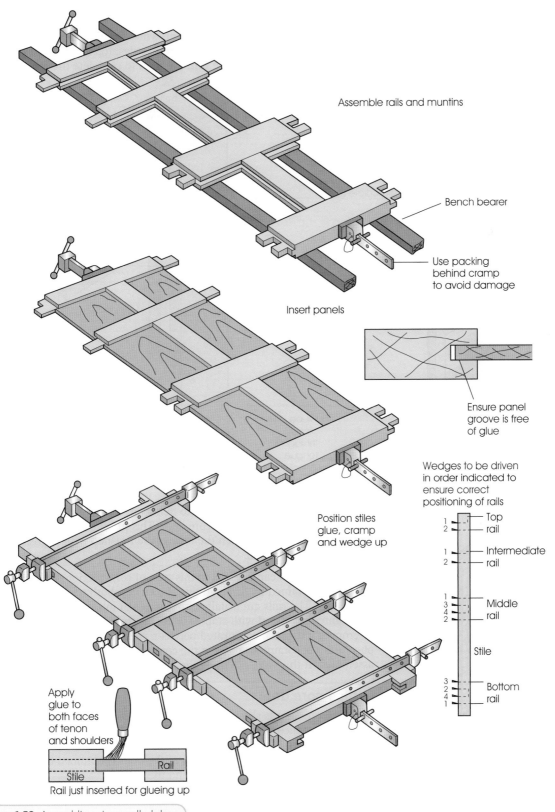

Assemble rails and muntins

Bench bearer

Use packing
behind cramp
to avoid damage

Insert panels

Ensure panel
groove is free
of glue

Wedges to be driven
in order indicated to
ensure correct
positioning of rails

Top
rail

Intermediate
rail

Middle
rail

Stile

Bottom
rail

Position stiles
glue, cramp
and wedge up

Apply
glue to
both faces
of tenon
and shoulders

Rail

Stile

Rail just inserted for glueing up

▲ **Figure 6.52** Assembling six-panelled door

Skeleton core door

Figure 6.53 shows a skeleton core door, which is suitable in one-off or limited production. It consists of 28 mm × 70 mm stiles, top and bottom rails; 20 mm × 28 mm intermediate rails are used to complete the framework.

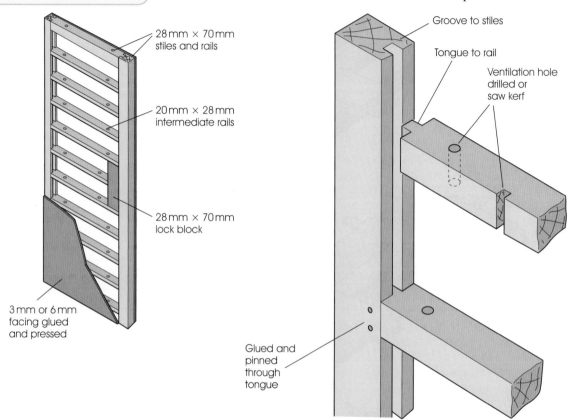

▼ **Figure 6.53** Skeleton core door construction

28 mm × 70 mm stiles and rails

20 mm × 28 mm intermediate rails

28 mm × 70 mm lock block

3 mm or 6 mm facing glued and pressed

Groove to stiles

Tongue to rail

Ventilation hole drilled or saw kerf

Glued and pinned through tongue

There is a tendency for the facing to deflect over time between the compartments of the skeleton core door. This produces a ripple effect, which is especially noticeable when the facings are finished with a gloss coating such as paint or varnish.

Very simple joints can be used in this type of construction, as its main strength is obtained by firmly gluing the facings both to the framework and to the core. The rails are usually either tongued into a groove in the stiles or butt jointed and fixed with staples or corrugated fasteners. Ventilation holes or grooves must be incorporated between each compartment. This is to prevent air becoming trapped in the compartment when the door is assembled. If this were not done the facings would have a tendency to bulge.

Lattice core door

A lattice core door is shown in Figure 6.54. It consists of 28 mm × 30 mm framework, which is simply stapled together. The core is made from narrow strips of hardboard slot jointed to produce the lattice. As with other types of flush door construction the strength of the door is obtained by firmly gluing the facings to both the framework and the core.

Honeycomb core door

Moulded panel doors, which are currently very popular, use a hollow core door construction with the application of facings that are pressed or moulded to give the impression of a traditional panelled door, see Figure 6.55. The core shown is typical of those that are used in mass production. It is formed from cardboard in a honeycomb pattern. The facings are moulded from hardboard for internal doors, and are normally impressed with a textured wood grain

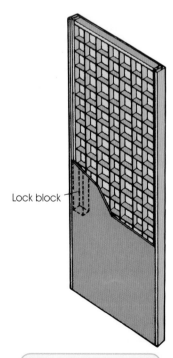

Lock block

▲ **Figure 6.54** Lattice core door construction

pattern, which is primed ready to receive a paint or wood stain finish. Doors with moulded plastic or metal facings are available for external use.

A **lock block** should be provided in all hollow core doors to accommodate a mortise lock or latch. The block also serves to provide a fixing for the lock furniture. The position of the lock block is normally indicated on the edge of the door.

Solid core door

Solid core flush doors are considered to be of a better quality than hollow flush doors. There are four main reasons for this view:

- ■ Their facing remains flat.
- ■ They have increased rigidity.
- ■ They have increased levels of sound and thermal insulation.
- ■ They have increased resistance to penetration by fire.

Laminated core door

Figure 6.56 shows a laminated core door. It consists of 25 mm strips, which are firmly glued together along with the two plywood facings. The strips should be laid with hearts facing alternately in order to balance any stresses.

Chipboard core door

Figure 6.57 shows a chipboard core door. It consists of the chipboard core, which is surrounded by a simple 30 mm framework. The framework is normally stapled together at the corners. Gluing the facings to either side completes the door. Extruded tubular core chipboard could be used as an alternative, where a lighter-weight door is required.

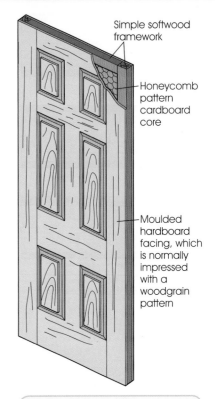

▲ **Figure 6.55** Moulded panel door with honeycomb core

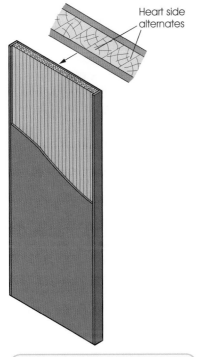

▲ **Figure 6.56** Laminated core door

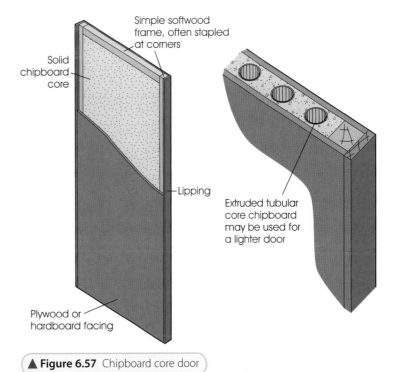

▲ **Figure 6.57** Chipboard core door

Lippings

These are narrow strips of timber, which are fixed along the edges of better-quality flush doors. Their purpose is to mask the edges of the facings and provide a neat finish to the door. External doors should have lippings fixed to all four edges for increased weather protection.

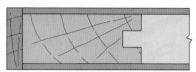

Plain

Tongued

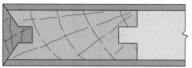

Splayed and tongued

▲ **Figure 6.58** Flush door lippings

Plain lipping is acceptable for internal doors, but for better-quality internal doors and external doors tongued lipping is preferred. Splayed and tongued lippings may be used for clear finished wood grained doors to avoid a visible stripe on the face, see Figure 6.58. Lipping should be glued in position and not fixed with panel pins, as these would cause problems when hanging.

Vision panels

These are often required in purpose-made flush doors. They provide a view of the other side, reducing the risk of collisions when the door is opened, as well as the transmission of light. Additional framing is required around the opening for hollow core doors, see Figure 6.59. Shaped blocking out pieces can be used to form circular vision panels (see Figure 6.60). This framing and blocking is normally done during the construction of the door, although in the case of simple rectangles, it may be possible to form the opening at a later stage. Vision panels in fire doors may be restricted in both size and shape, in order to comply with the certification.

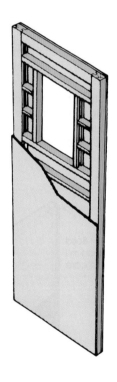

▶ **Figure 6.59** Flush door with vision panel

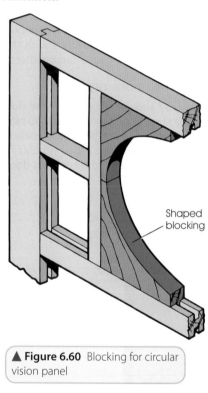

Shaped blocking

▲ **Figure 6.60** Blocking for circular vision panel

Glazing beads

These are used to hold the glass in place. Plain glazing beads are suitable for internal doors, but for increased protection from the elements on external doors they should be rebated (bolection) and the top surface weathered (see Figure 6.61). A better method for fixing external glazing beads is shown in Figure 6.62. As well as providing the increased weather protection of the previous method, it also provides far greater security as the beads cannot be removed from the outside.

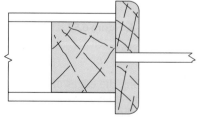

▶ **Figure 6.61** Internal and external glazing beads

Internal use only

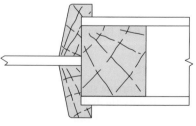

Internal and external use

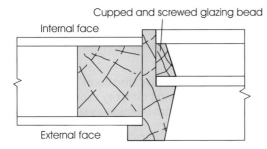

Internal face — Cupped and screwed glazing bead

External face

Glazing beads for doors are preferably fixed using countersunk brass screws and recessed cups on at least one face rather than nails or pins. This is to enable the glazing beads to be easily removed in the event of the glass needing replacement.

Door glazing

Safety glass should be used to glaze the critical areas of doors and other vulnerable low areas, particularly where children are present. The Building Regulations state that glass in these areas should either: break safely (does not produce pointed shards with razor sharp edges); be robust (adequately thick to prevent the likelihood of breaking); or be permanently protected (by screens).

The critical areas and Building Regulations requirements are shown in Figure 6.63. Toughened and laminated glass can meet the requirements. Where annealed (float) glass is used, it must be at least 6 mm thick and in panes not larger than 0.5 m^2 with a maximum width of 250 mm. Screens or railings on both sides, which are at least 800 mm high, should be used to permanently protect fixed annealed glass.

▼ **Figure 6.63** Glazing to doors, windows and screens

Safety glass required in shaded areas

Window

300mm   300mm

1500mm
800mm

Floor level

Door   Glazed screen

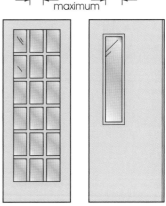

Use of annealed glass

250mm maximum

6mm annealed glass can be used in small panes up to 250mm wide and a maximum area of 0.5m^2

Fire-resisting doors

As stated previously, this type of door is not normally purpose made, as they must have approved fire-resistance certification. This entails submitting the door assembly, pre-hung in its frame and fitted with all its intended ironmongery, to an approved fire-testing organisation.

They will test the fire-resisting door assembly by reference to the following criteria:

- **Stability:** resistance to the collapse of the door.

■ **Integrity:** resistance to the passage of flames or hot gases to the unexposed face.

■ **Insulation:** resistance to the excessive rise in temperature of the unexposed face.

All fire-resisting doors are now called 'fire doors' (FD). Traditionally they were prefixed by their stability and integrity rating respectively. So a 60/45 fire-resisting door had a minimum 60 minutes stability rating and a 45 minutes integrity rating.

The more commonly used terms for fire doors were 'fire check' and 'fire resisting'. 'Fire check' was used to signify doors with a reduced integrity. For example 60/60 and 30/30 were known as 1 hour and $\frac{1}{2}$ hour fire-resisting doors; 60/45 and 30/20 were known as 1 hour and $\frac{1}{2}$ hour fire-check doors.

Today, all fire doors are termed by their integrity rating only (e.g. FD20, FD30 and FD60). A suffix 'S' may be added to a specification, to indicate a fire door that requires smoke seals fitted, for example, FD30S.

Fire doors should be identified by a permanent label or a colour-coded plug inserted into their hanging edge, see Figure 6.64: white for FD20; yellow for FD30 and blue for FD60. Plugs are further identified with a coloured central

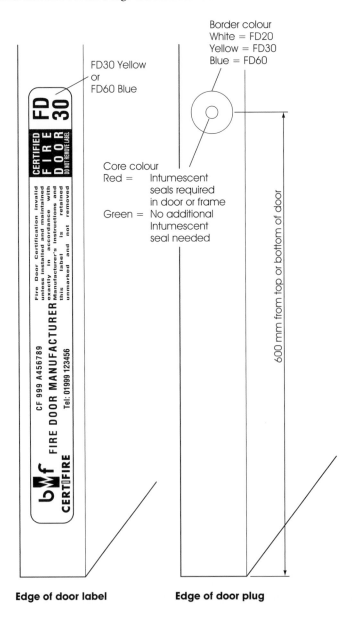

▶ **Figure 6.64** Identification of fire doors

core or spot. Red indicates that an intumescent seal must be fitted on site; green indicates that no additional intumescent seal is required. A blue core may also be used on a white background for FD20 doors without intumescent seals and FD30 doors with an intumescent seal.

The actual fire door required will vary depending on the relevant statutory regulations and the location of the door. For instance, it is significant if the door is providing access through a separating or compartment wall, opening on to a protective shaft, or separating a hazard (attached garage, boiler house, etc.). Therefore, consultation with the Local Authority's Fire Prevention Officer and the Building Regulations is advised in order to determine the specific requirements at the design stage.

Fire door seals

The weak point in fire door construction is the joint between the door and frame. This is where the fire and smoke will penetrate first. Intumescent strips or seals are required for many fire doors. They may be fitted into the door lipping, around the door frame or to both the door and the frame; however, this must always be in accordance with its test certification. When activated by high temperatures in the early stages of a fire, these strips expand, sealing the joint and prolonging the door's integrity, see Figure 6.65.

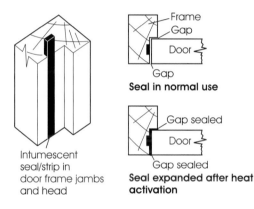

Figure 6.65 Fire door seals

Expanded strips also serve to lock the door into its frame, prolonging its stability. In the early stages of a fire, before the activation of the intumescent strips, the inhalation of smoke and fumes passing through the gap around the door presents the major hazard, particularly on designated escape routes. These doors should be fitted with smoke seals as a preventative measure. There are two main types of seal: compression seals are normally fitted into the doorstop, and wiping seals can be fitted into either the door edge or the frame itself, see Figure 6.66.

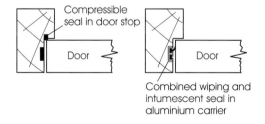

Figure 6.66 Smoke control door seals

Vision panels

These are permitted in fire doors provided they have been approved and tested. In fact vision panels in fire doors have several advantages. In addition to their general use, where they enable a person to see if the passage is clear on the other side, in the event of a fire they provide a visual warning before the door is opened. Georgian wired or a special fire-resisting safety glass is normally specified. This should be held in place using the specified method.

Untreated timber beads with the glass bedded in intumescent paste is normally suitable for lower-rated doors, although coating the beads with intumescent paint, or capping with a metal angle gives greater protection. Non-combustible glazing systems are recommended for higher-rated doors, as timber beads may not give the required levels of integrity. Figure 6.67 shows typical fire door glazing details. However, it must be remembered that any glazing to a fire door can only be undertaken if it conforms to its fire test certificate. Full details should be available from the door manufacturer.

▶ **Figure 6.67** Typical fire door vision panel glazing details

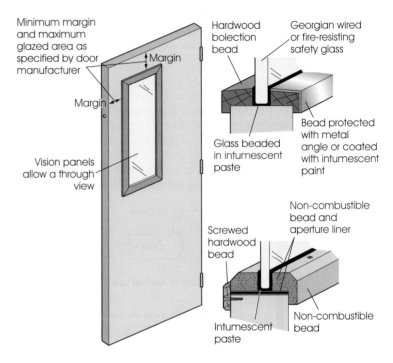

Methods of door operation

A door's method of operation will be determined by its location, construction and desired performance requirements. Figure 6.68 illustrates various methods of door operation.

Swinging or side hung doors

These are the most suitable doors for pedestrian use and the most effective for weather protection, fire resistance, sound and thermal insulation. Side hanging on hinges is the most common means of door operation, although pivoted floor springs are more effective where constant use is expected (e.g. shops and office reception areas and so on). The term **leaf** refers to the number of doors in an opening (single or double) and **swing or action** refers to the extent of travel. Single-action doors travel on one side of the door opening, whereas double-action doors can pass through the door opening and travel on both sides.

Sliding doors

These are mainly used either to economise on space where it is not possible to swing a door, or for large openings that would be difficult to close off with swinging doors.

Folding doors

These are a combination of swinging and sliding doors. They can be used as either movable internal partitions to divide up large rooms, or alternatively as doors for large warehouses and showroom entrances.

Related information

Also see Chapter 4 of Book 1, page 171, which gives details of schedules that are used to record repetitive design information. Door and ironmongery schedules when read in conjunction with floor plans and range drawings, identify the type of door, its size, numbers required, its method of operation and ironmongery to be fitted.

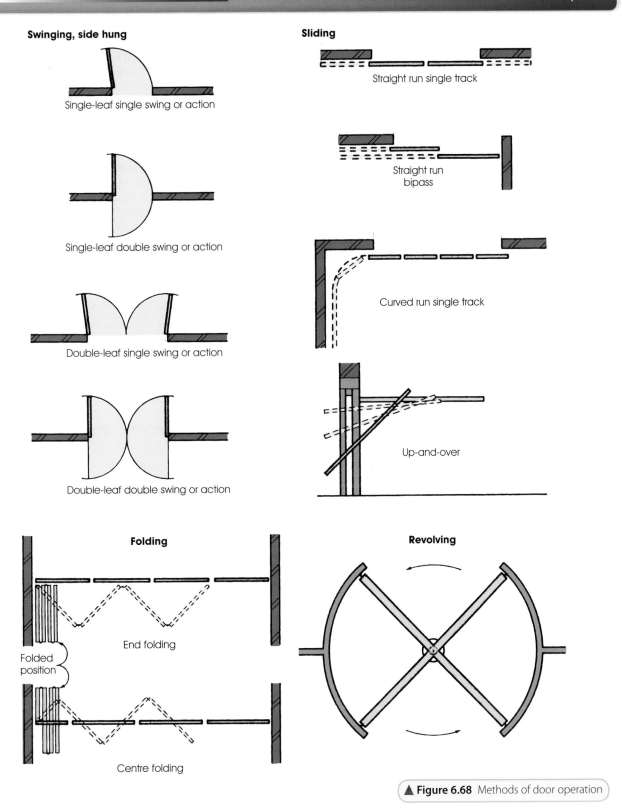

Figure 6.68 Methods of door operation

Revolving doors

These may be used as main entrances to hotels, offices, shops and so on. They allow a fairly constant flow of traffic while reducing draughts and heat loss, although they are often used adjacent to swinging doors in order to accommodate extra traffic and goods. A specialist firm normally carries out the design, construction and installation of these doors.

Door hanging

Door hanging is a second-fixing operation, which is carried out after the plastering is complete, but normally before skirting and architraves are fixed. Speed and confidence in door hanging can be achieved by following the procedures, see Figure 6.69.

1. Measure height and width of door opening.

2. Refer to door schedule(s) and select the correct door.

3. Locate and mark the top and hanging side of the opening and door. For flush and fire-check doors these should have been marked by the manufacturer. When the hanging side is not shown on the drawing, the door should open into the room to provide maximum privacy, but not onto a light switch.

4. Cut off the horns (protective extensions on the top and bottom of each stile) on panelled doors. Flush doors will probably have protective pieces of timber or plastic on each corner. These need to be prised off.

5. Where plain door linings are used, tack temporary stops either side at mid height to stop the door falling through the opening. These should be set back from the edge of the lining by the door thickness.

6. Shoot (plane to fit) in the hanging stile of the door to fit the hanging side of the opening. This should be planed at a slight undercut angle to prevent binding.

7. Shoot the door to width. Allow a 2 mm joint all around between door and frame or lining. Many woodworkers use a two pence coin to check. The closing side will also require planing to a slight angle to allow it to close. This is termed a leading edge.

8. Shoot the top of the door to fit the head of the opening. Saw or shoot the bottom of the door to give a 6 mm gap over the floor finish or to fit the threshold.

9. External doors may require rebating along the bottom edge, to fit over the water bar.

10. Mark out and cut in the hinges (see later hinge recessing procedure). Screw one leaf of each hinge to the door.

11. Offer up the door to the opening and screw the other leaf of each hinge to the frame.

12. Adjust fit as required. Remove all arrises (sharp edges) to soften the corners and provide a better surface for the subsequent paint finish. If the closing edge rubs the frame, the hinges may be proud and require the recesses to be cut deeper. If the recesses are too deep, the door will not close fully and tend to spring open, which is known as 'hinge bound'. In this case, a thin cardboard strip can be placed in the recess to pack out the hinge.

13. Fit and fix the lock or latch (see later lock-fitting procedure). This correctly positions the closing edge of the door flush with the face of the door frame or lining.

14. Remove temporary stops from plain linings and replace with planted stops pinned in position. Fix the head first, then the sides (see Figure 6.70). Rectangular section stops can be simply butted; moulded sections will require mitring. Allow a gap of up to 2 mm between the face of door and the edge of the stop. Otherwise it will bind after painting.

15. Finally, fit any other ironmongery, such as bolts, letter plates, handles and so on.

It is usual practice to only fit and not to fix the other ironmongery, such as handles and bolts and so on at this stage. They should be fixed later during the finishing stages, after all painting works are completed and the building is more secure.

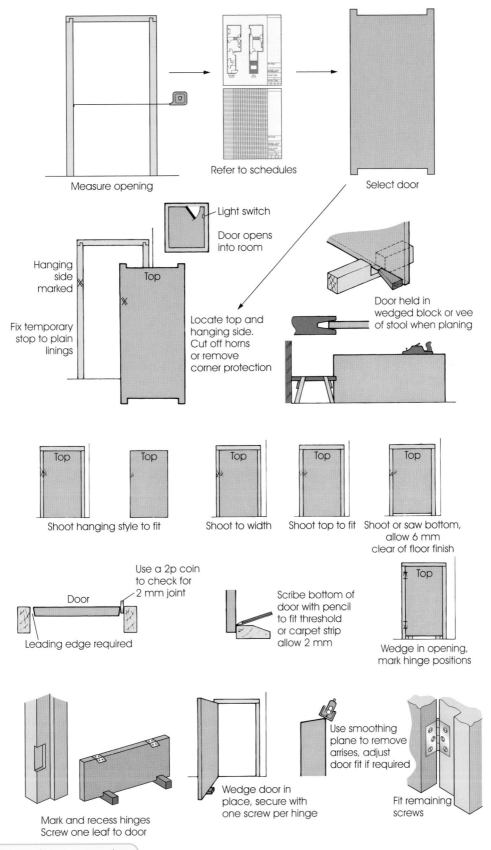

Measure opening

Refer to schedules

Select door

Light switch

Door opens into room

Hanging side marked

Top

Fix temporary stop to plain linings

Locate top and hanging side. Cut off horns or remove corner protection

Door held in wedged block or vee of stool when planing

Shoot hanging style to fit

Shoot to width

Shoot top to fit

Shoot or saw bottom, allow 6 mm clear of floor finish

Use a 2p coin to check for 2 mm joint

Door

Leading edge required

Scribe bottom of door with pencil to fit threshold or carpet strip allow 2 mm

Wedge in opening, mark hinge positions

Mark and recess hinges Screw one leaf to door

Wedge door in place, secure with one screw per hinge

Use smoothing plane to remove arrises, adjust door fit if required

Fit remaining screws

▲ **Figure 6.69** Door-hanging procedure

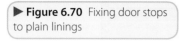
▶ **Figure 6.70** Fixing door stops to plain linings

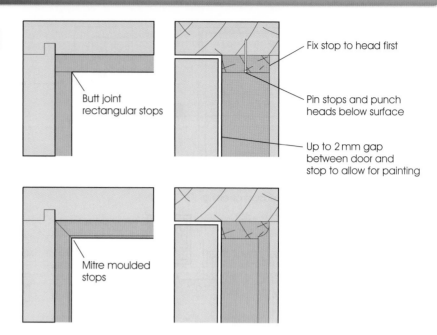

Butt joint
rectangular stops

Fix stop to head first

Pin stops and punch heads below surface

Up to 2 mm gap between door and stop to allow for painting

Mitre moulded stops

Weatherboard

Inward-opening external doors, in exposed positions, may require a weatherboard (see Figure 6.71). These are screwed to the bottom face of the door to throw rainwater clear of the water bar. The ends may be kept clear of the frame or let into it.

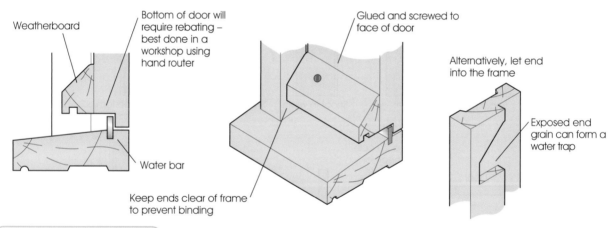

Weatherboard

Bottom of door will require rebating – best done in a workshop using hand router

Glued and screwed to face of door

Alternatively, let end into the frame

Exposed end grain can form a water trap

Water bar

Keep ends clear of frame to prevent binding

▲ **Figure 6.71** Weatherboard for inward-opening external doors

Hinge recessing

The leaves of a hinge can be recessed into the door and frame equally, termed half-and-half. See Figure 6.72. Alternatively, the hinge can be offset, with the front edge of both leaves being recessed into the door leaving a clean unbroken joint line. This all-in-the-door method is more popular for cabinet work than door hanging. The actual method used is dependent on personal preference as each is equally as good; check with the specification, foreman or customer.

The position of the knuckle in relation to the face of the door is optional, but is sometimes dependent on the hinge. Full-size room doors are mainly hung with the full knuckle protruding beyond the door face, to give increased clearance.

Some people like to see the leaf recessed up to the centre of the knuckle for a neater appearance. However, this is mainly used for cabinet doors. Once again, check the specification and personal preferences.

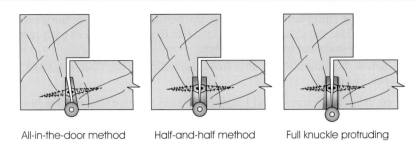

All-in-the-door method   Half-and-half method   Full knuckle protruding

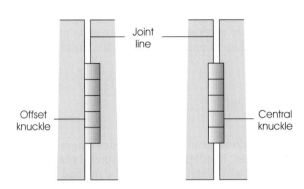

Joint line

Offset knuckle   Central knuckle

Hinge-recessing procedure

See Figure 6.73:

1. Decide on the hinge position. This will depend on the size and type of door being hung. (See Hinge size and positioning, page 179 for further details.)

2. Mark the hinge position on both the frame and the door or component to be hung.

3. Position the hinge on the edge of the door and mark the top and bottom of the leaf with a sharp pencil.

4. Repeat the process to mark the leaf position on the frame.

5. Set a marking gauge to the width of the hinge leaf and score a line on the edge of the door and frame, between the two pencil lines.

6. On frames and linings with a stuck rebated stop, a combination square can be set to mark the leaf width with a pencil.

7. Reset the marking gauge to the thickness of the hinge leaf. Gauge the door face and the frame edge at each hinge position.

8. Use a chisel held vertically and mallet to chop the ends of each hinge housing position to the recess.

9. Use the chisel bevel side down and a mallet to feather each hinge housing to depth.

▼ **Figure 6.73** Hinge recessing

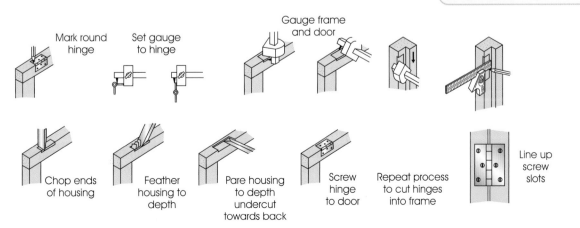

Mark round hinge   Set gauge to hinge   Gauge frame and door

Chop ends of housing   Feather housing to depth   Pare housing to depth undercut towards back   Screw hinge to door   Repeat process to cut hinges into frame   Line up screw slots

10. With the chisel bevel side up pare the feathered housing to the gauged depth. Use the chisel at a slight angle so that the bottom of each recess is slightly undercut towards the back.

11. Screw the hinges to the door first. Then offer up the door and screw to the frame.

When using brass screws and hinges, it is good practice to screw the hinges on initially with a matching set of steel screws. The brass screws, being softer, are easily damaged and may even snap off when being screwed in. The steel screws pre-cut a thread into the pilot holes for the brass screws to follow easily without risk of damage. Candle wax or petroleum jelly may be applied to lubricate the screw thread before insertion. If slot-head screws are being used, these should be lined up vertically. This gives an enhanced appearance and prevents the build up of paint or polish in the slots.

Door ironmongery

Door ironmongery is also termed door furniture and includes hinges, locks, latches, bolts, other security devices, handles and letter or postal plates. The actual items of ironmongery used will be determined by the method of operation, location, construction and desired performance requirements.

Door handing

The hand of a door must be established in order to select the correct items of ironmongery. Some locks and latches have reversible bolts, enabling either hand to be adapted to suit the situation. Traditionally, this has always been done by viewing the door from the hinge knuckle side; if the knuckles are on the left the door is left-handed, whereas if the knuckles are on the right, the door is right-handed. Doors may also be defined as either clockwise or anticlockwise closing when viewed from the knuckle side. When ordering ironmongery simply stating left-hand or right-hand, clockwise or anticlockwise can be confusing, as there may be variations between manufacturers and suppliers. The standard way now to identify handing is to use the following coding (see Figure 6.74):

◾ 5.0 for clockwise closing doors and indicating ironmongery fixed to the opening face (knuckle side).

◾ 5.1 for clockwise closing doors and indicating ironmongery fixed to the closing face (non-knuckle side).

◾ 6.0 for anticlockwise closing doors and indicating ironmongery fixed to the opening face.

◾ 6.1 for anticlockwise closing doors and indicating ironmongery fixed to the closing face.

▶ **Figure 6.74** Method for stating door handing

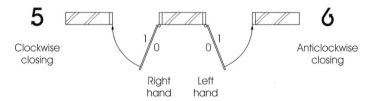

Hinge materials and types

Hinges are used to provide the opening and closing action required for most swinging doors. They are available in a variety of materials.

◾ **Pressed mild steel** hinges are the least expensive and are commonly used for internal painted doors. They are also available in different finishes such as electro-brass, bright zinc or chrome plated.

- **Solid brass and stainless steel** hinges are used for hardwood clear finished doors, both internally and externally. Steel hinges should not be used on hardwood or external doors, because of rusting and subsequent timber staining.
- **Cast iron** hinges are more suited to painted heavyweight doors. However, take care when fitting as they are brittle and easily broken if hit with a hammer.
- **Plastic and certain brass** hinges and other low-melting-point materials are not suitable for use on fire doors.

Always consult manufacturers' details regarding the suitability of hinges for a particular end use. Figure 6.75 shows a range of hinges, which are readily available in a variety of materials.

- **Butt hinge** is a general-purpose hinge suitable for most applications. It consists of two leaves, joined by a pin that passes through the knuckle formed on the edge of both leaves. As a general rule, the leaf with the greatest number of knuckles is fixed to the door frame.
- **Flush hinges** can be used for the same range of purposes as a butt hinge on both cabinets and full-size room doors. They are only really suitable for lightweight doors, but they do have the advantage of easy fitting, as they do not require 'sinking in'.
- **Loose pin butt hinges** enable easy door removal by knocking out the pins. For security reasons they should not be used for outward-opening external doors.
- **Lift-off butt hinges** also enable easy door removal, the door being lifted off when in the open position. These are available as right- or left-handed pairs. Each pair consists of one long pin hinge, which is fitted as the lower hinge. The upper hinge has a slightly shorter pin, to aid repositioning the door.
- **Washered butt hinges** are used for heavier doors to reduce knuckle wear and prevent squeaking.
- **Parliament hinges** have wide leaves to extend knuckles and enable doors to fold back against the wall, clearing deep architraves and so on.
- **Rising butts** are designed to lift the door as it opens to clear obstructions such as mats and rugs. They also give a door some degree of self-closing action. In order to prevent the top edge of the door fouling the frame as it opens and closes, the top edge must be eased. The hand of the door must be stated when ordering this item.
- **Soss hinges** are available in a range of sizes, which depend on the weight of the door. They are fitted by drilling and recessing into the door and frame. Their main feature is that they are invisible when the door is closed.
- **Strap hinges** are surface fixed, being screwed or bolted directly to the door and frame or post.
  - **Tee hinges** are used for ledged and braced doors and gates. Decorative versions are also available.
  - **Hook and band hinges** are suitable for more heavy-duty use.
- **Double and single action spring hinges** are designed to make a door self-closing. Their large knuckles contain helical springs, which can be adjusted to give the required closing action by using a tommy bar and moving the small pin at the top of the knuckles into a different hole. Where double-action hinges are used, they should not be cut into the door frame but screwed to a planted section the same width as the door and fixed to the frame.
- **Hawgood hinges** are another type of spring hinge and are suitable for industrial and heavy-duty double-swing doors. The spring housed in the cylinder is mortised and recessed into the door frame, while the moving shoe fits around both sides of the door. A twin-spring Hawgood hinge is available for use as the top hinge of very heavyweight doors.

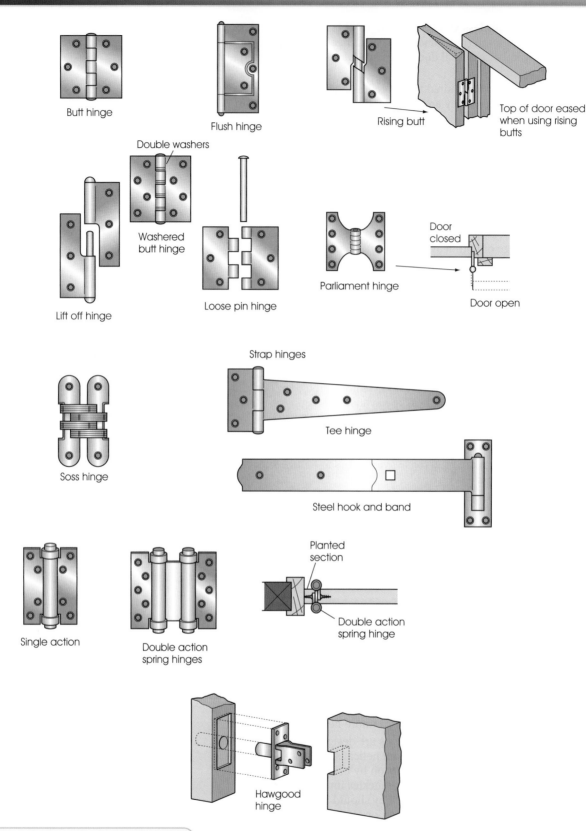

▲ **Figure 6.75** Range of hinges for doors

Hinge size and positioning

Lightweight internal doors are normally hung on one pair of 75 mm hinges; glazed, half-hour fire doors and other heavy doors need one pair of 100 mm hinges. All external doors and one-hour fire doors need three ($1\frac{1}{2}$ pairs) 100 mm hinges. The standard hinge positions for doors are shown in Figure 6.76. For flush doors, these are 150 mm down from the top, 225 mm up from the bottom and the third hinge where required, positioned centrally to prevent warping, or towards the top for maximum weight capacity. On panelled and glazed doors, the hinges are often fixed in line with the rails to produce a more balanced look. However, slight adjustment may be required to avoid the end grain of wedges.

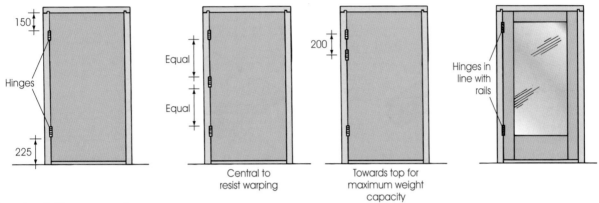

▲ **Figure 6.76** Hinge positioning

Pivoted floor springs

These provide a superior method of swinging and controlling both single- and double-action doors, although their use is restricted mainly to shops, offices and so on because of their expense.

Figure 6.77 shows single and double floor springs. Both consist of:

- a spring contained within a metal box, which is bedded into the floor screed
- a shoe, which is fixed to the bottom of the door and located over the pivot spindle
- a two-part top pivot, of which one part is fixed to the head of the frame and the other to the door.

It is essential for the smooth operation of the door that the loose box and floor spring is set into the floor at the correct position to the frame (see Figure 6.78). This is determined by the finished thickness of the door plus an allowance between the door in its opening position and frame of between 3 mm and 6 mm.

The closing action of a door fitted with a floor spring is shown in Figure 6.79. A stand-open position, normally at 90°, can be included to hold the door in the open position, although the stand-open device is not suitable for fire doors.

The delayed closing action over the first few degrees can delay the closing of the door for up to one minute before it reverts to its normal closing speed and is therefore suitable where goods, trolleys and so on have to pass through. Turning the closing adjustment screw in the floor spring will vary the door's closing action.

The single-action doors may be removed from the frame by unscrewing the head of the frame plate, tilting the door slightly outwards to clear the frame and lifting it off the bottom pivot spindle.

Double-action doors are removed by first retracting the top pivot pin. This is achieved by turning the adjustment screw in the head of the frame plate. The door can then be tilted slightly outwards and lifted off the bottom pivot.

Brush or rubber seals may be required around the edges of double-action doors to prevent the passage of smoke and draughts. These may also include intumescent strips for fire doors (see Figure 6.80).

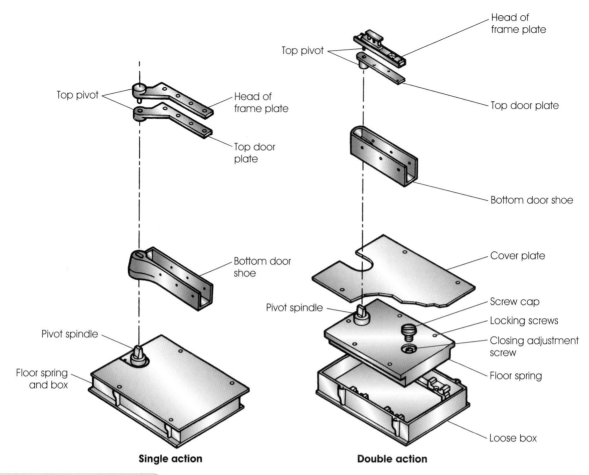

Single action   **Double action**

▲ **Figure 6.77** Pivoted floor springs

▶ **Figure 6.79** Action of door with floor spring

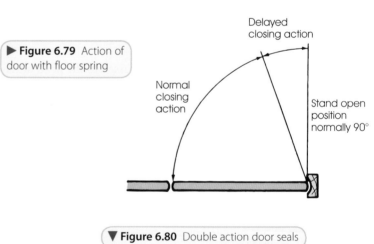

Delayed closing action

Normal closing action

Stand open position normally 90°

▼ **Figure 6.80** Double action door seals

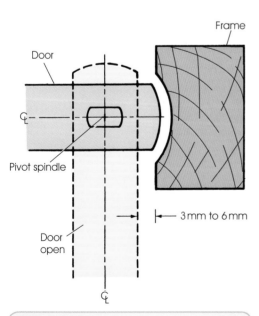

▲ **Figure 6.78** Floor spring loose box positioning

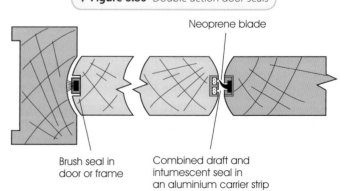

Neoprene blade

Brush seal in door or frame

Combined draft and intumescent seal in an aluminium carrier strip

Sliding door ironmongery

Various types of sliding gear are available for straight sliding, bypassing and around-the-corner arrangements. In general, top-sliding gear is used for lightweight doors and bottom-sliding gear is used for heavier doors.

Figure 6.81 shows typical details and sections of an internal sliding door that is top hung. The bottom of the door is controlled by a small nylon guide that runs in a channel set in the bottom of the door. Sliding doors require special locks and latches with hook-shaped bolts, which prevent the door being slid open. The furniture is usually flush fitting so that the doors are able to slide without fouling.

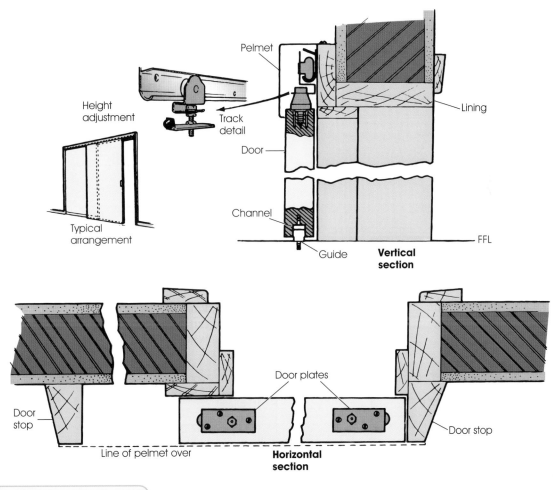

▲ **Figure 6.81** Internal sliding door

Folding-door ironmongery

Folding doors may be either end folding or centre folding. Top-hung or bottom-roller gear is available for both types.

End-folding units should consist of an even number of leaves, up to six, hinged on either jamb of the frame (Figure 6.82). For extra wide openings, additional 'floating' units of four or six leaves not hinged to either jamb may be added. An extra leaf can be added if a swinging access door is required. A leaf width of between 600 mm and 900 mm is recommended. These must have solid top and bottom rails so that a firm fixing for the edge-fitting hangers and guides can be achieved. Figure 6.82 shows the arrangement and sections of an external end-folding top-hung unit.

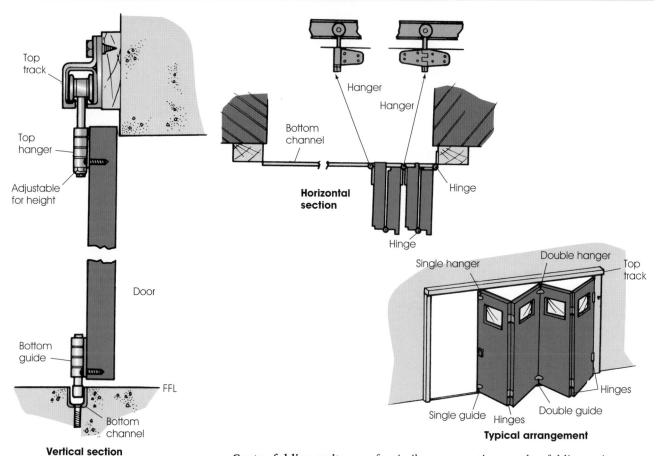

Figure 6.82 End-folding doors

Centre-folding units are of a similar construction to edge-folding units except that there will always be a half-leaf hinged to the frame to permit the centre folding arrangement. From $1\frac{1}{2}$ to $7\frac{1}{2}$ leaves may be hinged together on either jamb of the frame. An extra leaf can be added if a swinging access door is required. Figure 6.83 shows details and sections of an internal centre-folding bottom-running partition.

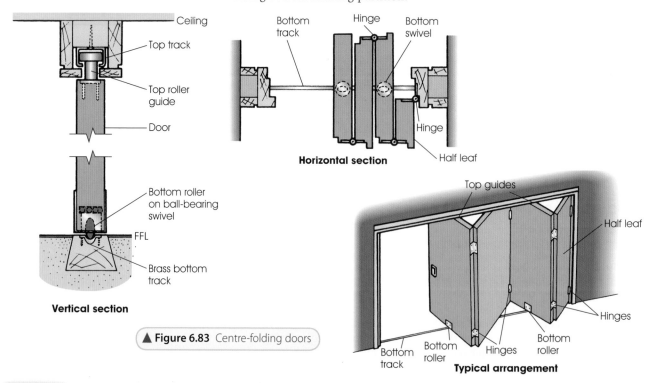

Figure 6.83 Centre-folding doors

Locks, latches and other door furniture

Always consult manufacturers' details regarding the suitability of locks, latches and other door furniture for a particular end use. Figure 6.84 illustrates a range of commonly used items, which are readily available in a variety of materials and finishes.

See Table 6.1 for typical materials used to manufacture ironmongery and their common abbreviations.

Locks and latches

- **Cylinder rim night latches.** Mainly used for entrance doors to domestic property, but as they are only a latch, provide little security on their own. When fitted, the door can be opened from the outside with the use of a key and from the inside by turning the handle. Some types have a double-locking facility that improves their security. Double-locking types are essential when used for glazed doors, as the inside handle will not turn even if an intruder breaks a pane to reach inside for it.

- **Mortise deadlocks.** Provide a straightforward key-operated locking action and are often used to provide additional security on entrance doors where cylinder rim latches are fitted. They are also used on doors where simple security is required (e.g. storerooms). The more levers a lock has, the better it is. A five-lever lock is more difficult to pick than a three-lever one. For public buildings, shops and offices, a *master keyed system* is available. The 'grand master key' will operate all locks in a system and sub-masters will operate only a specific range of locks in a system. 'Servant' keys will only operate individual locks.

- **Mortise latches** are used mainly for internal doors that do not require locking. The latch, which holds the door in the closed position, can be operated from either side of the door by turning the handle.

- **Mortise lock/latches** are available in two main types. The horizontal one is little used nowadays because of its length, which means that it can only be fitted to doors with substantial stiles. The vertical type is more modern and can be fitted to most types of doors. It is often known as a narrow-stile lock/latch. Both types can be used for a wide range of general-purpose doors in various locations. They are, in essence, a combination of the mortise deadlock and the mortise latch. A further variation is the Euro pattern mortise lock/latch, which uses a cylinder lock to operate the dead bolt. Also, bathroom privacy locks are available which use a turn button on the inside of the door.

- **Rebated mortise lock/latches** should be used when fixing a lock/latch in double doors that have rebated stiles. The front end of this lock is cranked to fit the rebate on the stiles. A conversion kit is also available for use with a standard mortise lock/latch.

- **Rim deadlocks** are surface-fixed locks, traditionally used in the place of a mortise deadlock, now rarely used.

- **Rim lock/latches** are surface-fixed versions of the mortise lock/latch. Mainly used today for garden gates and sheds. Knob furniture is used to operate the latch.

Table 6.1 Ironmongery finishes

Abbreviation	Material finish
Al	Aluminium
An Al	Anodised aluminium
B	Brass
BJ	Black japanned
BMA	Bronze metal antique
CP	Chrome plated
EB	Electroplated brass
NP	Nickel plated
PA	Polished aluminium
PB	Polished brass
PC	Powder coated
PNP	Polished nickel plated
SAA	Satin anodised aluminium
SB	Satin brass
SC	Satin chrome
SS	Stainless steel
ZP	Zinc plated

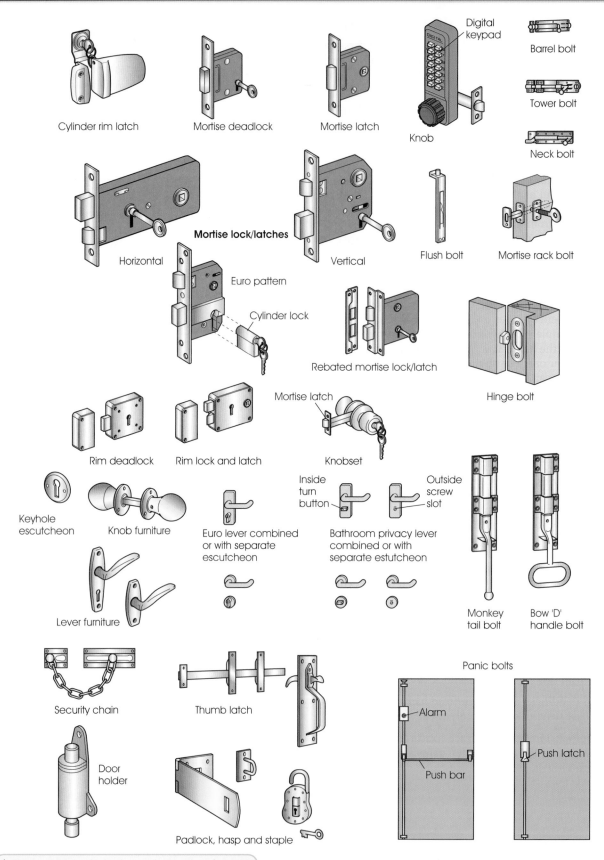

▲ **Figure 6.84** Locks, latches and other door furniture

Lock accessories

- **Knobsets.** These consist of a small mortise latch and a pair of knob handles that can be locked with a key so that it can be used as a lock/latch in most situations both internally and externally. Knobsets can also be obtained without the lock in the knob for use as a latch only.
- **Knob furniture** is for use with rim and horizontal mortise lock/latches. It should not be used with vertical mortise lock/latches, as the knob is very close to the edge of the door, hand injuries may be caused when closing the door (from the closing face). Versions are available with a digital key pad *entry system* for controlled access.
- **Keyhole escutcheon** *plates* are used to provide a neat finish to the keyhole of both deadlocks and horizontal mortise lock/latches. A bathroom privacy version has a screw slot on the outside for use in an emergency.
- **Lever furniture** is available in a wide range of patterns, for use with the mortise latches and mortise lock/latches. A bathroom privacy version has a screw slot on the outside for use in an emergency.
- **Thumb latches** are also known as Norfolk or Suffolk latches. They are used to latch doors and gates in the closed position and may be installed either side of the door.
- **Padlock, hasp and staple.** Used on external doors, such as sheds and gates, to provide security. The hasp is fixed to the door and the staple to the frame. Larger versions are fixed by bolting through the door rather than screwing, to provide greater protection.

Bolts and chains

- **Barrel and tower bolts** are used on external doors and gates to secure them from the inside. Two bolts are normally used, one at the top of the door and the other at the bottom. Heavier versions, such as the monkey tail bolt and the D-handle bolt, are used for garage and industrial doors.
- **Flush bolts** are flush fitting and, therefore, require recessing into the timber. They are used for better-quality work on the inside or edge of external doors to provide additional security and also on double doors and French windows to bolt one door in the closed position. Two bolts are normally used, one at the top of the door and the other at the bottom.
- **Mortise rack bolts** are a fluted key-operated dead bolt, which is mortised into the edge of a door at about 150 mm from the top and bottom.
- **Hinge bolts** help to prevent a door being forced off its hinges, particularly on outward-opening external doors where the hinge knuckle pin is vulnerable.
- **Security chains** can be fixed on front entrance doors, with the slide to the door and the chain to the frame. When the chain is inserted into the slide, the door will only open a limited amount until the identity of the caller is checked.
- **Panic bolts and latches** are used on the inside of emergency exit doors to bolt them closed. Pressure applied to the push bar will disengage the bolts. They can be fitted with an audible alarm to deter unauthorised use.
- **Door holders** are foot-operated door stops to hold doors and gates in the open position. They are particularly useful for holding side-hung garage doors open while driving into or out of the garage.

Positioning locks, latches and door furniture

The position of door ironmongery or furniture will depend on the type of door construction, the specification and the door manufacturer's instructions. Figure 6.85 shows the recommended fixing height for various items.

- The standard position for mortise locks and latches is 990 mm from the bottom of the door to the centreline of the lever or knob furniture spindle. However, on a panelled door with a middle rail, locks/latches may be positioned centrally in the rail's width.

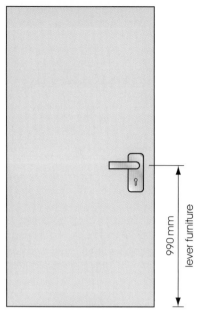

Mortise locks and latches

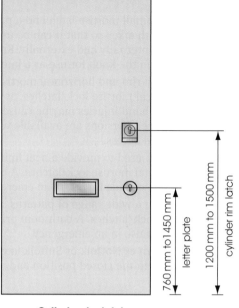

Cylinder rim latch and letter plate

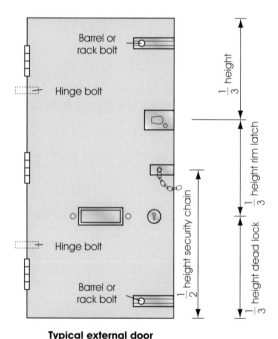

Typical external door ironmongery

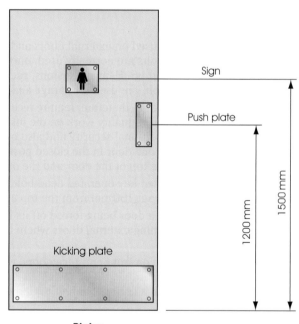

Plates and signs

▲ **Figure 6.85** Fixing heights

- Cylinder rim latches are positioned in the door's stile between 1200 mm and 1500 mm from the bottom of the door and the centre-line of the cylinder.
- Before fitting any locks/latches, the width of the door stile should be measured to ensure the lock/latch length is shorter than the stile's width, otherwise a narrow stile lock may be required.
- On external doors using both a cylinder rim latch and a mortise dead lock, the best positions for security is one third up from the bottom for the dead lock and one third down from the top for the cylinder rim latch.
- Security chains are best positioned near the centre of the door in height.
- Hinge bolts should be positioned just below the top hinge and just above the bottom hinge.

Miscellaneous door furniture

- **Letter plates** are normally positioned centrally in a door's width and between 760 mm and 1450 mm from the bottom of the door to the centre line of the plate. Again, on panelled doors, letter plates may be positioned centrally in a rail and sometimes even vertically in a stile.
- **Information signs** are normally positioned centrally in a door's width and 1500 mm from the bottom of the door to the centre of the sign.
- **Kicking plates** are fixed to the bottom face of the door for protection, keeping an even margin.
- **Push or finger plates** are positioned near the closing edge of a door at a height of 1200 mm to the centre of the plate. On panelled and glazed doors, they should be fixed centrally in the stile's width.

Before starting work, always read the job specification carefully as exact furniture positions may be stated. Also read both the door manufacturer's and the ironmongery manufacturer's instructions to ensure the intended position is suitable to receive the item, for example, the positioning of the lock block on a flush door, and to ensure the correct fixing procedure.

Door closers

These are mainly used in offices, shops and industrial premises to give these normally larger and heavier doors a controlled self-closing action, and also to enable the door to be held in the open position when required (Figure 6.86). Turning the adjustment screw, which is normally under the main cover, will alter the speed of closing. Certain types of closer are fitted with a temperature-sensitive device that automatically closes the door from the held-open position in the event of a fire.

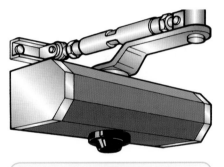

▲ **Figure 6.86** Overhead door closer

Other door closers are available with the following actions:

- Delayed closing action as with floor springs.
- Snap action over the last few degrees of travel to ensure that the door closes fully into the latch.
- Back check action, which controls the last few degrees of opening travel, preventing the door from slamming into a return wall adjacent to the opening.

A concealed spring door closer is shown in Figure 6.87. The cylinder, which contains the spring, is mortised into the edge of the door and the plate recessed flush. The anchor plate is recessed in and screwed to the frame. Adjustment to the closing action is achieved by inserting the metal plate near the cylinder plate to hold the chain, unscrewing the anchor plate and turning it either clockwise to speed the action or anticlockwise to slow the closing action.

◀ **Figure 6.87** Concealed door closer

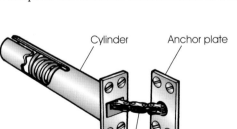

Cylinder   Anchor plate

Cylinder plate   Chain

Metal plate to hold chain for adjustment

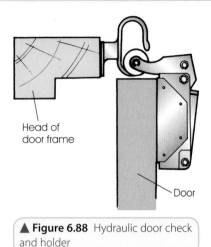

Head of
door frame

Door

▲ Figure 6.88 Hydraulic door check
and holder

The **hydraulic door check and holder** shown in Figure 6.88 can be used in conjunction with spring door closers or spring hinges to control the last few degrees of closing travel. This prevents slamming and holds the door firmly closed against strong winds.

In order to ensure that self-closing double doors with rebated stiles close in the correct sequence, a door selector must be fitted (see Figure 6.89).

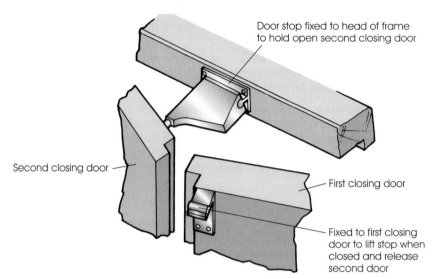

Door stop fixed to head of frame
to hold open second closing door

Second closing door

First closing door

Fixed to first closing
door to lift stop when
closed and release
second door

▶ Figure 6.89 Double-door selector

Installation of door ironmongery

Installation of mortise dead lock, latch or lock/latch

See Figure 6.90. Always consult manufacturers' instructions.

1. Wedge the door in the open position. Use the lock as a guide to mark the mortise position on the door edge at the pre-marked height.
2. Set a marking gauge to half the thickness of the door. Score a centre line down the mortise position to mark for drilling.
3. Select a drill bit the same diameter as the thickness of the lock body. An oversize bit will weaken the door; use an undersize one and additional paring will be required. A hand brace and bit or power drill and spade bit may be used. Mark the required drilling depth on the bit using a piece of masking tape. Drilling too deep will again weaken the door as well as risk the drill breaking out right through the stile in panel and glazed doors.
4. Drill out the mortise working from the top, each hole slightly overlapping the one before.
5. Use a chisel to pare away waste between holes to form a neat rectangular mortise.
6. Slide the lock into the mortise and mark around the faceplate.
7. Remove the lock and use a marking gauge to score deeper lines along the grain, as this helps to prevent the fine edge breaking out or splitting when chiselling out the housing.
8. Use a chisel to form the housing for the faceplate (let in).
9. The surface can first be feathered (as when recessing hinges) and finally cleaned.
10. Hold the lock against the face of the door, with the faceplate flush with the door edge and lined up with the faceplate housing. Mark the centre positions of the spindle and keyhole as required with a bradawl.

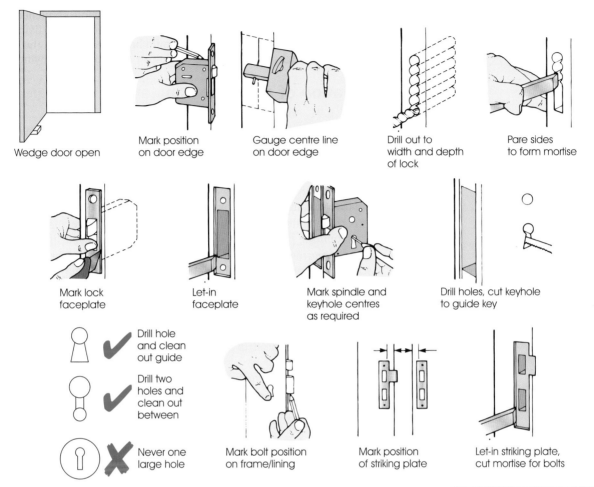

Wedge door open   Mark position on door edge   Gauge centre line on door edge   Drill out to width and depth of lock   Pare sides to form mortise

Mark lock faceplate   Let-in faceplate   Mark spindle and keyhole centres as required   Drill holes, cut keyhole to guide key

Drill hole and clean out guide

Drill two holes and clean out between

Never one large hole   Mark bolt position on frame/lining   Mark position of striking plate   Let-in striking plate, cut mortise for bolts

▲ **Figure 6.90** Fitting procedure for mortise locks and latches

11. Use a 16 mm drill bit to drill the spindle hole, working from both sides to avoid breakout, or clamp a waste piece to the back of the door.

12. Use a 10 mm drill bit to drill the keyhole, again working from both sides. Cut the key slot with a padsaw and clean out with a chisel to form a key guide. Alternatively, use a 6 mm drill bit to drill out a second hole below the 10 mm hole and chisel out waste to form a key guide. Never drill a larger hole for the key as it will not give a guide when inserting the key, making it a hit-and-miss affair.

13. Insert the lock, check the spindle and keyholes, and align from both sides. Secure the faceplate with screws. Re-check that the key works.

14. With the dead bolt out, close the door against the frame and mark the bolt and latch positions on the edge of the frame. Square these positions across the face of the frame.

15. Set the adjustable square from the face of the door to the front edge of the latch or dead bolt. Use it to mark the position on the face of the frame.

16. Hold the striking plate over the latch or dead bolt position and mark around the striking plate. Gauge vertical lines to prevent breakout when chiselling.

17. Chisel out to let in the striking plate. Again you may find it easier to feather first before finally cleaning out. The extended lead-in or lip for the latch may require a slightly deeper bevelled housing or recess.

18. Check for fit and screw the striking plate in place. Select a chisel slightly smaller than the striking plate bolt-holes.

19. Chop mortises to accommodate both the latch and dead bolt. Some striking plates have boxed bolt-holes. These must be cut beforehand.

20. Finally, fit the lever furniture, knob furniture or keyhole escutcheon plates as appropriate and check for smooth operation. When fixing keyhole escutcheon plates, the key should be passed through the plate and into the lock and centralised on the key shaft before screwing.

Installation of a cylinder rim latch

See Figure 6.91. Always consult manufacturers' instructions.

1. Wedge the door in an open position. Use the template supplied with the latch at the pre-marked height to mark the centre of the cylinder hole.
2. Use a 32 mm auger bit in a brace or a spade bit in a power drill to drill the cylinder hole. Drill from one side until the point just protrudes. Complete the hole from the other side to make a neat hole, avoiding breakout or, alternatively, by cramping a block to the door.
3. Pass the cylinder through the hole from the outside face and secure it to the mounting plate on the inside with the connecting machine screws.
4. Ensure that the cylinder key slot is vertical before fully tightening the screws. For some thinner doors these machine screws may require shortening before use with a hacksaw. If required, secure the mounting plate to the door with woodscrews.
5. Check the projection of the flat connection strip. This operates the latch and is designed to be cut to suit the door thickness. If necessary, use a hacksaw to trim the strip so that it projects about 15 mm past the mounting plate.
6. Align the arrows on the backplate of the rim latch and the turnable thimble.
7. Place the rim latch case over the mounting plate ensuring that the connection strip enters the thimble.
8. Mark out and let-in the rim latch lip in the edge of the door if required.
9. Secure the rim latch case to the door or mounting plate with wood or machine screws as required. Check both the key and inside handle for smooth operation.
10. Close the door and use the rim latch case to mark the position of the keep (striking plate) on the edge of the door frame.
11. Open the door and use the keep to mark the lip recess on the face of the frame. Chisel out a recess to accommodate the keep's lip. Secure the keep to the frame using woodscrews.
12. Finally, check from both sides to ensure a smooth operation.

▼ **Figure 6.91** Fixing a cylinder rim night latch

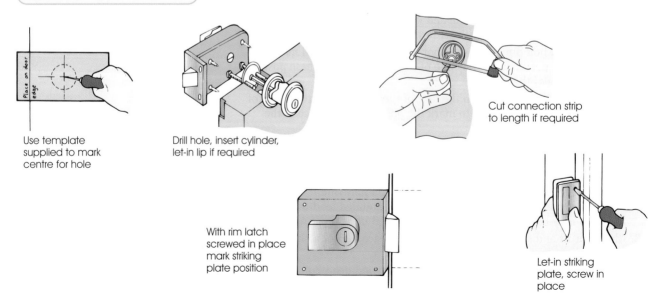

Use template supplied to mark centre for hole

Drill hole, insert cylinder, let-in lip if required

Cut connection strip to length if required

With rim latch screwed in place mark striking plate position

Let-in striking plate, screw in place

Installation of a letter plate

See Figure 6.92. Always consult manufacturers' instructions.

1. Wedge the door in the open position. Mark the centre line of the plate on the face of the door.
2. Position the plate over the centre line and mark around it.
3. Measure the size of the opening flap and mark the cut-out on the door. Allow about 2 mm larger than the flap, to ensure ease of operation.
4. Mark the position of the holes for the securing bolts.
5. If the door is easily removed, cramp an off-cut of timber to the back of the door to prevent damage from drill breakout. Alternatively, the holes can be drilled from both sides.
6. Drill holes for the fixing bolts and at each corner of the flap cut-out.
7. Use a jigsaw or padsaw to saw from hole to hole.
8. Neaten up the cut-out using glass paper if required. Remove the arris from the inside edges.
9. Position the letter plate and secure, using the fixing bolts.
10. Check the flap for ease of operation and adjust if required.

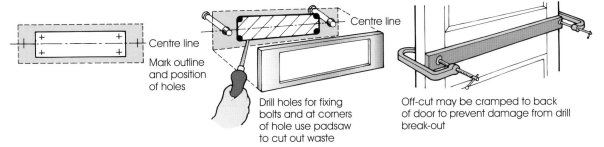

Centre line

Mark outline and position of holes

Centre line

Drill holes for fixing bolts and at corners of hole use padsaw to cut out waste

Off-cut may be cramped to back of door to prevent damage from drill break-out

▲ **Figure 6.92** Installing a letter plate

Installation of barrel and tower bolts

See Figure 6.93. Always consult manufacturers' instructions.

1. Place the bolt in the required position. Mark one of the screw holes through the backplate with a bradawl or pilot drill.
2. Insert a screw, ensuring the bolt is parallel or square to the edge of the door, and insert a screw at the other end of the bolt.

▼ **Figure 6.93** Installing barrel and tower bolts

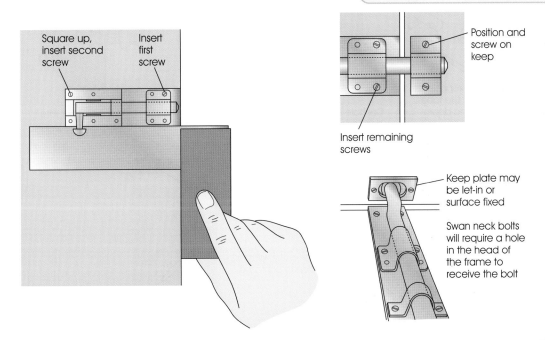

Square up, insert second screw

Insert first screw

Position and screw on keep

Insert remaining screws

Keep plate may be let-in or surface fixed

Swan neck bolts will require a hole in the head of the frame to receive the bolt

3. Move the bolt to the locked position and slide the keep over the bolt.
4. Mark the screw holes in the keep and screw in place.
5. Check the bolt works smoothly before inserting the remaining screws.

Cranked or swan-necked bolts will require a hole drilling in the head of the door frame to receive the bolt.

1. Position and secure the bolt as before.
2. Slide the bolt to the locked position and mark around the bolt.
3. Use an auger bit slightly larger than the width of the bolt to drill a hole in the marked position. Ensure the bolt can slide to its full length.
4. Check for smooth operation.

Installation of a mortise rack bolt

See Figure 6.94. Always consult manufacturers' instructions.

1. Use a marking gauge to score a centre line on the edge of the door.
2. Mark the centre of the bolt-hole using a try square and pencil. Transfer the line onto the face of the door that the bolt will operate from.

▼ **Figure 6.94** Installing a mortise rack bolt

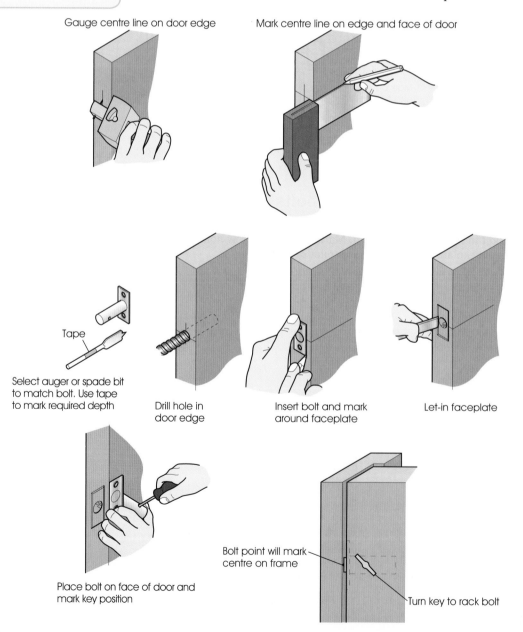

Gauge centre line on door edge

Mark centre line on edge and face of door

Tape

Select auger or spade bit to match bolt. Use tape to mark required depth

Drill hole in door edge

Insert bolt and mark around faceplate

Let-in faceplate

Place bolt on face of door and mark key position

Bolt point will mark centre on frame

Turn key to rack bolt

3. Select an auger or spade bit slightly larger than the bolt barrel. Drill a hole in the edge of the door. Use a piece of tape wrapped around the bit as a depth guide.

4. Insert the bolt and then turn the faceplate so that it is parallel to the door edge. Mark around the faceplate with a pencil.

5. Remove the bolt and gauge the parallel edges. Using a sharp chisel let-in the faceplate.

6. Place the bolt on the face of the door. Use the bradawl to mark the key position.

7. Use a 10 mm bit to drill the keyhole through the face of the door, into the bolt-hole.

8. Insert the bolt into the hole then check that the faceplate finishes flush and that the keyhole lines up. Insert the fixing screws.

9. Insert the key through the keyhole plate into the bolt. Position the keyhole plate centrally over the key and insert the fixing screws.

10. Close the door and turn the key to rack the bolt to its locked position. The point on the bolt will mark the clearance hole centre position on the frame.

11. Drill the clearance hole for the bolt in the frame on the marked centre.

12. Close the door. Rack the bolt and check for the correct alignment of the bolt clearance hole.

13. Let-in the keep centrally over the clearance hole and insert the fixing screws.

14. Check for smooth operation.

Door frames and linings

The timber surround to the wall opening on which doors are hung may be termed as either a door frame or lining. The main differences between the two are explained below and shown in Figure 6.95.

- **Position:** door frames are mainly used for external doors, while door linings are mainly used for internal doors. An exception to this would be for heavy-weight internal doors such as fire doors, which are hung in a frame.

- **Sectional size:** door frames are of a bigger section, typically 95 × 58 mm, in order to provide sufficient strength and security; door linings are of a lighter section, typically 25 mm to 38 mm thick and 115 mm or 138 mm wide, to cover the full finished thickness of the internal wall.

- **Profile:** door frames for external use normally have a rebated section with stuck-on solid stops to accommodate the door rather than a plain section with planted-on door stops, giving them increased security and weathering properties. Internal door frames can have a plain or rebated section; door linings normally use planted-on doorstops, however, stuck-on rebated stop linings are also available and termed as door casings.

- **Weather and draft proofing:** external door frames will normally include anti-capillarity measures, such as a mortar key, throats and capillary grooves, to prevent rainwater entering the building; horizontal members will be weathered on their upper surface to shed rainwater and grooved on the undersurface to make rainwater drip off clear of the building. Weathering measures do not prevent drafts, so existing door frames can be fitted with a surface-mounted compressible seal. More modern door frames are supplied complete with a grooved-in PVC or foam draft seal.

Frames

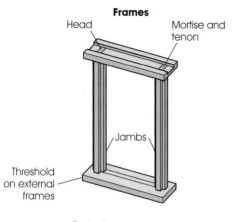

Head

Mortise and tenon

Jambs

Threshold on external frames

Linings

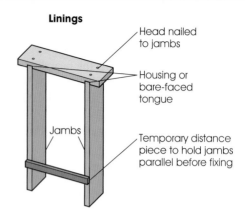

Head nailed to jambs

Housing or bare-faced tongue

Jambs

Temporary distance piece to hold jambs parallel before fixing

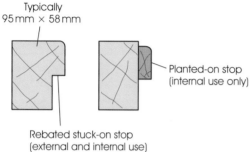

Typically 95 mm × 58 mm

Planted-on stop (internal use only)

Rebated stuck-on stop (external and internal use)

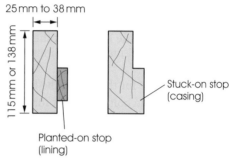

25 mm to 38 mm

115 mm or 138 mm

Planted-on stop (lining)

Stuck-on stop (casing)

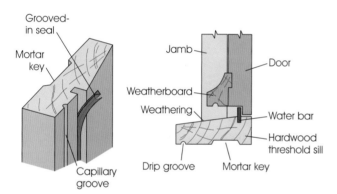

Grooved-in seal

Mortar key

Capillary groove

Jamb

Door

Weatherboard

Weathering

Water bar

Hardwood threshold sill

Drip groove

Mortar key

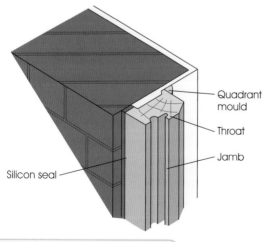

Quadrant mould

Throat

Jamb

Silicon seal

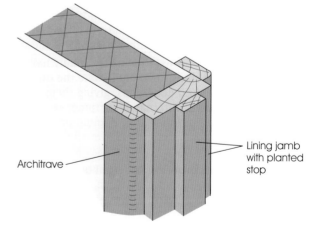

Architrave

Lining jamb with planted stop

▲ **Figure 6.95** Door frames and linings

- **Assembly:** door frames for good-quality work are normally fully assembled in the joinery workshop, while it is common practice to produce linings in a knockdown (flat pack) form to be assembled on site by the carpenter. Cheaper-quality door frames, mass produced or off-the-shelf. may also be supplied as a flat pack for site assembly. The use of flat packs reduces both transport costs and expensive joinery works manufacturing overheads but places greater quality-control issues on site operatives.

- **Installation detail:** external door frames are normally positioned slightly back from the outside face brickwork, with the outer joint between frame and brickwork sealed with silicone or acrylic to prevent rainwater penetration and the inner gap between the frame and plasterwork covered with a quadrant mould or scribed architrave to mask the effects of shrinkage. Architraves are fixed around door linings to cover the joint between the plaster and the lining. Internal frames can be fixed centrally in the reveal or flush with the wall on one side and use both architraves and quadrant moulds.

- **Materials:** softwood is widely used for both frames and linings that are to be painted. The threshold of external frames should be of hardwood to provide durability and protection from wear. Hardwood frames and linings for clear finished work are normally purpose made. Linings are also manufactured using 30 mm MDF. These can be ready primed as a protection from moisture for a paint finish. Veneered MDF can be used to give the appearance of a 'solid hardwood' lining and are suitable for clear finish: Melamine-faced MDF (MFMDF) also termed 'foil wrapped' is pre-finished, giving the appearance of wood grains or solid colour. The use of moisture-resistant MDF (MRMDF) is preferable, especially in high-moisture areas, such as kitchens and bathrooms, which could result in standard MDF soaking up moisture and swelling, particularly at joints and the feet of the jambs.

Standard door frames

Door frames consist of a head, two jambs and, where required, a threshold and transom (see Figure 6.96). These component parts are jointed together with draw-pinned mortise and tenon joints. Draw pins are used in preference to wedges, as they will hold the joint even if the horn is cut off. In addition, the use of draw pins has the advantage of not requiring cramps to pull up the joint during assembly. By off-setting the hole in the tenon slightly towards the shoulder, the joint will be drawn up tight as the pin is driven in (see Figure 6.97). The draw pin can simply be cut-off flush with the frame after driving in, or have a wedge added.

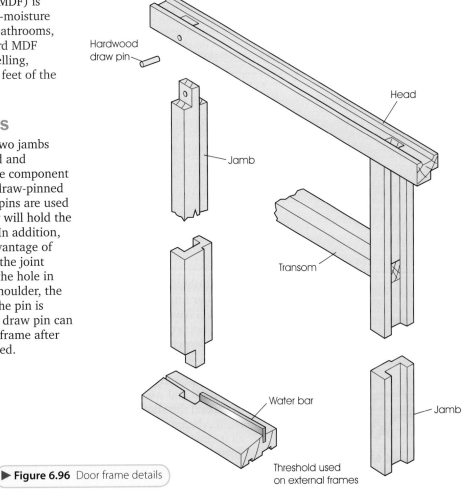

Hardwood draw pin

Head

Jamb

Transom

Water bar

Jamb

Threshold used on external frames

▶ **Figure 6.96** Door frame details

▶ **Figure 6.97** Draw pinning (joint cut away to show detail)

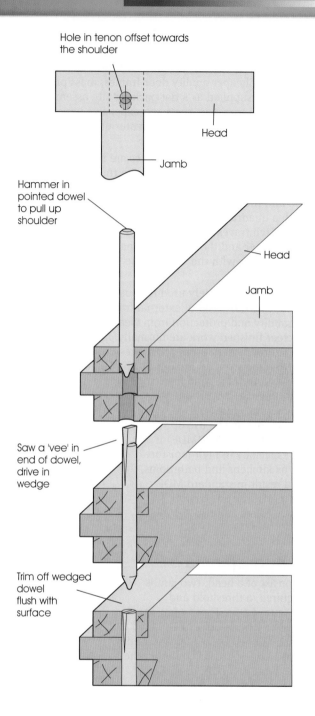

Hole in tenon offset towards the shoulder

Head

Jamb

Hammer in pointed dowel to pull up shoulder

Head

Jamb

Saw a 'vee' in end of dowel, drive in wedge

Trim off wedged dowel flush with surface

▲ **Figure 6.98** Proprietary metal weatherboard and seal

The threshold of external frames should have a galvanised metal or plastic water bar fitted to prevent the passage of driving rain under the door. The door will have to be rebated along its bottom edge to fit over the bar. As an additional protection against rainwater running down the face of the door and underneath, a weatherboard can be fitted to throw it clear.

As an alternative to the traditional water bar and timber weatherboard, two-piece proprietary metal systems are available that incorporate strips to act as both draft and weather seals. Various types are available but basically consist of a metal threshold section screwed to the timber sill and a metal weatherboard section screwed to the face and bottom of the door. As with all ironmongery and proprietary fittings, read the manufacturer's instructions before fitting as some types require bedding in silicone before screwing in place while others do not. Figure 6.98 illustrates this alternative threshold detail.

Storey-height door frames

These are used in frames either to provide glazing above a door termed a fanlight or borrowed light, or provide stability to thin non-load-bearing internal blockwork partitions. They are known as storey-height frames as they normally extend from floor to ceiling (one storey).

External storey-height door frames (see Figure 6.99) are used to provide additional light into an entrance, particularly where solid doors are used. The glazed area, termed a fanlight, is formed by extending the jambs and inserting a transom to act as the door head. Traditionally, the transom extends beyond the face of the jamb for weathering purposes and is jointed using a double mortise and tenon, with the overhanging edge of the transom housed across the face of the jambs for improved durability. The fanlight can be directly glazed into the rebates using beads or by adding a fixed or opening casement, as shown in Figure 6.100.

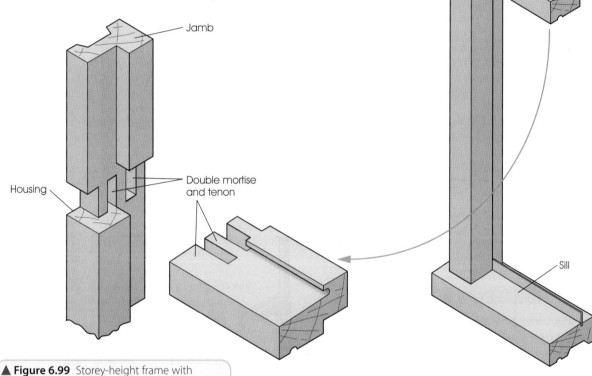

Jamb

Housing

Double mortise and tenon

Head

Transom

Sill

▲ **Figure 6.99** Storey-height frame with fanlight for external use

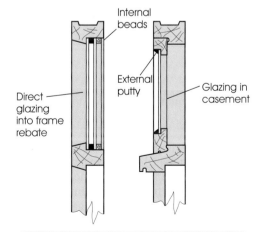

Internal beads

Direct glazing into frame rebate

External putty

Glazing in casement

▲ **Figure 6.100** External fanlight either directly glazed or with a casement

Heritage Link

Traditionally, **fanlights** were a semi-circular window above an entrance door or other window, with radiating glazing bars resembling a woman's open fan. Early examples use wooden glazing bars, but the demand for more elaborate shapes led to the introduction of composite metal fanlights, consisting of lead mouldings soldered onto a metal strip. These were a feature of many Georgian buildings, particularly those of Scottish architect Robert Adams.

Internal storey-height door frames are used either to simply provide a borrowed light above a door frame (Figure 6.101) or to provide stability for blockwork partition walls at door openings, with or without a borrowed light (Figure 6.102). Both types are best constructed using mortise and tenon joints. Single joints are suitable for plain frames; double joints may be required for rebated and wider section frames. The jambs of frames used to provide wall stability extend past the head to provide a point of fixing to the floor or ceiling joists above. Both the jambs and head are grooved out on their back face to receive the building blocks. In addition, the jambs above the head are cut back to finish flush with the blockwork and allow for the finished thickness of plaster or plasterboard.

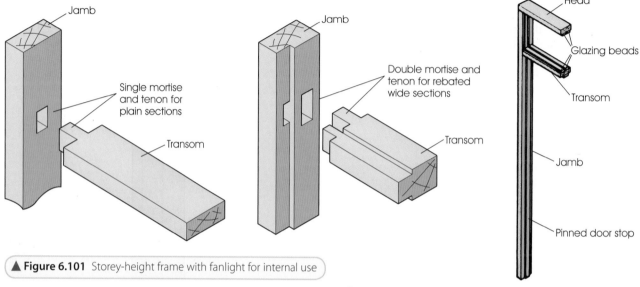

▲ **Figure 6.101** Storey-height frame with fanlight for internal use

Heritage Link

Traditionally, storey-height frames can also be termed clerestory or clear-storey frames, which can be applied to any window or light in the upper part of a wall, that gives light to the inner space of a large building.

The term is particularly applied to the lights or windows above the nave of a church, where the walls rise above the roofline of the lower aisles.

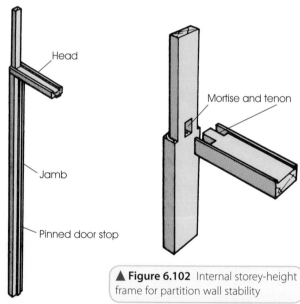

▲ **Figure 6.102** Internal storey-height frame for partition wall stability

The term **borrowed light** can be used to refer to any glazing in an internal wall, which connects a day lit room to an internal room or space that has no natural light. This borrowed daylight from an adjacent room with an external window can provide adequate illumination for internal circulation spaces, storage areas and so on and so is seen as energy efficient as it makes the use of artificial light unnecessary during daylight hours (see Figure 6.103).

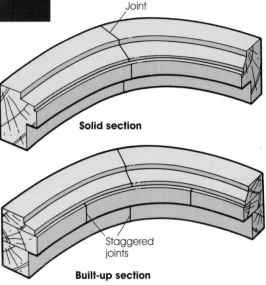

▼ **Figure 6.104** Forming curved members

Joint

Solid section

Staggered joints

Built-up section

Curved head frames

It is normal to form the head of curved door frames from solid sections, which are joined together to produce the required shape. Alternatively, the curved member may be built up using a number of layers with end joints staggered and glued together (see Figure 6.104).

A third method of forming the curved head is by glulam construction. However, this is not often used as it is the most expensive, requires a cramping jig and the grain pattern resulting from the thin laminations may not be suitable for high-quality clear finished work.

The shape of the head is used to classify curved head door frames. Four typical shapes, along with suitable jointing methods, are shown in Figure 6.105.

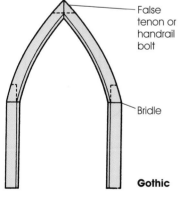

False tenon or handrail bolt

Bridle

Gothic

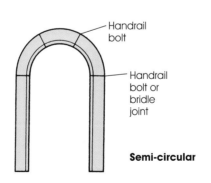

Handrail bolt

Handrail bolt or bridle joint

Semi-circular

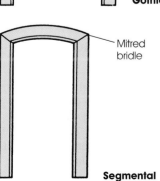

Mitred bridle

Segmental

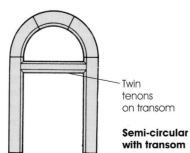

Twin tenons on transom

Semi-circular with transom

> **Related information**
>
> Also see Chapter 4, page 112, which covers the principles of glulam construction.

◀ **Figure 6.105** Curved head door frames

The main methods used to join the members are: bridle joints pulled up and secured with draw pins; butt joints pulled up and secured using handrail bolts, with the addition of dowels to prevent twisting; or false tenons pulled up and secured with draw pins (see Figure 6.106). Where a transom is required near the springing point, it should be jointed to the jamb using two tenons in order to minimise short graining problems. Traditionally, hammer-headed tenons and keys were also used to join the curved members together. They are rarely used today, as they are not suitable for machine production methods.

▶ **Figure 6.106** Curved head frame joints

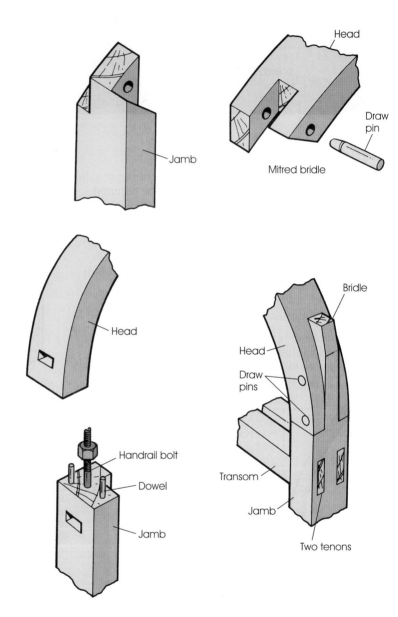

Combination frames, glazed screens and entrance lobbies

Combination frames that incorporate both a door and side glazing (also known as a side light) are often used for front entrances. The general arrangement of members for an external combination frame is shown in the elevation, vertical and horizontal sections (see Figure 6.107). The middle rail of the door has been continued across the side light as a feature. However, it also serves to break up the area of glass and provide protection at a vulnerable point.

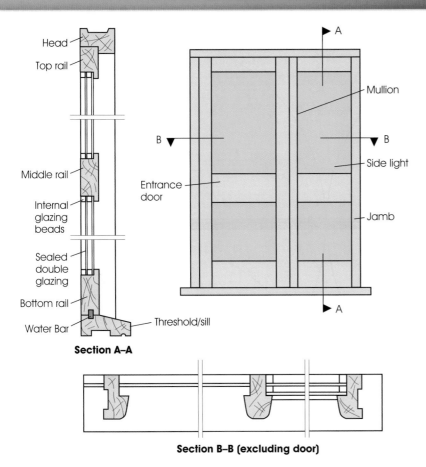

Section A–A

Head
Top rail
Middle rail
Internal glazing beads
Sealed double glazing
Bottom rail
Water Bar
Threshold/sill

A
Mullion
B
Side light
B
Entrance door
Jamb
A

Section B–B (excluding door)

Figure 6.107 External combination frame with side light

> **Related information**
>
> Also see Chapter 7, page 221, which gives further information concerning combination frames.

Glazed screens or partitions are used internally to divide up space and form a borrowed light. Details of a glazed screen that incorporates a double-action door are shown in Figure 6.108. Although Figure 6.108 shows a door, this could be omitted and the same detail used to construct a glazed partition. When the frame is fixed the mullion and plinth on the side screen must be secured with the aid of galvanised steel dowels grouted into the floor.

Entrance lobbies or vestibules were traditionally a common feature at the front entrance of public buildings, banks, offices and so on. Figure 6.109 shows a plan of such a lobby. It consists of a pair of single-action

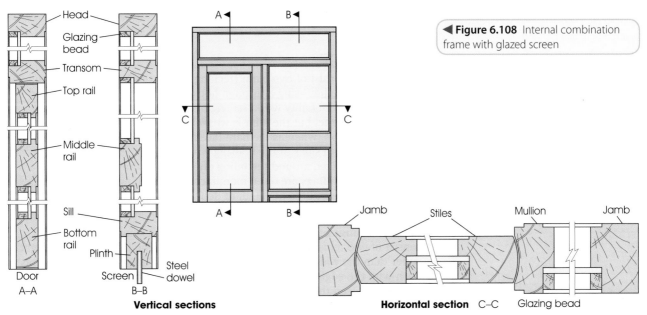

Head
Glazing bead
Transom
Top rail
Middle rail
Sill
Bottom rail
Plinth
Door A–A
Screen B–B
Steel dowel
Vertical sections

A
B
C
C
A
B

Jamb
Stiles
Mullion
Jamb
Horizontal section C–C
Glazing bead

Figure 6.108 Internal combination frame with glazed screen

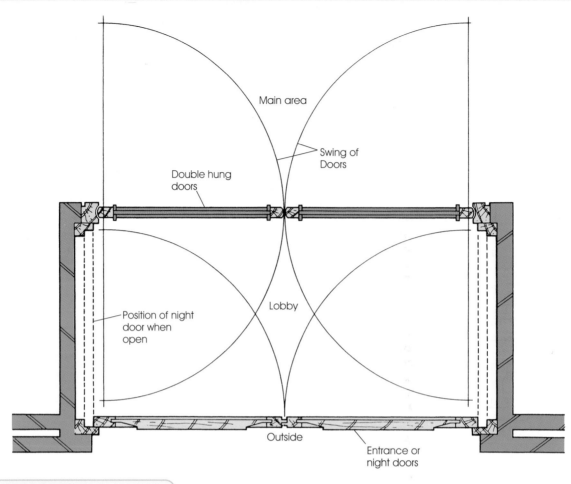

Main area

Swing of Doors

Double hung doors

Lobby

Position of night door when open

Outside

Entrance or night doors

▲ **Figure 6.109** Plan of entrance lobby

entrance doors leading to a pair of double-action doors. The entrance doors are only intended to be closed when the building is locked up out of business hours, so they are often referred to as night doors. During the day they are kept open to give the impression of a panelled lobby. A pictorial sketch of this lobby with the entrance doors in the open position is shown in Figure 6.110. The actual walls of the lobby may be constructed in various ways, including brickwork, as shown in the plan, or studwork framing.

Door linings

Plain linings consist of two plain (rectangular section) jambs and a plain head joined together using a through housing or a bare-faced tongue and housing for better-quality work, see Figure 6.111. Through housings are simpler to construct but the joint is weakened when the horn is removed before installation, whereas the bare-faced tongue is stronger as it is still supported by both sides of the housing when the horn is removed. The planted stop is fixed around the lining after the door has been hung. Mass-produced linings are often manufactured with housings on both sides of the head to suit different standard door widths. This reduces costs as the supplier does not have to stock two different width linings.

Rebated linings or door casings are used for better-quality work. They consist of two rebated jambs and a rebated head. The rebate must be slightly wider than the width of the door so that when the door is hung it will finish flush with the edges of the lining, but still have a working clearance of 1.5 mm between the door and the stop in order to allow for finishing and prevent the door from binding or rubbing, see Figure 6.112.

▲ **Figure 6.110** Entrance lobby with night doors open

▼ **Figure 6.111** Plain lining

▼ **Figure 6.112** Rebated lining

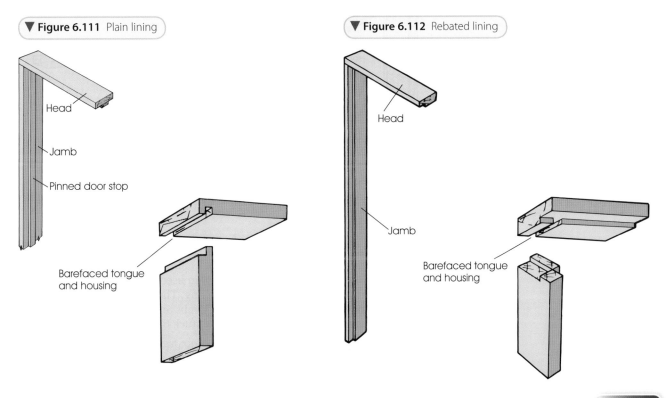

Head

Jamb

Pinned door stop

Barefaced tongue
and housing

Head

Jamb

Barefaced tongue
and housing

Storey-height linings may be constructed from either plain or rebated sections. In both types the transom is normally through housed into the jamb. The glazing can either be bedded into the rebate and held in place with pinned glazing beads or a twin set of beads (normally of the same section as the doorstop) can be used, depending on the type of lining, see Figure 6.113.

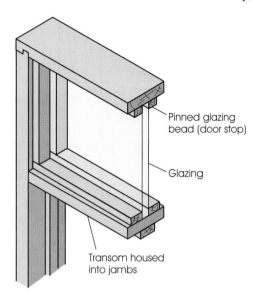

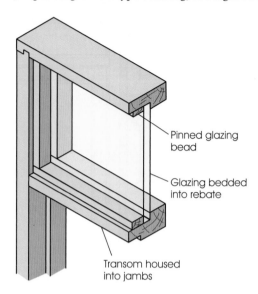

▲ **Figure 6.113** Storey-height linings (glazing detail)

Skeleton lining are used for deeper reveals, where the internal division wall (often load-bearing) is too thick for a normal lining to be used. They consist of a basic framework that is stub tenoned together. A stopped bare-faced tongue and housing is used between the head and the jambs. A plywood, MDF or solid timber lining is used to cover the framework and form the rebate to receive the door, see Figure 6.114.

▶ **Figure 6.114** Skeleton lining

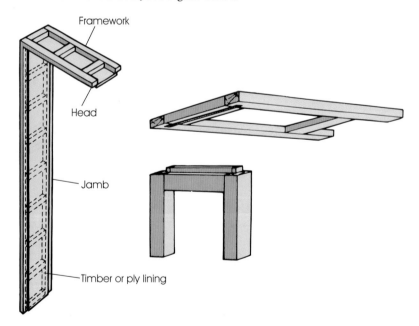

Related information

Also see Chapter 10, page 312, which covers the construction of wall panelling.

Panelled linings (see Figure 6.115) are again used in thick walls and particularly in conjunction with wall panelling. Their construction is similar to skeleton linings except that the framed jambs and head are grooved out to receive the panels and the stop is normally stuck on the solid.

Splayed linings (see Figure 6.116) are mainly used in conjunction with a door frame to finish deep reveals. They give a lead into the opening and create a distinctive feature. The linings, which may be plain, rebated,

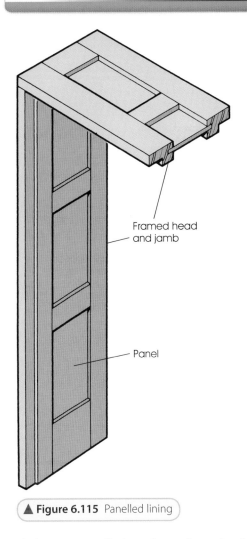

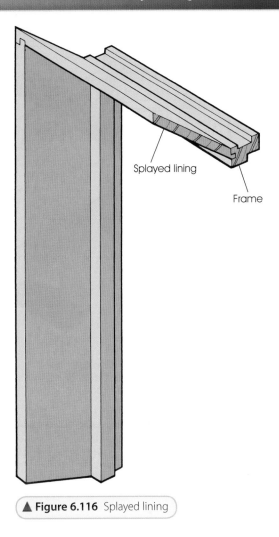

Splayed lining

Frame

Framed head and jamb

Panel

▲ **Figure 6.115** Panelled lining

▲ **Figure 6.116** Splayed lining

skeleton or panelled, are housed or mitred together at their head and tongued into the frame. In order to construct the lining, the true shape and bevels will have to be geometrically developed (see Book 1).

Assembly and fixing of door frames and linings

It is standard practice to assemble purpose-made door frames in the joiners shop, mass-produced frames may be supplied in flat pack form, for on-site, assembly by the carpenter. The majority of linings are also supplied as flat packs.

Extra care must be taken when assembling frames and linings on site to avoid damage. Assembly on a bench is ideal. Alternatively, levelled bearers (to prevent winding) or stools are suitable, but never directly on the floor.

Frames with a threshold are normally assembled with temporary braces between the jamb and head for squaring purposes; these are simply nailed in place at an angle of approximately 45°. Frames without a threshold require a temporary distance piece (also termed a stretcher) and braces to hold the jambs parallel and square (see Figure 6.117). Both the braces and the distance piece should remain in place until the frame is securely fixed. To check for square the diagonal measurements from the head to threshold or jamb feet are compared. Measurements should be equal if the frame is square. If not, apply pressure across the longest opposite corners until the diagonal measurements are equal.

> ### Related information
>
> Also see Chapter 9 in Book 1, page 396, which covers the geometrical development of splayed linings.

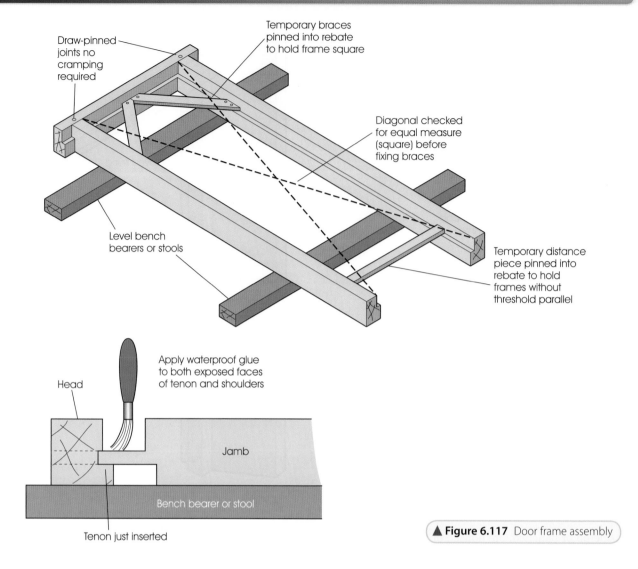

Draw-pinned joints no cramping required

Temporary braces pinned into rebate to hold frame square

Diagonal checked for equal measure (square) before fixing braces

Level bench bearers or stools

Temporary distance piece pinned into rebate to hold frames without threshold parallel

Apply waterproof glue to both exposed faces of tenon and shoulders

Head

Jamb

Bench bearer or stool

Tenon just inserted

▲ **Figure 6.117** Door frame assembly

Linings: apply glue to the housing joints and tongue shoulders, secure with screws or round head nails in a dovetail pattern. Temporary braces, distance piece and squaring procedures are the same as for frames. Two distance pieces may be used particularly on wide linings, one on either side of the jambs. This helps to keep the jambs parallel across their width and prevent them from twisting during fixing. When squaring, the diagonal may be measured from the distance piece, but ensure it has been positioned parallel to the head; say about 150 mm up from the feet of the jambs (see Figure 6.118).

Fixing of door frames and linings

Door frames can be either 'built in' or 'fixed in'.

▪ Built-in frames are fixed into a wall or other element by bedding in mortar and surrounded with the walling components.

▪ Fixed-in frames are inserted into a ready-formed opening after the main building process at the first-fixing stage before the walls are plastered.

Built-in frames

The bricklayer fixes the majority of frames by building in as the brickwork proceeds. Prior to this, the frame has to be prepared, accurately positioned, plumbed, levelled and temporary-strutted by the carpenter. Temporary struts are used to hold the frame upright. The feet of the door frame jambs, in the

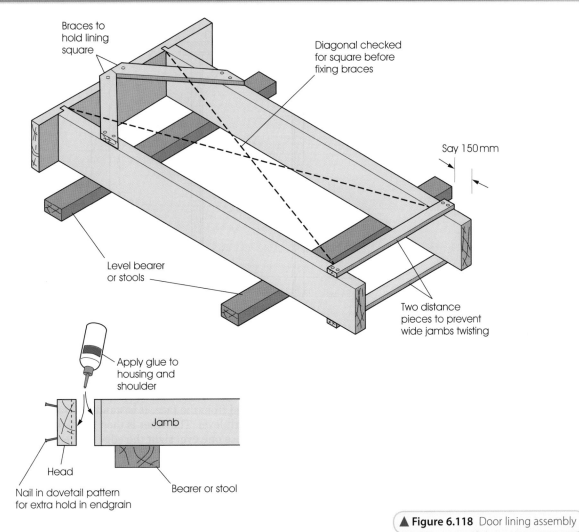

Braces to hold lining square

Diagonal checked for square before fixing braces

Say 150 mm

Level bearer or stools

Two distance pieces to prevent wide jambs twisting

Apply glue to housing and shoulder

Jamb

Head

Nail in dovetail pattern for extra hold in endgrain

Bearer or stool

▲ **Figure 6.118** Door lining assembly

absence of a threshold, are held in position by galvanised metal dowels, which are drilled into the end of the jambs and are grouted into the concrete (see Figure 6.119).

Temporary braces and distance pieces are left in place to keep the frame square and the jambs parallel during the building-in process.

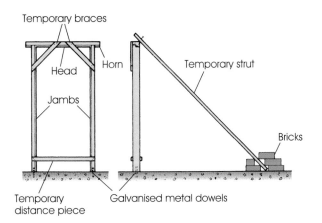

◀ **Figure 6.119** Building in a frame

Temporary braces

Head

Horn

Temporary strut

Jambs

Bricks

Temporary distance piece

Galvanised metal dowels

The frame's head should be checked for level and packed up as required; frames with thresholds are normally bedded level using bricklayer's mortar (see Figure 6.120).

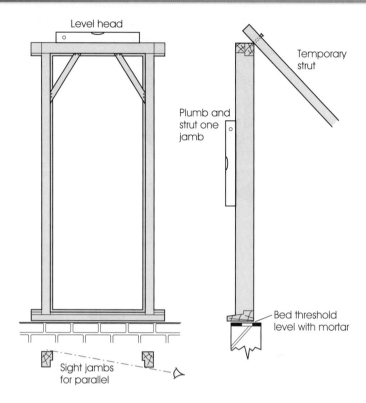

► **Figure 6.120** Building in a frame with threshold

Level head

Temporary strut

Plumb and strut one jamb

Bed threshold level with mortar

Sight jambs for parallel

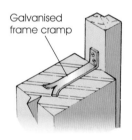

Galvanised frame cramp

▲ **Figure 6.121** Attaching frame cramps

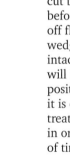

Head

▲ **Figure 6.122** Horn cut back ready for building in

Jambs should be plumbed from the face. It is standard practice to plumb and strut the first using a spirit level. The other is then sighted parallel: stand to the side of the frame, close one eye, sight the edge of the plumbed jamb with the edge of the other and adjust if required until both jambs are parallel. The frame is said to be out of wind if jambs are sighted parallel and in-wind or winding when they are not sighted parallel.

As the brickwork proceeds, galvanised metal frame cramps should be screwed to the back of the jambs and built into the brickwork, see Figure 6.121. Four cramps should be evenly spaced up each jamb.

The horns of the frame should be cut back as shown in Figure 6.122 before building in, rather than cut off flush. This not only keeps wedged mortise and tenon joints intact, but the horns, when built in, will help secure the frame in position. After trimming the horns it is essential that the cut ends are treated with paint or preservative in order to reduce the possibility of timber rot.

Storey-height frames used to provide stability in thin block work partitions are positioned before building the blockwork (see Figure 6.123).

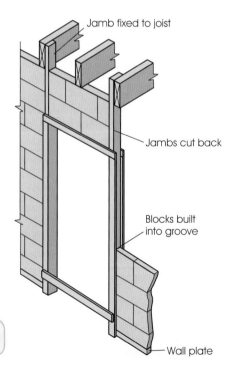

Jamb fixed to joist

Jambs cut back

Blocks built into groove

Wall plate

► **Figure 6.123** Building in a storey-height frame

The storey frame should be fixed in position by nailing or screwing, at the bottom to the wall plates and at the top to the joists. As with other built-in frames, ensure the head is level and the jambs are both plumb and parallel to each other. Three or four equally spaced frame cramps should be used to secure each jamb to the blockwork.

Fixed-in frames

These are installed into prepared openings in the wall and apply mainly to expensive hardwood frames and pre-finished timber, metal or uPVC door sets. This is to protect them from possible damage or discoloration during the building process. In addition, frames that were not available during the building process or replacement frames will have to be fixed in.

Four fixings are required for each jamb, say 100 mm from both ends, and the other two equally spaced between. Various methods can be used to fix the frame to the wall, see Figure 6.124.

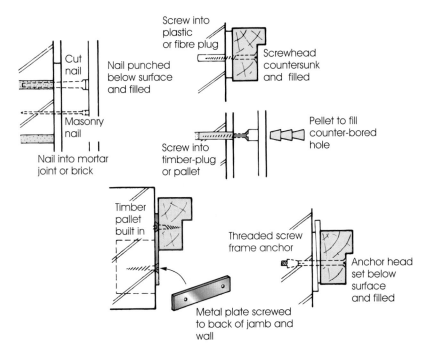

◀ **Figure 6.124** Methods of fixing frames

- Nailing through the jamb using cut nails into the blockwork or brickwork mortar joints or masonry nails into the actual brick.
- Screwing through the jamb using either plastic plugs and screws or special masonry screws on their own. Screwheads in softwood frames may be countersunk below the surface and filled. Screwheads in hardwood frames should be concealed by counter-boring and pelleting.
- Metal plates used as fixing lugs screwed at intervals to the back of jamb before the frame is put in the opening. The lugs are screwed and plugged to the brick or block reveals.
- Frame anchors – proprietary fixing consisting of a metal or plastic sleeve and matching screw. The jamb and reveal are drilled out to suit the sleeve, which is inserted in position and screwed up tight.

Sequence of operations to fix a frame as shown in Figure 6.125:

1. Assemble frame if supplied as a flat pack.
2. Saw off horns flush with the back of the jambs, as they are not required as a fixing on fixed-in frames. Remember to paint or preservative treat the cut ends.
3. Insert metal dowels into the end grain of the jambs feet if required as a fixing into the concrete floor, or later laid cement screed.

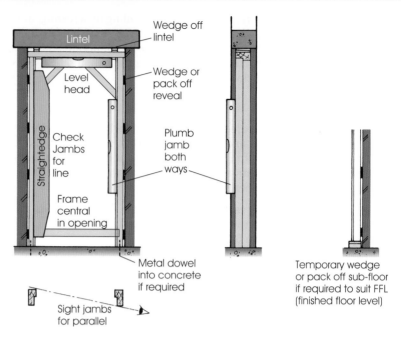

► **Figure 6.125** Positioning fixed-in frames

4. Place the frame centrally in the prepared opening, temporarily holding it in position with the aid of folding wedges off the lintel. The position of the jamb on the reveal will depend on the frames location, normally just back from the outside face for external work and either flush with one face of the finished wall or centrally in the reveal for internal work. Check specification and drawings for specific information.
5. Check the head or sill for level and adjust the wedges as required. Wedges or packers may be required at the feet of the jambs where a cement screed is to be later laid over a concrete sub-floor.
6. Plumb one jamb both ways using a spirit level and 'sight in' the other.
7. Check jambs for line using a long straightedge. Use packers or folding wedges to adjust if required.
8. Fix the top and bottom of one jamb first, followed by the two intermediate fixings. Ensure fixings will not interfere with the later positioning of hinges, lock striking plates and so on. Fixings should be through the packers to prevent them being dislodged at a later stage. Where folding wedges are used, fixings should be positioned just below them. After any adjustment they can be secured by nailing into them through the jamb.
9. Check the jambs are still parallel, adjust if required and fix second jamb as in step 8.
10. Re-check frame for level, plumb, line and winding.

Fixing door linings

Door linings are normally fixed in place. The opening in the wall to receive the lining is formed while the wall is being built and the lining is fixed at a later stage. Fixings may be:

■ **Nailing to twisted wooden plugs** (see sequence of operations)
■ **Nailing or screwing to timber pallets** that have been built into the brick joints by the bricklayer, or into the door stud of a stud partition. Folding wedges are used as packings down the sides of the jambs.
■ **Nailing directly into the blockwork** reveal or brickwork mortar joint. Folding wedges or other packings will be required.
■ **Using plugs and screws** or other proprietary fixing.

Sequence of operations to fix a lining (using twisted timber plugs) as shown in Figure 6.126:

1. Assemble lining. This is normally done by skew nailing through the head into the jambs.

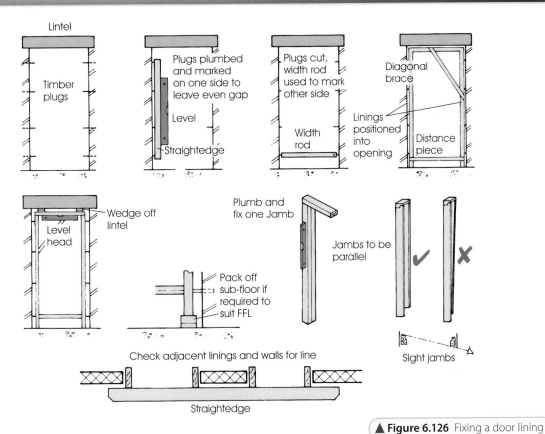

▲ **Figure 6.126** Fixing a door lining

2. Fix a distance piece near the bottom of the jambs and, when required, diagonal brace or braces at the head.

3. Rake out brickwork joints and plug (see Figure 2.127).

4. There should be at least four fixing points per jamb. Omit this stage if the bricklayer has built in wooden pallets or pads into the brickwork.

5. Offer lining into opening and mark where the plugs need to be trimmed. The plugs should project equally from both reveals.

6. Cut the plugs and check the distance with a width rod. The ends of the plugs should be in vertical alignment. Check with a straightedge and spirit level.

7. Fix lining plumb and central in the opening by nailing or screwing through the jambs into the plugs. Before finally fixing check head for level, wedge off lintel, ensure the lining is out of wind: check by sighting through the jambs. When fixing to unplastered walls, check adjacent linings and wall surfaces are lined up.

8. Ensure lining jambs are packed up off a concrete sub-floor to suit the finished floor level (FFL) if required.

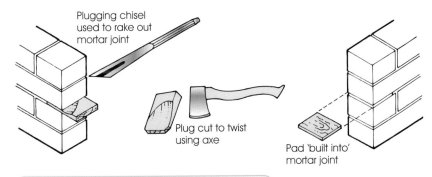

▲ **Figure 6.127** Plugs or pads used for fixing linings

Door sets

Door sets consist of a pre-hung door and frame or lining, complete with all ironmongery (Figure 6.128). Loose-pin or lift-off hinges are normally used for easy door removal. Architraves are normally fixed to one side of the set and loose pinned on the other. All work is undertaken in the joinery shop and supplied to site ready for fixing into its opening. The fixing of door sets should not be carried out until at least the second fixing stage. This is to avoid the possibility of any damage during the building process. The main advantages of using doorsets are the joiner's shop quality and the considerable saving in on-site time to install. Door sets are normally installed at the second-fixing stage, after the building has been plastered and allowed to dry out.

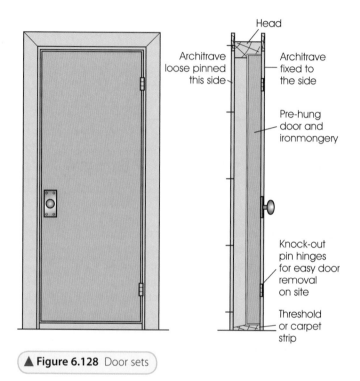

▲ **Figure 6.128** Door sets

7

Chapter Seven

Windows

This chapter covers the work of both the site carpenter and bench joiner. It is concerned with principles of constructing and installing windows. It includes the following:

- Classification of windows.
- The performance requirements of windows.
- Methods of window construction.
- Installation of windows.
- Types of window ironmongery.
- Single and double glazing.
- Fixing of windows and window boards.
- Window ventilation.

Windows are glazed areas in walls and roofs to allow natural lighting into buildings and to give the occupants an outside view. In addition, they may also be used to provide rapid natural ventilation. They consist of glass panes held into a timber, metal or plastic frame, which in turn is fixed to the wall or roof structure.

The frame material is used as the first classification of windows, followed by whether or not they open. Windows that do not open are termed fixed or direct glazing; others are termed by their method of opening and whether or not they project from the main building line.

General classification of windows

The majority of windows come under one more of the following groups, see Figure 7.1.

- **Fixed glazing.** Glazing that is fixed directly into an opening or sub frame; also termed **direct glazing**.
- **Casements.** Normally opening windows that are either top- or side-hung on hinges to a main frame, but may also be fixed.
- **Pivot-hung windows.** A type of casement that is either horizontally or vertically hung and includes what is termed tilt-and-turn windows.
- **Sliding sashes.** Opening windows that slide in a main frame either horizontally or vertically.
- **Bay windows.** These project beyond the main building line, often giving an increased floor area in the room. Bays that project from an upper storey only are known as **oriel windows**.

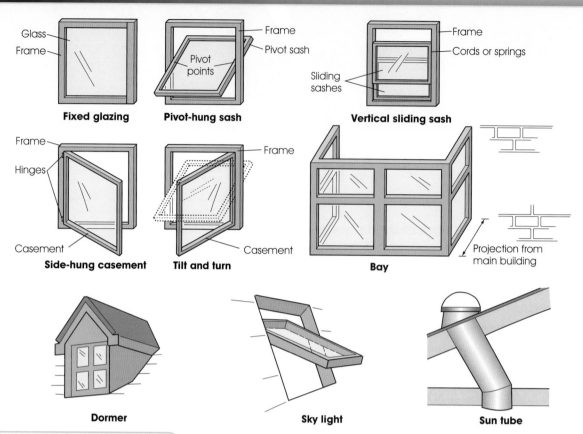

Fixed glazing   **Pivot-hung sash**   **Vertical sliding sash**

Glass

Frame

Frame

Pivot points

Pivot sash

Frame

Cords or springs

Sliding sashes

Frame

Hinges

Casement

Side-hung casement

Frame

Casement

Tilt and turn

Casement

Bay

Projection from main building

Dormer   **Sky light**   **Sun tube**

▲ **Figure 7.1** Classification of windows

■ **Roof windows.** These include dormer windows (normally casements that project above the main roof line) and skylights that follow the main roof slope. Both can be installed in new buildings to provide natural daylight or installed at a later stage as part of an extension or refurbishment project.

■ **Sun or solar tubes.** Although not windows in the true sense, these can be used to provide daylight to hallways and other areas. They capture sunlight on the roof through a small dome and reflect the light down a tube, which runs through the roof space. The light comes through a ceiling-mounted diffuser resembling a recessed light fitting; often LED lights are incorporated to provide dual function after dark.

The recommended method of indicating the type of window and its opening arrangement is shown in Figure 7.2.

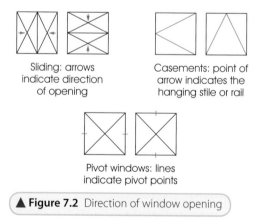

Sliding: arrows indicate direction of opening

Casements: point of arrow indicates the hanging stile or rail

Pivot windows: lines indicate pivot points

▲ **Figure 7.2** Direction of window opening

Design features

The design of windows requires careful consideration. In addition to the main functions, they have other performance requirements, similar to those of external doors. These are weather protection (including thermal insulation), security, ease of operation, durability and sound insulation. The extent to which these apply and their order of importance will vary from job to job. However, in every case the following points should be borne in mind. They can be considered the principles of window design, and apply mainly to timber windows.

- Use only durable, preservative-treated timber with suitable moisture content.
- Ensure all framing joints are tight fitting and use a waterproof adhesive.
- Prime or seal the whole assembly, preferably directly after manufacture.
- External horizontal surfaces should be weathered.
- The horizontal underside of projecting sills and transoms must be provided with a drip edge.
- The opening joint must be a tight fit for draught proofing, but must open out within rebate to avoid capillarity.
- Anti-capillary grooves and throatings should be incorporated around opening parts.
- External glazing in painted work is best secured with weathered face putty and sprigs.
- External glazing in clear finished work and internal glazing should be bedded in a suitable material but secured with cupped and screwed glazing beads from the inside of the building.
- Rectangular window openings should have well-balanced proportions related to their height and width.
- A 'golden rectangle', having sides in the proportion of approximately 5:8, can be used to determine the size of rectangular window openings with well-balanced proportions, see Figure 7.3. The method used is to first draw a square with sides equal to the smallest dimension, divide it in half to form two equal-sized rectangles and then swing the diagonal of one rectangle down to give the longest side. Golden rectangles are equally suitable for the proportions of other items of joinery where a pleasing, balanced effect is required.

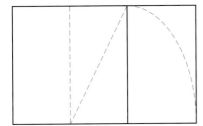
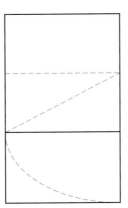

▲ **Figure 7.3** Forming golden rectangles

The majority of timber windows in use are made in a variety of standard styles and sizes by joinery manufacturers using volume production techniques to produce a standard product of known quality at an economical price. A typical example from a timber window manufacturer's range is shown in Figure 7.4. Rather than the mass production, the carpenter and joiner will be more involved with window construction from the following points of view: the like-for-like replacement of rotting windows in maintenance work; the one-off or limited batch of windows for an extension to match those existing; the production of limited batch runs to a purpose-made design or size.

▶ **Figure 7.4** Example from window manufacturer's range of casements

Window type		Frame dimensions width × height	
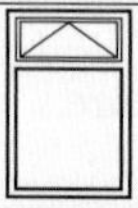	V	634 × 921 634 × 1073 634 × 1226 634 × 1378	
	V	921 × 921 921 × 1073 921 × 1226	
	V	1221 × 1073 1221 × 1226	Side hung
	CV AS CV OPP	1221 × 1073 1221 × 1226	
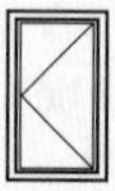	C AS C OPP	634 × 1073	
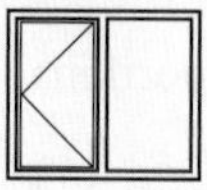	CD AS CD OPP	1221 × 1073	
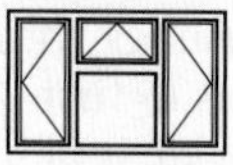	CVC	1808 × 1073 1808 × 1226	Fanlight opening

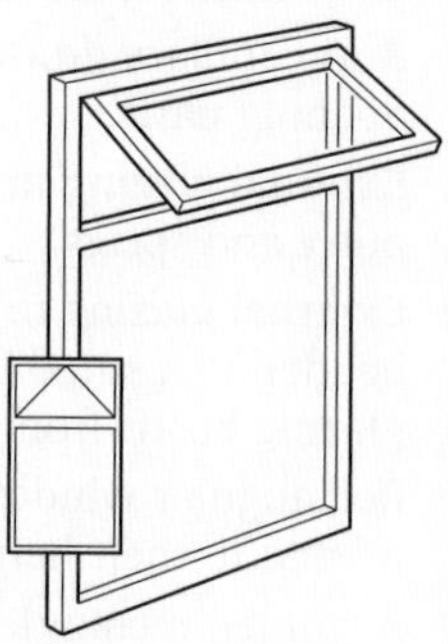

Notes: **V** = Ventlight   **C** = Casement   **OPP** = Opposite hand
D = Fixed light   **AS** = Hand as shown

Casement windows and fixed lights

Casement windows are comprised of two main parts:
- **The frame**, which consists of head, sill and two jambs. Where the frame is subdivided, the intermediate vertical members are called mullions and the intermediate horizontal members are called transoms.
- **The casement**, which can be fixed or opening, consists of top rail, bottom rail and two stiles. Where the casement is subdivided, both the intermediate vertical and horizontal members are called glazing bars.

Opening casements above the transom are known as **fanlights** or **vent lights**.

Fixed glazing is called a dead light and glazing at the bottom of a window, normally below a casement is a sub-light. Where glass is bedded in the mainframe itself, it is known as direct glazing or direct lights.

The elevation of a four-light casement window is shown in Figure 7.5 with all of the component parts named. The 'four' refers to the number of glazed openings or lights in the window frame, in this case the casement has also been sub-divided into four lights using glazing bars.

Casement windows can further be divided into two types, traditional and storm-proof, depending on their design and methods used in construction. See Figure 7.6.

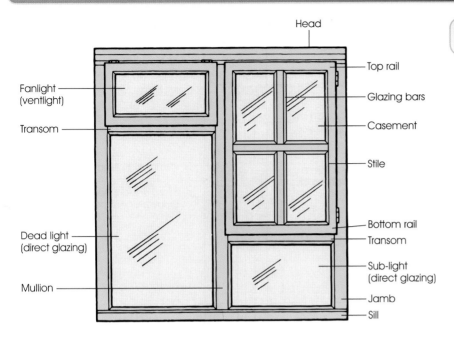

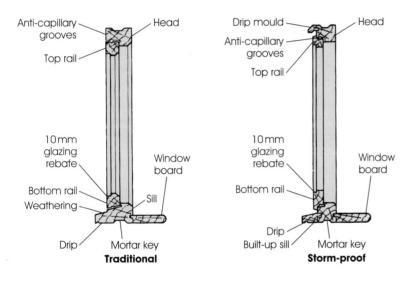

Traditional casements

These are fully rebated into and sit flush with the main frame. Early forms had a casement sash in each section of the main frame, whether or not they opened to create a balanced effect. However, due to cost implications more modern windows tend to only use casement sashes in the opening sections. A vertical section through a traditional casement window is shown in Figure 7.7. Anti-capillary grooves are incorporated into the frame and the opening casements in order to prevent the passage of water into the building. Drip grooves are made towards the front edges of the transom and sill to stop the water running back underneath them.

A mortar key groove is run on the outside face of the head, sill and jambs. The sill also has a plough groove for the internal windowsill or board to tongue into. Both the transom and sill incorporate a throat to check the penetration of wind-assisted rain. In addition, this feature may be continued up the jambs. Finally, the fronts of the transom and sill are weathered: they have a 9° slope for the rainwater to run off.

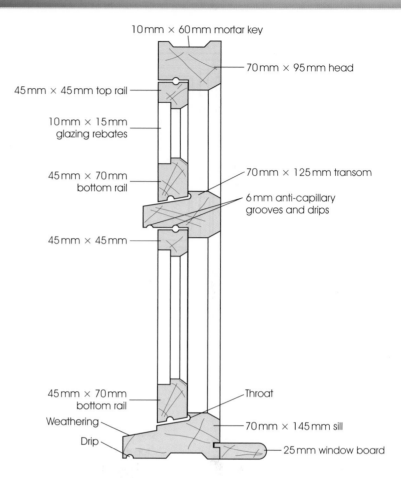

10 mm × 60 mm mortar key

70 mm × 95 mm head

45 mm × 45 mm top rail

10 mm × 15 mm glazing rebates

45 mm × 70 mm bottom rail

70 mm × 125 mm transom

6 mm anti-capillary grooves and drips

45 mm × 45 mm

45 mm × 70 mm bottom rail

Throat

Weathering

70 mm × 145 mm sill

Drip

25 mm window board

Figure 7.8 shows part of a horizontal section through a traditional casement window. It also shows the sizes and positions of the rebates, grooves and moulding in the jambs, mullion and the casement stiles.

► **Figure 7.8** Traditional casement window (part horizontal section)

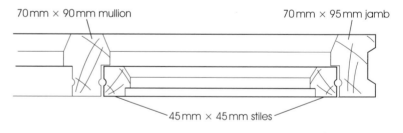

70 mm × 90 mm mullion

70 mm × 95 mm jamb

45 mm × 45 mm stiles

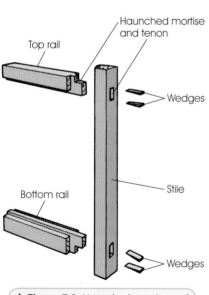

Haunched mortise and tenon

Top rail

Wedges

Stile

Bottom rail

Wedges

▲ **Figure 7.9** Haunched mortise and tenon joint secured with wedges

All joints used in traditional casement window construction are mortise and tenons. Standard haunched mortise and tenons are generally used for the actual casements, although a sash haunch is preferable where smaller sections are used, see Figure 7.9. As a matter of good practice, the depth of the rebates should be kept the same as the depth of the mouldings. This simplifies the jointing, as the shoulders of the tenons will be level.

The jointing of head, jamb and sill of the main frame are mortise and tenons, see Fig 7.10. These joints are normally wedged, although the use of draw pins or star dowels are acceptable and even preferable where the horn is to be later cut off.

Figure 7.11 shows an exploded isometric view of the joint between the transom and jamb. In order to make a better weatherproof joint, the front edge of the transom is housed across the jamb.

Figure 7.12 shows types of seal, which can be fitted around the rebate of opening sashes to further improve the window's draught-proofing qualities.

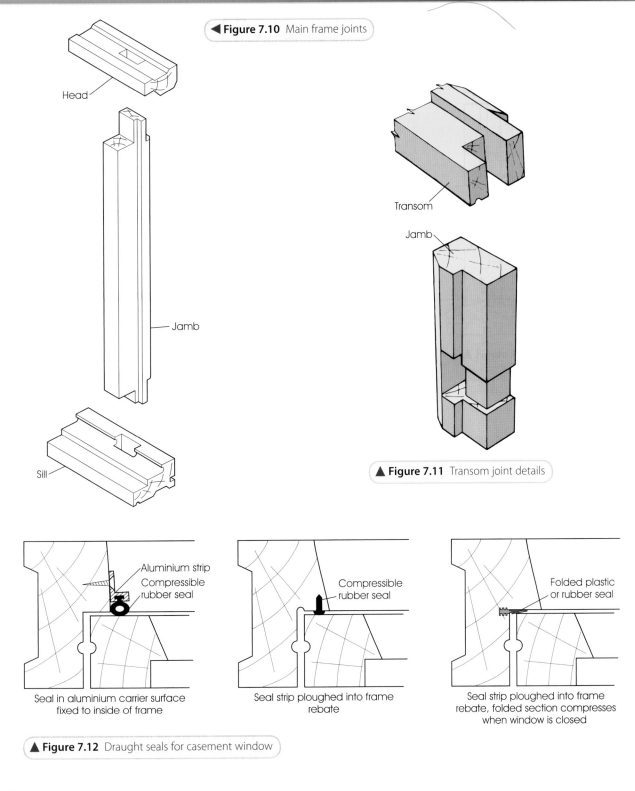

◀ **Figure 7.10** Main frame joints

Head

Jamb

Sill

Transom

Jamb

▲ **Figure 7.11** Transom joint details

Aluminium strip
Compressible rubber seal

Seal in aluminium carrier surface fixed to inside of frame

Compressible rubber seal

Seal strip ploughed into frame rebate

Folded plastic or rubber seal

Seal strip ploughed into frame rebate, folded section compresses when window is closed

▲ **Figure 7.12** Draught seals for casement window

Storm-proof casements

These are partially rebated into and sit proud of the main frame. Storm-proof casement windows incorporate two rebates, one around the main frame and the other around the casement. These rebates, in conjunction with the drip, anti-capillary grooves and throat, make this type far more weatherproof than traditional casements, see Figure 7.13. Also shown is an alternative sill detail, which is used in conjunction with a stone, concrete or tiled sub-sill, when the window is set well back from the face of the wall.

French doors

French casements are often referred to as French doors. Figure 7.21 shows a French casement window with fixed sidelights and double opening casements. The construction of the frame is similar to that of the traditional casement frame or an external doorframe with a threshold. The opening casements and the fixed sidelights are constructed in a similar way to external glazed doors.

▶ **Figure 7.21** French casements

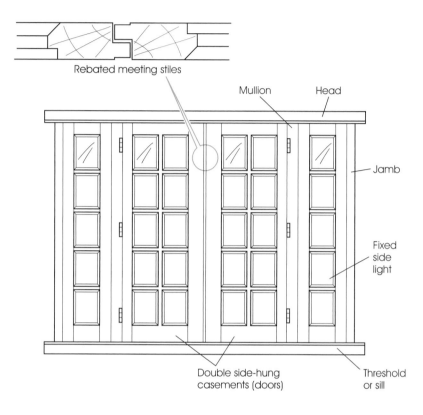

The hanging procedure for double casement doors is similar to that used for single doors, except that only the hanging stiles are planed to fit the frame. The meeting stiles are normally rebated together to provide an overlapping seal rather than a straight joint. The doors should be offered into the opening, keeping a 2 mm gap between the rebates and the hanging stiles marked accordingly. Any surplus timber should be removed evenly from both hanging stiles.

Bay windows

Bay windows are normally incorporated into a building for one or a combination of three reasons:

- To increase the amount of daylight and ventilation admitted into a room.
- To increase room size.
- To provide an architectural feature for the building.

The majority of bay windows are constructed using outward-opening casements. The main construction details vary little from the standard casement windows covered above.

Bay windows are classified according to their shape on plan, see Figure 7.22:

- Square or rectangular bay.
- Cant or splayed bay.
- Segmental bay or bow window.
- Combination bay, square and cant.

Heritage Link

Oriel windows are bay windows, normally in an upper storey, that project from the wall but do not extend down to the ground.

They can often be seen as a feature of buildings in the Victorian gothic revival style.

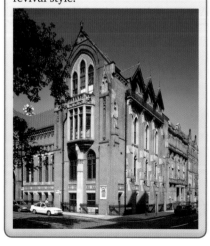

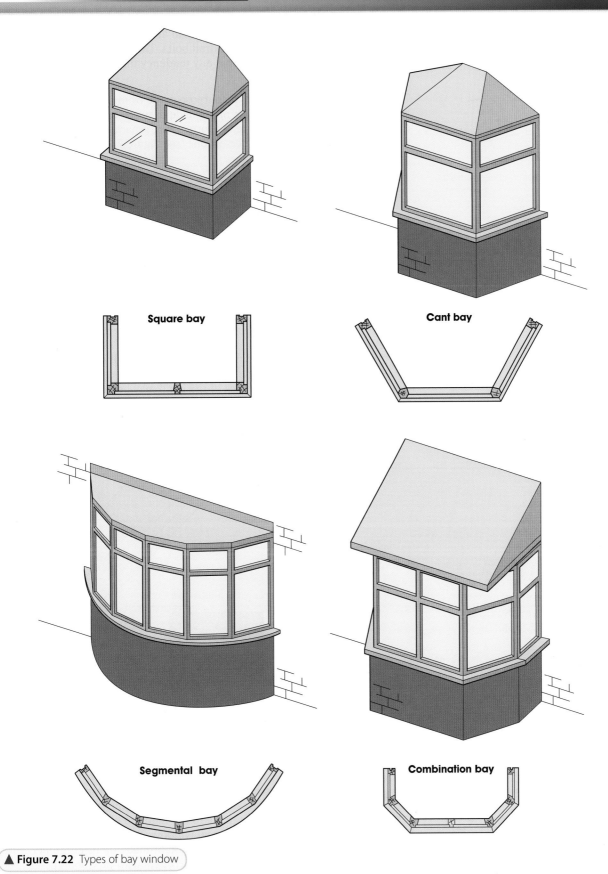

Square bay

Cant bay

Segmental bay

Combination bay

▲ **Figure 7.22** Types of bay window

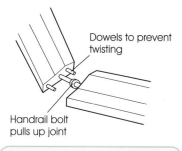

Dowels to prevent twisting

Handrail bolt pulls up joint

▲ **Figure 7.23** Joining head and sill

The angled intersections of the head and sill are normally mitred. These mitres are drawn up tight using handrail bolts. Two hardwood dowels are also used for each joint to overcome any tendency that the joint has to twist (see Figure 7.23).

The mullions or angle posts can be purpose made from solid timber, although for ease of manufacture these are often made in two sections. Alternatively, where standard windows are used to form a bay, separate corner posts can be used to join the jambs at the intersections (see Figure 7.24).

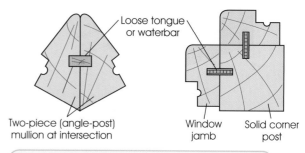

Loose tongue or waterbar

Two-piece (angle-post) mullion at intersection

Window jamb

Solid corner post

▲ **Figure 7.24** Mullion and corner post intersections

The preferred method of 'building in' the bay at the wall junction, is to keep the wall jambs square and angle the brickwork off to meet the frame, see Figure 7.25. Alternatively, the back of the jamb may be built up and splayed off to meet the face brickwork.

▶ **Figure 7.25** Building in a bay window

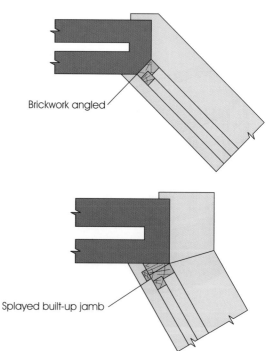

Brickwork angled

Splayed built-up jamb

▼ **Figure 7.26** Bay window template

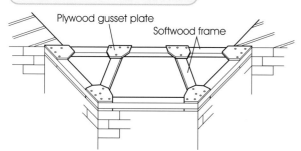

Plywood gusset plate

Softwood frame

A template should be provided with a bay window as an aid to the bricklayer when building the supporting wall. This should be made to the required brickwork sizes and not those of the actual window. Templates may be either cut from a sheet of plywood or made up using timber battens reinforced at the corners with plywood gussets, see Figure 7.26. The method used will depend on the material available and the window size.

Bow window

Segmental bay windows are often mistakenly called bow windows, but they are really a series of flat sections joined together to give a curved effect. A true bow window, often called a Georgian bow is shown in Figure 7.27. Shaped cantilevered brackets may be required for support where the brickwork does not follow the window curve.

Related information

Also see Chapter 4, page 112, which gives further details on making glulam components.

◀ **Figure 7.27** Bow window

Typical sections for a direct glazed bow window are shown in Figure 7.28. The head and sill may be both cut and jointed from a solid section or vertically glulaminated to the required curve. Jambs and vertical glazing bars are mortised and tenoned into the head and sill. The horizontal glazing bars can be jointed to the jambs and vertical glazing bars using short stub tenons.

Where the horizontal bars have a large curvature, short graining will weaken the tenons. In such cases they may be simply housed into the vertical members or, alternatively, they may be also vertically glulaminated. The glazing rebate in the head, sill and horizontal bars is worked flat to avoid the necessity of using expensive special-order curved glass. In cheaper-quality work, often only the head and sill are curved leaving the horizontal bars straight.

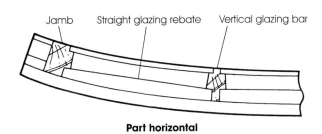

Part horizontal

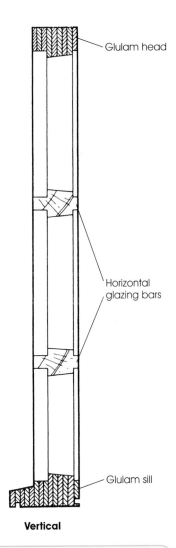

Vertical

▲ **Figure 7.28** Bow window sections

Roofs over bay and bow windows

These may be pitched and covered with tiles or slates to match the main roof or flat and covered with built-up felt or sheet metal.

The construction of the roofs follows the methods covered in Chapter 2 of this book. Figure 7.29 shows a range of bay window roof configurations and details. Pitched roofs are all lean-to and may have hipped or gable ends.

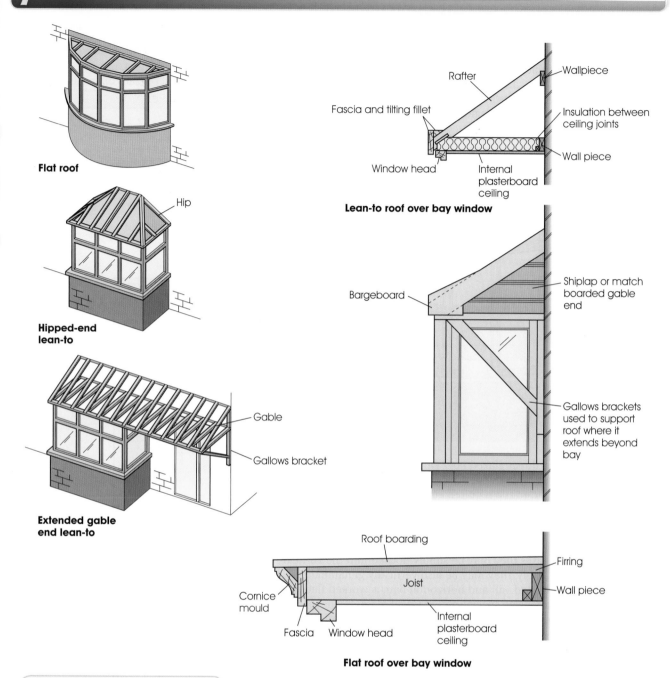

Flat roof

Hipped-end lean-to

Extended gable end lean-to

Gable

Gallows bracket

Rafter

Wallpiece

Fascia and tilting fillet

Insulation between ceiling joints

Window head

Wall piece

Internal plasterboard ceiling

Lean-to roof over bay window

Bargeboard

Shiplap or match boarded gable end

Gallows brackets used to support roof where it extends beyond bay

Roof boarding

Firring

Joist

Wall piece

Cornice mould

Fascia   Window head

Internal plasterboard ceiling

Flat roof over bay window

▲ **Figure 7.29** Roofs over bay windows

Where the roof is extended past the window, possibly to form a porch over a front entrance, a timber gallows bracket may be required for support. Also available are one-piece GRP (glass fibre reinforced plastic) flat roofs, made to suit a standard range of bay windows.

Curved-head windows

Where required, these can be constructed using the same method as that used for curved-head doors and frames. Curved members are built up from a number of pieces, to avoid short grain, and jointed together with handrail bolts, false tenons, hammer-headed key and tenons or bridle joints.

A typical semi-circular-headed window with direct glazing is shown in Figure 7.30. Glazing bars are used to subdivide the area into seven lights. These bars are cross-halved and scribed together where they intersect, then stub tenoned to the main frame. The radiating bars are stub tenoned at one end to a semicircular boss that is cut over and dowel-jointed to the cross-glazing bar.

Related information

Also see Chapter 2 in this book, which covers both flat and pitch roof construction and eaves/verge finishing in detail.

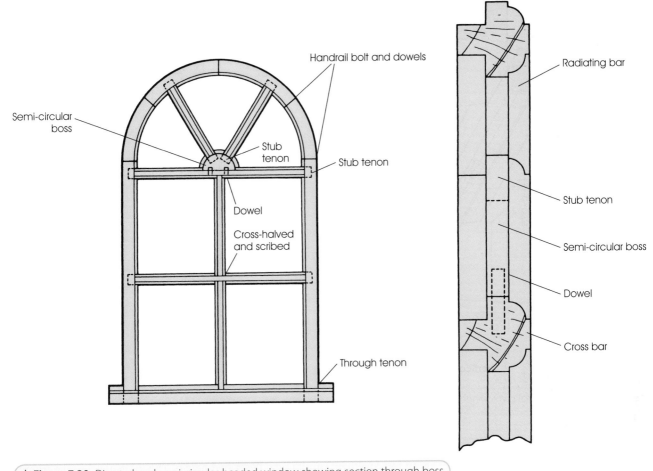

▲ **Figure 7.30** Direct glazed semi-circular-headed window showing section through boss

Pivot windows

Centre-hung pivot windows

These are often used for taller buildings, as both sides of the glass can be cleaned from the inside of the building with ease. The main disadvantage of this type is that, when opened, the top of the sash interferes with the curtains.

Figure 7.31 shows the vertical and horizontal sections of a traditional pivot window. Both the surrounding frame and the pivoting sash are constructed using mortises and tenons. The sash is hung on pivot pins about 25 mm above its centre line height to give it a self-closing tendency. The pivot pin usually is fixed to the frame and the socket to the sash. The planted stops, which form the rebates, serve to weatherproof the window. Those above the pivots are nailed or screwed to the frame on the outside and the sash on the inside, whereas those below are fixed to the frame on the inside and the sash on the outside.

The actual positions of intersection of the beads needs to be precisely determined, especially where the sash is required to be removable without taking off any beads. The sash stile and its planted top bead must be grooved as shown to allow removal of the sash. The head and top rail is splayed to give sufficient opening clearance. This is provided at the sill by its weathering.

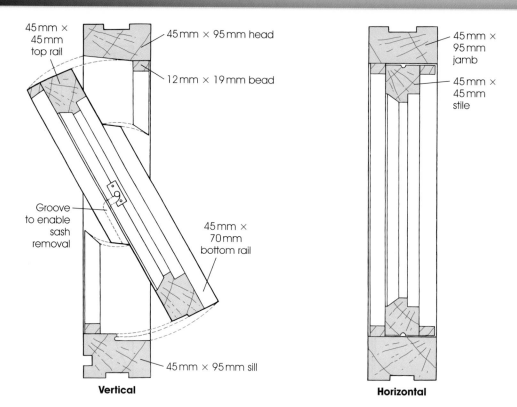

45mm × 45mm top rail

45mm × 95mm head

12mm × 19mm bead

45mm × 95mm jamb

45mm × 45mm stile

Groove to enable sash removal

45mm × 70mm bottom rail

45mm × 95mm sill

Vertical

Horizontal

▲ **Figure 7.31** Traditional pivot window sections

Storm-proof pivot window

Figure 7.32 shows details of a storm-proof centre-hung pivot window, which is an improvement on the type described above. The joints used in the frame are mortise and tenons. Comb joints are used for the sash. Face-fixing friction pivots or back-flap hinges are used for hanging the sash.

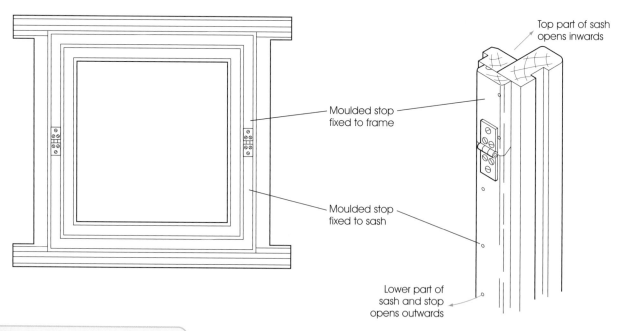

Top part of sash opens inwards

Moulded stop fixed to frame

Moulded stop fixed to sash

Lower part of sash and stop opens outwards

▲ **Figure 7.32** Stormproof pivot window

The moulded stop, which is mitred around the frame and cut on the pivot line, is glued and pinned to the top half of the frame and the bottom half of the sash.

Vertically hung pivot windows may be constructed using similar details to those of the horizontal type. Although, as the bottom pivot takes all of the sash weight, a special pivot is required that normally incorporates some form of ball race mechanism.

Bull's eye windows

These are circular windows consisting of a main frame, which can be glazed, direct or contain a sash. The sash itself may be either fixed or pivot hung.

Figure 7.33 shows an elevation and section of a bull's eye window with a pivot-hung sash that has been subdivided with glazing bars into four lights.

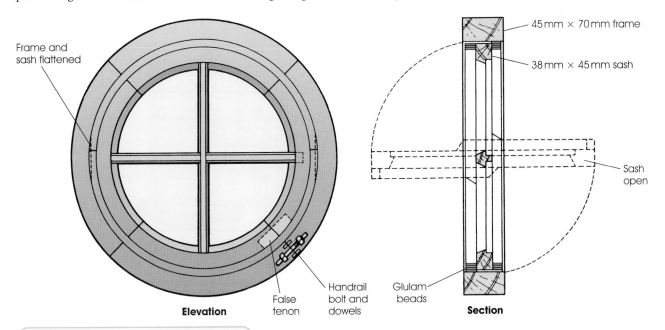

Elevation   False tenon   Handrail bolt and dowels   Glulam beads   **Section**

Frame and sash flattened

45 mm × 70 mm frame

38 mm × 45 mm sash

Sash open

▲ **Figure 7.33** Bull's eye window (pivot hung)

There are various ways (already described in the section on curved work with doors) in which the frame and sash can be made. In this case they have both been formed from four curved segments, joined with handrail bolts and dowels on the main frame with false tenons used for the sash. The planted beads, which are fixed partly to the frame and partly to the sash, have been glulamed using a number of thin laminations bent around a suitable former. Both the frame and sash will require flattening around the pivot positions on the inside to enable it to pivot without binding. Threaded centre screw pivots are normally used for bull's eyes, as these allow easy removal of the sash.

Figure 7.34 shows a section of a direct glazed bull's eye window. In this case it has been built up in two layers with end joints staggered. The joint between the layers, which has been positioned to coincide with the rebate, is glued, screwed and pelleted. The glulamed glazing bead run around the inside of the frame is secured with recessed cups and screws.

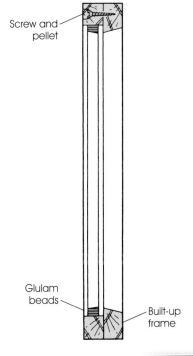

Screw and pellet

Glulam beads

Built-up frame

▶ **Figure 7.34** Bull's eye window section (direct glazed)

Window ventilation

Ventilation to rooms is a requirement of the Building Regulations. Opening lights in the window normally provide rapid ventilation equal to 1/20th of the floor area, at least part of which must be at least 1.75 m above floor level. Background ventilation of 8000 mm^2 for habitable rooms and 4000 mm^2 for kitchens and bathrooms is also required and again this must be at least 1.75 m above floor level.

Vents

Background ventilation is normally provided by trickle type permanent vents incorporated into the window by one of the means shown in Figure 7.35:

- In the head of the main frame.
- In the top rail of the casement or sash.
- Within the glazed area itself.

▶ **Figure 7.35** Window vents

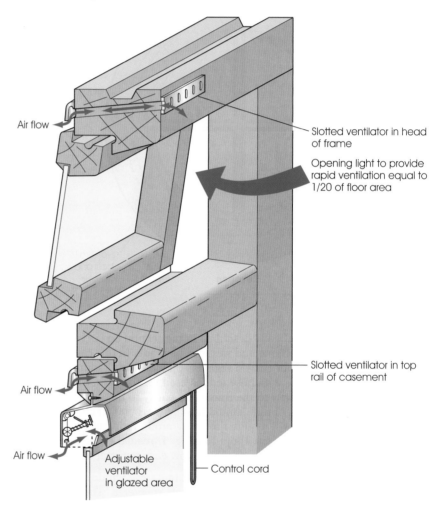

Air flow

Slotted ventilator in head of frame

Opening light to provide rapid ventilation equal to 1/20 of floor area

Slotted ventilator in top rail of casement

Air flow

Air flow

Adjustable ventilator in glazed area

Control cord

Ventilators

These are used where permanent ventilation only is required and not daylight or a view. They consist of a main frame into which louvre slats are housed. The slats, which are normally set at 45°, should project beyond the face of the frame to throw rainwater clear. The top and bottom edges of adjacent slats should overlap to prevent through vision and restrict the passage of driving rain.

The elevations and sections of the two most common ventilators are shown in Figure 7.36. The triangular frame may be jointed using dovetails or pinned bridles and the circular one built up using any of the methods covered previously. The true shapes of the louvre slats and the housings in the frame must be determined geometrically. This is because the slats are inclined and do not intersect with the frame at right angles.

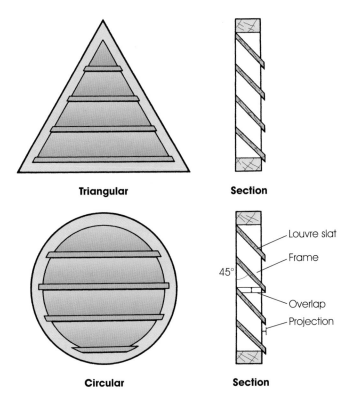

Triangular   **Section**

Circular   **Section**

Louvre slat
Frame
45°
Overlap
Projection

◀**Figure 7.36** Louvre ventilators

Assembly procedure for window frames and opening sashes

This can be carried out for all the above types of frames and opening sashes using the following assembly procedure, see Figure 7.37.

1. Dry assemble to check fit of joints, overall sizes, square and winding.
2. Clean up inside edges of all outer framing components and both edges of inner ones.
3. Glue, assemble, cramp up and wedge. Re-check for square and winding.
4. Clean up remainder of item and prepare for finishing.
5. Hang, where used, opening casement sashes and fanlights.

Hanging casement sashes

Stormproof casements fit on the face of the frame and normally require no fitting at all. The only operation necessary to hang the casement is the screwing on of the hinges. Traditional casements fit inside the frame and require both fitting and hanging, see Figure 7.38.

The hanging of casement sashes follows a similar procedure to that used for doors, see Figure 7.39.

1. Mark the hanging side on both the sash and the frame.
2. Cut off the horns.
3. Shoot in (plane to fit) the hanging stile.

Related information

Also see Book 1, Chapter 9, Applied geometry, page 390, which deals with the geometry of louvre ventilators.

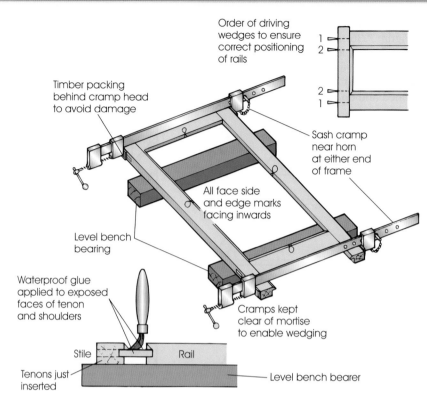

► **Figure 7.37** Assembling casement sash

Order of driving wedges to ensure correct positioning of rails

Timber packing behind cramp head to avoid damage

Sash cramp near horn at either end of frame

All face side and edge marks facing inwards

Level bench bearing

Waterproof glue applied to exposed faces of tenon and shoulders

Cramps kept clear of mortise to enable wedging

Stile

Rail

Tenons just inserted

Level bench bearer

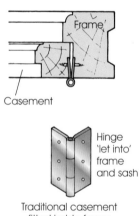

Frame

Casement

Hinge 'let into' frame and sash

Traditional casement fitted inside frame

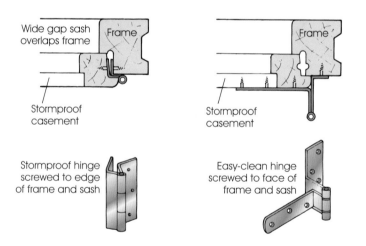

Wide gap sash overlaps frame

Frame

Stormproof casement

Stormproof hinge screwed to edge of frame and sash

Frame

Stormproof casement

Easy-clean hinge screwed to face of frame and sash

▲ **Figure 7.38** Hanging of casements sashes

4. Shoot the sash to width.
5. Shoot in the top and bottom of the sash.
6. Mark out and cut in the hinges.
7. Screw one leaf of the hinges to the sash.
8. Offer up the sash to the opening and screw the other leaf to the frame.
9. Adjust fit if required and fix any other ironmongery.

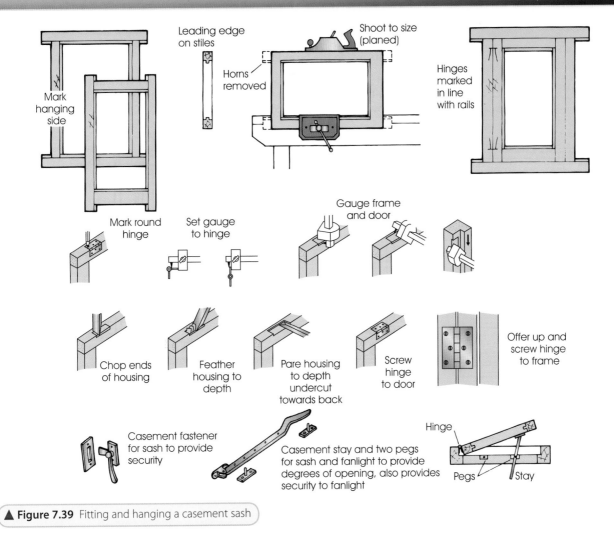

▲ **Figure 7.39** Fitting and hanging a casement sash

Sliding sash windows

These may be classified into two groups according to their direction of opening:

■ Vertical sliding sash window.
■ Horizontal sliding sash window.

Vertical sliding sashes

These consist of two sashes, which slide up and down in a main frame. They are also known as double-hung sliding sash windows. There are two different forms of construction for these types of window:

■ Those with boxed frames.
■ Those with solid frames.

Boxed frames

This type of window is the traditional pattern of sliding sashes and for many years has been superseded by casements and solid frame sash windows. This was mainly due to the high manufacturing and assembly costs of the large number of component parts. An understanding of their construction and operation is essential, as they will frequently be met in renovation and maintenance work.

The double-hung boxed window consists of two sliding sashes suspended on cords or chains, which run over pulleys and are attached to counterbalanced weights inside the boxed frame.

Figure 7.40 shows an elevation and horizontal and vertical sections of a boxed-frame sliding sash window. It shows the make-up of this type of window and names the component parts.

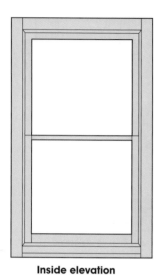

Inside elevation

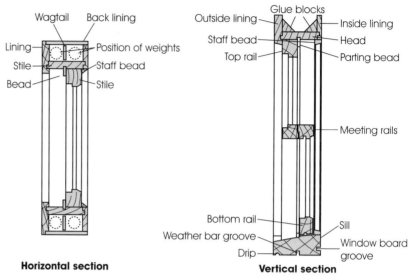

Horizontal section

Vertical section

▲ **Figure 7.40** Boxed-frame sliding sash window details

The pulley stiles are housed into the head and sill. These are simply skew nailed at the head and wedged at the sill, see Figure 7.41. The sill will require cutting away on both ends, to accommodate the linings. Wagtails, which separate the weights to prevent them fouling in use, are simply pinned into slots cut in the head.

▶ **Figure 7.41** Boxed-frame joint details

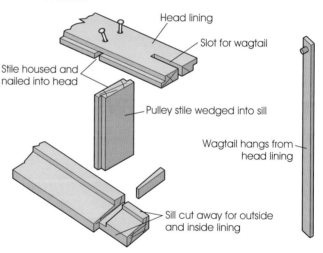

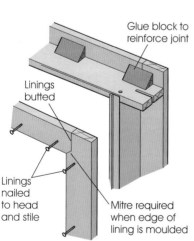

▲ **Figure 7.42** Boxed-frame head details

The inside and outside linings, which are butt jointed at the corners, are tongued and nailed to the head and pulley stiles. Glue blocks are used to strengthen the joints at the head, but must not be used in the stiles, as they will interfere with the sash weights, see Figure 7.42. The outer linings, which extend past the pulley stiles and head, often have a bevelled edge, requiring a mitre to be incorporated in the butt joint.

■ Figure 7.43 shows how the outside lining and parting bead at sill level are cut away in order to prevent water and dirt being trapped at this point.

■ Figure 7.44 shows an alternative detail, which can be used at sill level. It uses a deeper draughtboard tongued into the sill replacing the standard staffbead. This allows the bottom sash to be partly opened to provide ventilation at the meeting rails without causing a draught.

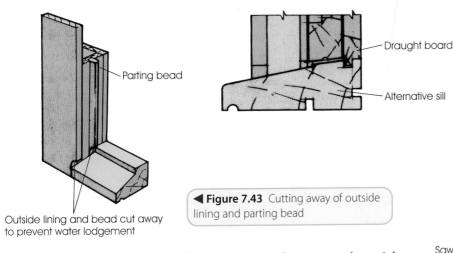

Parting bead

Draught board

Alternative sill

Outside lining and bead cut away
to prevent water lodgement

◀ **Figure 7.43** Cutting away of outside
lining and parting bead

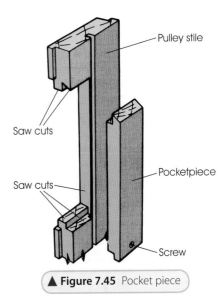

Pulley stile

Saw cuts

Pocketpiece

Saw cuts

Screw

▲ **Figure 7.45** Pocket piece

- A pocket piece is cut in the pulley stiles to provide access to the weights. It is cut out by making saw cuts, see Figure 7.45, and then firmly tapping out from the back of the stile.

- Figure 7.46 shows the mortise and tenon joint used between the sash stile and rail. The sash haunch or reverse franking is used instead of a normal haunch. If this were not done, the joint would be seriously weakened by the sash cord groove that runs into the edge of the stile.

- Figure 7.47 shows two alternative methods used for the joints between the stiles and meeting rails. Where the horn is left on, it is usually moulded and a mortise and tenon joint can be used. A dovetail joint should be used where no horn is required.

Where a window is more than one sash wide, boxed double mullions, see Figure 7.48, are introduced to accommodate the four weights. These have a wide bulky appearance, so they were often concealed on the outside with ornamental stone mullions or pilasters.

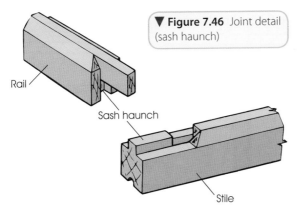

▼ **Figure 7.46** Joint detail
(sash haunch)

Rail

Sash haunch

Stile

▼ **Figure 7.47** Alternative meeting rail joint details

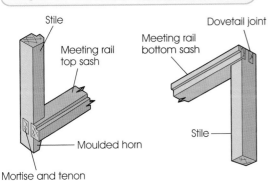

Stile

Meeting rail
top sash

Meeting rail
bottom sash

Dovetail joint

Moulded horn

Stile

Mortise and tenon

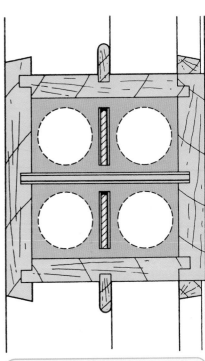

▲ **Figure 7.48** Boxed double mullion
(plan view)

Venetian windows

Venetian windows do not require boxed mullions as normally only the middle sashes slide, with the narrow side sashes being screw-fixed permanently in the closed position, see Figure 7.49.

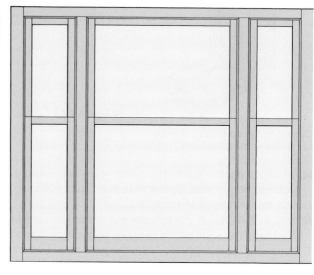

▶ **Figure 7.49**
Venetian window (inside elevation)

■ Figure 7.50 shows a section through the solid mullion of a Venetian window and the method used to hang the centre sashes. The pulleys are fixed to the pulley stiles as high as possible. The cords fixed at one end to the sashes pass over the mullion pulleys, across the heads of the side sashes, then finally over the side pulleys to the weights.

■ Figure 7.51 shows how the cords are concealed across the top. The top rail of the side sash is grooved out to receive the outer cord. A grooved cover mould that also incorporates the top staff bead conceals the inner cord. Both the cover mould and the fixed top sash must be made removable for cord replacement.

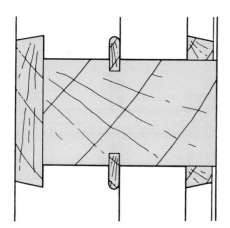

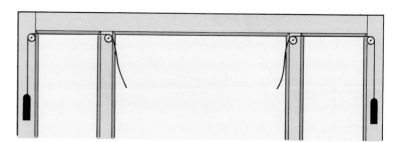

▲ **Figure 7.50** Solid mullion section and method used to hang centre sashes

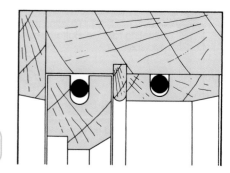

▶ **Figure 7.51** Concealing sash cords across head

Curved-head boxed-frame sashes

The inside elevation of a semi-circular-headed vertical sliding sash window is shown in Figure 7.52. The construction details of this sash window are the same as those of a standard sash up to the springing line. Above this a number of alternative methods may be used to construct the frame and sash.

Figure 7.53 shows the curved head built up in three layers with joints well staggered and glued together. The middle layer, often of plywood, also forms a fixed parting bead. Below the springing line a removable parting bead is used. Also illustrated is how the sash stile to curved rail joint (handrail bolt with dowels) is positioned beyond the springing to enable a shoulder to be formed. The shoulder stops on the projecting edge of the built-up head when the sash is lifted. This prevents jamming and impact at the crown, so avoiding the risk of sash damage and breaking glass.

An alternative method of forming the curved head is shown in Figure 7.54. In this case the head has been glulamed from thin solid laminates or thin plywood. Blocks are glued around the head to provide a fixing for the curved inner and outer linings, which, unlike the straight ones, are merely butt jointed and pinned, not tongued and grooved. The parting bead in this case would be shaped from ply or solid timber and grooved into the head.

Balancing of sashes

For all forms of boxed sash windows, two cast iron or lead weights are required to balance each sash. To determine the size of these weights the fully glazed sashes can be weighed with a spring balance. The weights for the top sash should each be half the sash weight plus

▲ **Figure 7.52** Semi-circular-head sash (inside elevation)

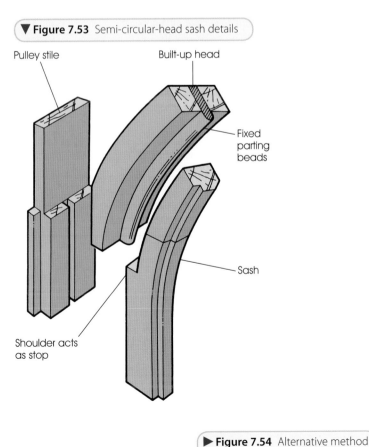

▼ **Figure 7.53** Semi-circular-head sash details

Pulley stile

Built-up head

Fixed parting beads

Sash

Shoulder acts as stop

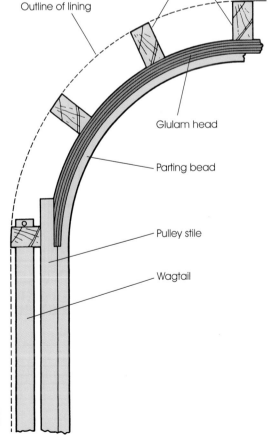

Fixing blocks

Outline of lining

Glulam head

Parting bead

Pulley stile

Wagtail

▶ **Figure 7.54** Alternative method of forming a curved head about 250 g and the bottom sash weights half the sash weight less 250 g. Using this principle, both sashes are counterbalanced, but will have the tendency to remain closed in their respective positions, firmly against the head or sill.

Solid-frame sash windows

Sash balances are manufactured for use in place of sash cords and weights. They do away with the need for boxed frames, thereby, reducing the number of component parts and simplifying the construction of sliding sash windows.

Figure 7.55 shows the elevation, horizontal and vertical sections of a solid-frame sliding sash window. It can be seen from this that apart from the solid jambs, the arrangement of the component parts is similar to that of the boxed-frame sash window. The frame and sashes can be jointed using

▼ **Figure 7.55** Solid-frame sash window sections

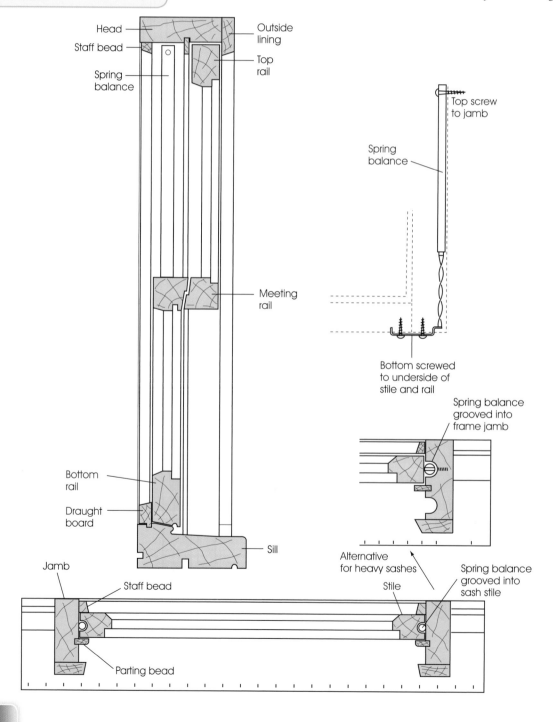

mortise and tenon joints, although modern mass-produced windows are now exclusively manufactured using comb joints with metal star dowels. For lightweight domestic sashes, the spring balances are accommodated in grooves run in the back of the sash stiles. The spring balances for heavyweight industrial sashes are accommodated in grooves run in the actual jamb of the frame.

Two balances are required for each sash. Where the sashes are the same size, the longer balances are for the bottom sash and the shorter balances are for the top. The balances are fixed to the tops of the jambs with a screw. The plate on the other end is attached to the bottom rail of the sash. As each sash is raised or lowered the spring is tensioned. The sash is then supported or balanced by the tension of the springs.

After glazing the sashes, the springs should be adjusted to balance the sash and ensure a smooth operation. If the sash is not balanced both springs on the sash should be adjusted. Unscrewing the sash plate and twisting it one turn in a clockwise or anticlockwise direction will adjust the tension of the springs. The plate should then be re-screwed to the sash and the operation of the sash checked to see if it is correct.

Horizontal sliding sashes

The elevation and sectional details of a window with horizontally sliding sashes is shown in Figure 7.56. The main frame is simply constructed from rectangular section material using mortises and tenons. It has planted staff beads and grooved parting beads. The sashes, which are also framed using mortises and tenons, have rebated meeting stiles to prevent draughts. Both sashes may be made to slide on a track or waxed hardwood runners.

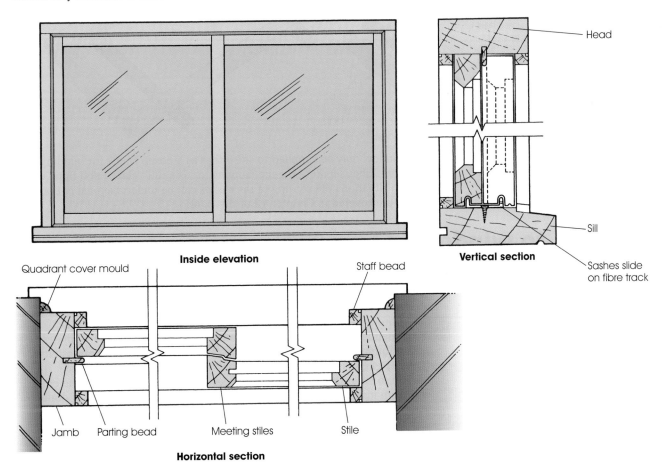

Inside elevation

Vertical section

Head

Sill

Sashes slide on fibre track

Quadrant cover mould

Staff bead

Jamb Parting bead Meeting stiles Stile

Horizontal section

▲ **Figure 7.56** Horizontal sliding sash window

Larger three or four sash windows can be constructed, using similar details to those shown. Normally only the middle sash or sashes slide, while the outer ones are fixed to the main frame.

Assembly procedure for boxed frames

The sashes are normally made up first as they can be used when checking the boxed frame for size. After preparing and cleaning up all of the components and fitting the pulleys, the following assembly procedure can be used for the boxed frame, see Figure 7.57:

1. Assemble basic frame: glue and wedge pulley stiles to sill; glue and nail head to pulley stiles; place on level bench bearers and check for square and wind.
2. Glue and pin the outer linings to the stiles and head. Fix glue blocks to reinforce the head joints.
3. Turn the frame over and repeat Step 2, this time for the inner linings.
4. Check sashes for size and shoot to fit if required, so that they slide easily. A working clearance of 2 mm on both sides should be sufficient to allow for subsequent finishing. Remove sashes for glazing and weighing.
5. Mitre the parting beads and dry position, insert the head before the sides.
6. Mitre the staffbeads and dry nail in position, leaving the nail heads protruding.

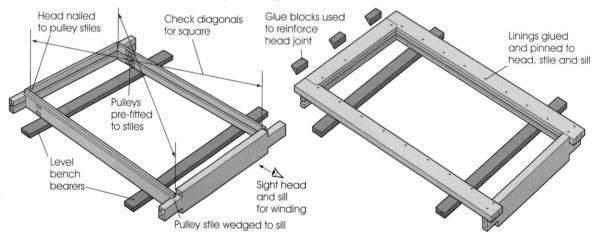

Head nailed to pulley stiles

Check diagonals for square

Glue blocks used to reinforce head joint

Linings glued and pinned to head, stile and sill

Pulleys pre-fitted to stiles

Level bench bearers

Sight head and sill for winding

Pulley stile wedged to sill

▲ **Figure 7.57** Assembly of a boxed frame

The window may be supplied to site either as a boxed frame for building in, with the sashes separate for site hanging, or as a fully glazed and hung unit. In both situations the cording of the sashes can be undertaken using the procedure covered in Chapter 12 of this book.

Glazing to windows

Windows are single, double or triple glazed using clear or patterned (obscure) glass. Glazing to windows below 800 mm from the internal floor level should be done using a laminated or toughened safety glass that will not, if broken, produce pointed shards with razor sharp edges. It is also recommended that environmental control glass be used. In cool climates such as ours, where thermal insulation is most important, **low emissivity (Low E) glass** is used. This is glass that has a coating that allows heat from the sun to enter the building but reduces the amount of heat loss the other way from inside the building. It is used mainly as the inner pane of double-glazed windows with the coated side facing into the gap or air space. In warmer climates, a **solar control glass** may be used to limit the amount of heat gain from the sun into the building.

Heat loss through materials can be compared by their U values. The higher the U value the greater the heat loss. Single pane windows have a typical U value of 5.6; sealed double glazed units using ordinary glass are typically

2.8, which can be reduced to around 1.9 if low E glass is incorporated, and still further if the air space is filled with an inert gas such as argon. A further reduction in U values can be achieved using triple glazed units.

Glass with self-cleaning properties has a coating bonded to its outside surface that absorbs the ultraviolet light given off by the sun. The absorption process causes a reaction on the glass's surface that breaks down and loosens any dirt. When it rains, the coating causes the rain water to sheet off the surface of the glass, washing away the loosened dirt particles and also preventing the formation of water droplets and subsequent streaks.

Single glazing has been traditionally used for windows, 4mm or 6mm thickness glass being the norm, which is held in position with either putty or glazing beads, see Figure 7.58.

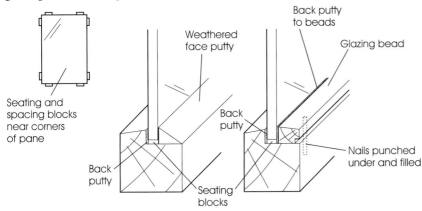

◀ **Figure 7.58** Single glazing

> ### Related information
> Also see Chapter 12, page 414, which deals with glazing procedures.

Double and triple glazing

This is now preferred as it offers higher levels of both thermal and sound insulation. In addition, as the temperature of the inner pane is warmer, the risk of condensation on the room side is reduced.

Sealed double-glazing units

These are used to double glaze windows mainly for thermal insulation purposes. These consist of two panels of glass with a gap of up to 25mm between them. The panes are held apart by a spacer that runs around the perimeter and is sealed to provide an airtight joint. The units are assembled in a clean dry atmosphere or an inert gas is used to fill the gap to avoid any possibility of condensation forming between the panes, see Figure 7.59.

In situations where obscure or patterned glass is required, such as in a bathroom for privacy purposes, only the internal pane is normally obscure with the external one being in standard glass. This minimises costs but also standard glass is far easier to clean.

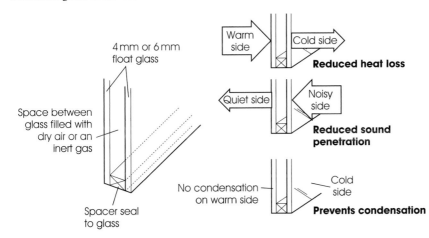

◀ **Figure 7.59** Sealed double-glazing units showing advantages

Windows glazed using sealed units can also be given a mock Georgian or leaded light look by incorporating glazing bars or lead strips between the two panes, see Figure 7.60.

In addition, the greater mass of sealed units compared to single glazing, will reduce the passage of sound to some extent.

Wider rebates are normally required around the edges of casements and sashes to accommodate sealed units. However, where the rebate in existing windows is too small to take the extra thickness of glass, stepped units can be used. Figure 7.61 shows how normal and stepped sealed glazed units are bedded and held in position. Bedding can be done using either self-adhesive strips or a non-setting glazing compound. Spacer blocks are required along the bottom of the rebate to support the unit and at the back of the rebate all around the perimeter when using a glazing compound to prevent it squeezing out as the unit is pushed home. Beads are the preferred method of securing the units, either screwed or nailed as appropriate.

▲ **Figure 7.60** Mock Georgian and leaded light effect windows

▶ **Figure 7.61** Fixing sealed double-glazing units

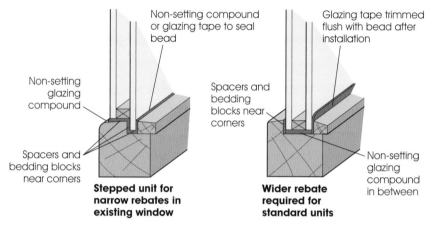

Stepped unit for narrow rebates in existing window

Wider rebate required for standard units

Figure 7.62 shows a method of double glazing that is similar in effect to sealed units. This is achieved by fixing a second pane of glass on the inside of the sash or casement with the aid of rebated beads. A distinct disadvantage of this method is the very high risk of condensation occurring between the two panes of glass.

Sealed triple-glazing units

With triple glazing an extra pane of glass and an additional air or inert gas filled gap are added to the sealed unit, see Figure 7.63. This gives both increased thermal and sound insulation levels compared to double glazing, as there are now three panes and two air gaps to resist heat and sound transfer. As with double glazing, triple-glazed units can also be made using low E, self-cleaning, toughened or laminated safety glass and can also incorporate glazing bars or lead strips between the panes.

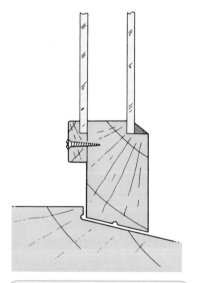

▲ **Figure 7.62** Alternative to sealed double-glazing units

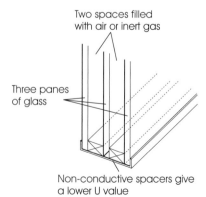

▶ **Figure 7.63** Sealed triple-glazed unit

Secondary double glazing

While sealed units do provide a certain amount of sound insulation, this is often insufficient for noisy locations, for example, near busy roads or airports. In general, the sound insulation properties of a window increase as the air space between the two windows increases.

For efficient sound insulation, the air space should be between 150 mm and 200 mm. In order to achieve this air space, a second window can be built on the inside of the reveal behind the existing window. This type of double glazing is known as secondary double glazing.

Figure 7.64 shows secondary double glazing detail. As a further improvement, the reveals between the two frames can be lined with a sound absorbent material, such as strips of insulation board.

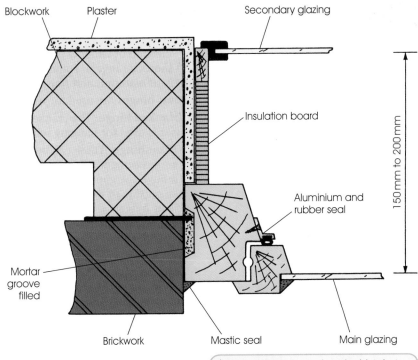

▲ **Figure 7.64** Secondary double glazing

Linked casement windows

These provide an alternative method of double glazing, which is popular in Europe (see Figure 7.65). The two casements are hinged and screwed together. The outer one is normally hinged to the frame. Where the casements are screwed together, the fixings should be accessible, as the inner faces of the glass will require cleaning occasionally. This can be reduced to a minimum if the two casements are sealed together with an air-permeable but dust-excluding sealing strip.

Condensation can occur between the two panels of glass where this space is not ventilated. However, ventilation has the effect of reducing the window's insulating value.

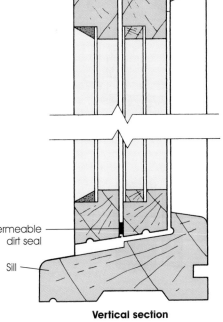

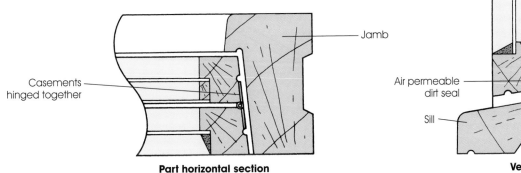

Part horizontal section

Vertical section

▲ **Figure 7.65** Linked casements

Window ironmongery

The selection of window fittings will depend on its type and location. Figure 7.66 shows a range of commonly used items that are readily available in a variety of materials and finishes.

- **Butt hinges** are used for traditional side-hung casements.
- **Cranked hinge.** The cranked hinge is also known as the *storm-proof hinge* as it is used for hanging rebated storm-proof casements.
- **Offset hinge.** These can be used for hanging both traditional and storm-proof casement windows. When open, they allow an arm to be passed through the gap between the frame and the casement in order to clean the outside of the window from the inside, hence the popular name 'easy-clean' hinge.
- **Window pivot.** These pivots are recessed into the edge of the window jamb and stile. They are used for the concealed hanging of pivot windows.
- **Friction pivot.** This is a type of hinge used for hanging pivot windows. It screws on to the face of the window and also contains a friction action, allowing the window to remain open in any position.

▼ **Figure 7.66** Window ironmongery

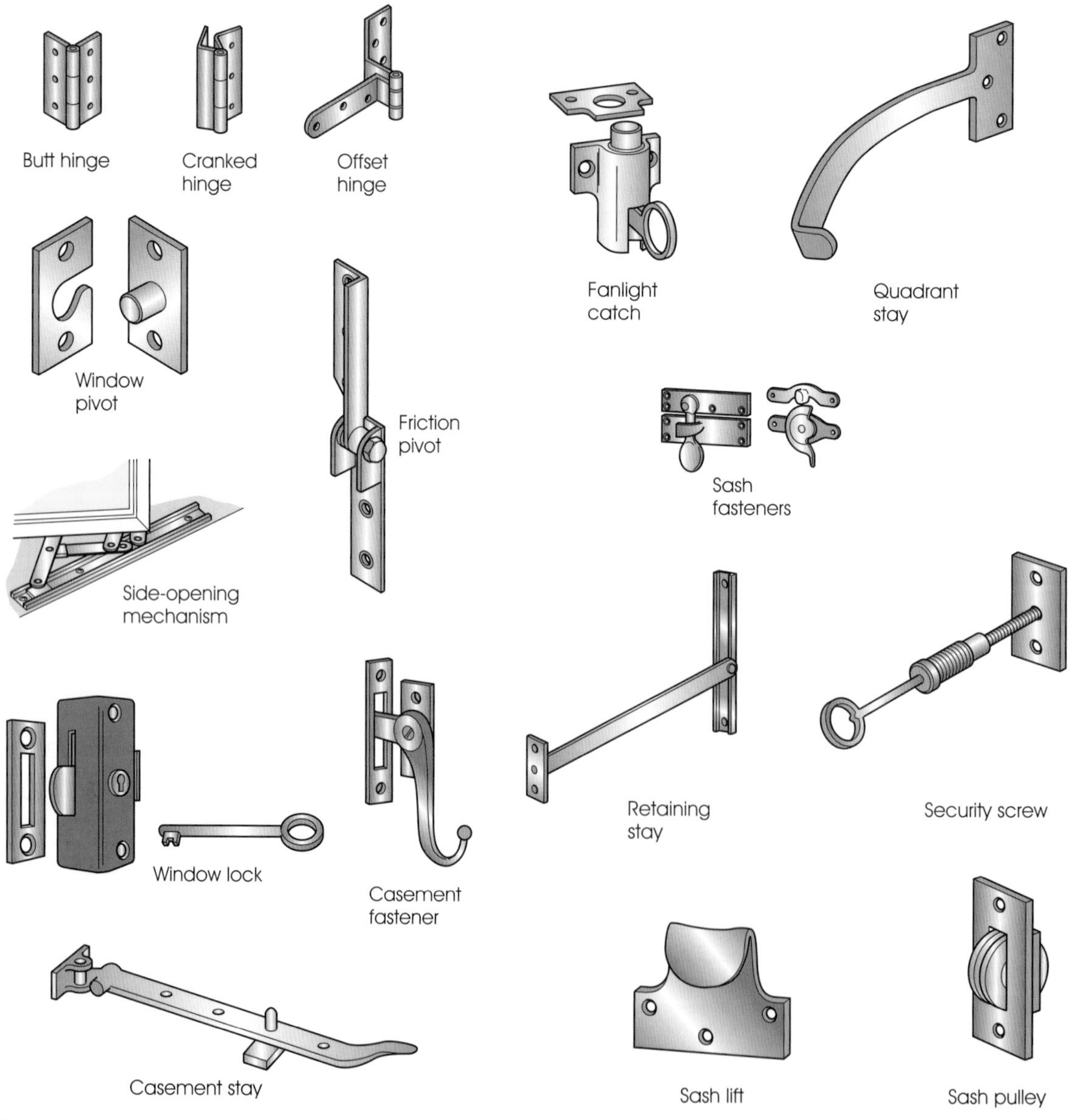

Butt hinge

Cranked hinge

Offset hinge

Fanlight catch

Quadrant stay

Window pivot

Friction pivot

Sash fasteners

Side-opening mechanism

Window lock

Casement fastener

Retaining stay

Security screw

Casement stay

Sash lift

Sash pulley

- **Side-opening mechanism.** These pivot and slide the sash from the side, creating a gap at the jamb for ease of cleaning the outside. Versions are available that enable the sash to tilt as well as turn.
- **Casement stay.** This is also known as a *peg stay*. It is used to hold the window in various open positions.
- **Casement fastener.** This is also known as a *cock spur* and is used to secure a casement window in the closed position. Versions are also available with a key-operated lock for additional security.
- **Window lock.** A key-operated lock fitted to casements and fanlights for additional protection.
- **Fanlight catch.** Fitted to the top rail of horizontal pivots and inward-opening fanlights to secure them in the closed position. For high-level windows the ring pull may be operated with a window pole or cord system.
- **Quadrant stay.** Used in pairs to limit travel of inward-opening fanlights.
- **Sash fasteners.** These are available in a number of types and are used to hold the meeting rails of a sash window together, thereby securing the sashes in a closed position. It is recommended that two are fitted to ground floor windows for additional security.
- **Retaining stay.** Used to limit the travel of horizontal pivots and fanlights.
- **Security screw.** Fitted mainly to the meeting rails of sliding sash windows for extra security. The threaded screw passes through the socket drilled into the inner meeting rail and locates in the plate fixed to the outer meeting rail. It can also be used to secure casement windows by passing through the casement stile and into the jamb.
- **Sash lift.** One or two are fixed to the bottom rail of a vertical sliding sash to act as a finger grip when lifting or lowering the sash.
- **Sash pulley.** Various patterns are available to take different-sized cords and chains. They are mortised and recessed flush into the top of the pulley stile.

Fixing windows

In common with door frames, widows may be fixed in position by either 'building in' during the main building process or by being 'fixed in' at a later stage. The actual method used will depend on the window construction, the material used (hardwood and plastic windows are usually fixed in), and the availability of the windows at the wall construction stage.

Fixing casement window frames

Built in

The bricklayer fixes the majority of timber frames by building in as the brickwork proceeds, see Figure 7.67.

1. The frame sill is bedded level using bricklayer's mortar and the head is checked for level and packed up as required.
2. Temporary struts are used to hold the frame upright.
3. Jambs should be plumbed from the face. It is standard practice to plumb and strut the first using a spirit level. The other is then sighted parallel: stand to the side of the frame, close one eye, sight the edge of the plumbed jamb with the edge of the other and adjust if required until both jambs are parallel.
4. As the brickwork proceeds, galvanised metal frame cramps should be screwed to the back of the jambs and built into the brickwork. Three or four cramps should be evenly spaced up each jamb.

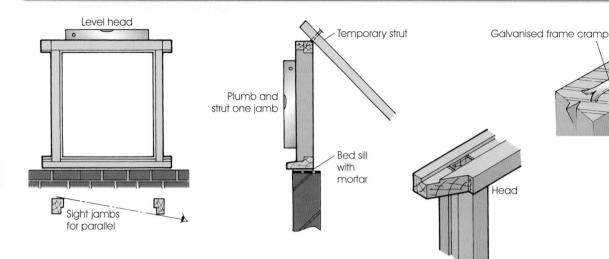

▲ Figure 7.67 Built in frames

5. The horns of mortise and tenoned frames should be cut back for building in, rather than cut off flush, which helps to fix the frame in position. The horns of comb-jointed frames are often cut off flush with the jambs. After trimming the horns it is essential that the cut ends be treated with paint or preservative in order to reduce the possibility of timber rot.

Fixed in

Where window frames are fixed in at a later stage, the carpenter is often required to make a timber template or profile frame. These are made up either on site or in the workshop using 50 mm × 75 mm softwood and plywood corner gussets, see Figure 7.68.

▶ Figure 7.68 Profile frame for windows that are to be fixed at a later stage

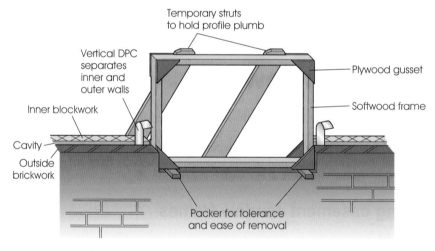

The profile is made about 6 mm wider than the actual window to give a fixing tolerance. The vertical tolerance is achieved by setting the profile on 6 mm packing blocks to aid removal. After the wall has set the profiles may be removed, leaving the prepared opening ready to receive the frame.

1. The frame is placed in the prepared opening by temporarily holding it in position with the aid of folding wedges (see Figure 7.69). The actual position of the frame in the reveal will depend on the type of frame and presence of a sub-sill. Standard casement windows are fixed with about a 25 mm setback from the face of the wall; those with flush sills that are intended to be used in conjunction with a stone, concrete or tiled sub-sill are set back about 110 mm.

2. Check the sill for level and the jambs for plumb; adjust the wedges as required.

3. Fix the frame to the wall reveal by the use of normal plugs and screws or proprietary sleeved frame-fixing plugs and screws. As an alternative fixing for timber windows, metal plates may be prefixed to the back of the jambs before positioning the window and subsequently secured to the wall by plugs and screws. A minimum of three fixings per jamb will be required, whichever method is used.

4. Seal the gap around the frame with a silicone or acrylic frame sealant. These are normally supplied in cartridges and gun applied. The seal should be forced as far back into the gap as possible, finally finishing off with a narrow angled face bead. Any slight unevenness in the face bead can be smoothed using a small paintbrush dipped in water.

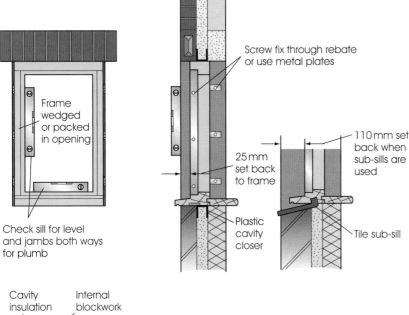

◀ **Figure 7.69** Fixing windows in prepared openings

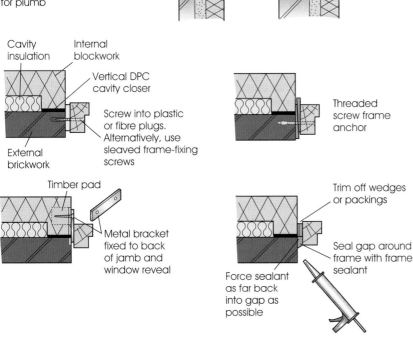

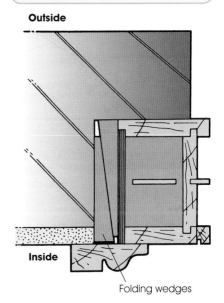

▼ **Figure 7.70** Fixing boxed-frame sash windows

Fixing sash windows

Solid-frame sash windows are usually built in in the same manner as casements, using galvanised frame cramps. Boxed-frame sash windows are normally fixed into a prepared opening in the wall by driving wedges in behind the jambs (see Figure 7.70).

Window boards

A window board is normally used to finish the top of the wall internally at sill level. They may be formed from solid timber, plywood, MDF or plastic and should be fitted at the first fixing stage before the internal walls have been plastered. Where the window board fits up to a timber window, it is normally tongued into the sill groove. If the window is metal or plastic (UPVC), the window board is only butted to the sill, with the gap being filled with caulking or covered with a small bead, see Figure 7.71.

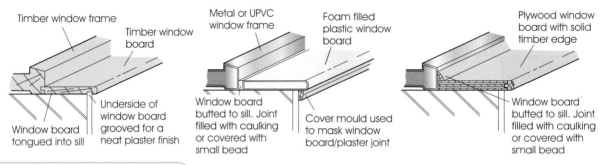

▲ **Figure 7.71** Window board details

Many window boards have a groove on the underside; this provides a finish for the plaster and masks any gap that may result later from shrinkage. Alternatively, a cover mould may be used to mask the joint.

Cutting and fixing window boards

See Figure 7.72:

1. Cut a length of board about 100 mm longer than the sill.
2. Mark out each end and cut out a portion of the board to fit the window reveal.
3. Return the front edge nosing profile around the ends, using a block plane and finish with an abrasive paper.
4. Use an off-cut of window board and a boat level to check for front to back level. Packing may be required under the board where it is not level; proprietary plastic shims available in a range of thicknesses are the best. However, a piece of hardboard or several pieces of damp-proof material will also do the job. These should be positioned about 50 mm from each end, with intermediates spaced at about 450 mm centres.
5. Place the window board in position and fix the front edge to the wall, using either cut nails, masonry nails or plugs, screws and pellets. Fixings should preferably go through any packers so the board is not pulled down out of line and also to ensure they are not misplaced later.

Alternative fixing methods

A batten or ground may be pre-fixed to the wall at the correct level for the window board, using cut or masonry nails, the window board in turn being fixed directly to the ground using ovals nails. Plastic and other pre-finished window boards may be bedded and fixed to the wall using beads of a gap-filling 'no nails' type of adhesive, or frame anchors screw-fixed to the underside of the window board and face of the wall.

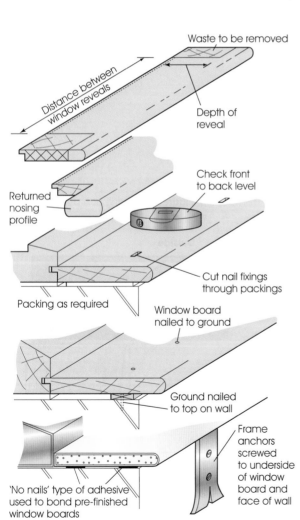

◀ **Figure 7.72** Fixing window boards

Chapter Eight

Mouldings and trim

This chapter covers the work of the site carpenter and joiner. It is concerned with the recognition and installation of mouldings and trim and it includes the following:

- Timber mouldings and trim.
- Materials and profiles.
- Installation of architraves.
- Installation of skirting.
- Installation of dado and picture rails.

Introduction to timber mouldings and trim

Mouldings are the ornamental contours or shapes applied to angles or features of building materials for decorative purposes. They are variously named according to their profile, see Figure 8.1.

The cutting and fixing of timber mouldings and trim is carried out by the carpenter after the building has been plastered and is classified as a

▼ Figure 8.1 Mouldings

Arrised	Pencil rounded	Rounded	Ovolo	Barefaced ovolo
Splayed	Chamfered	Sunk chamfered	Splayed and rounded	Weathered 9° min
Throat	Rebate — Barefaced tongue — Grooves	Quirk — Bead	Astragal — Fillet	Ogee (cyma-recta)

'second fixing' operation. Joiners may also be involved in the cutting and fixing of trim in the joinery works when producing items such as door sets.

Traditionally, these profiles were 'stuck' moulds on timber produced by the carpenter using a range of hand-held wooden moulding planes. Today stuck moulds are normally produced on a spindle moulder, multi-head planer/moulder or a hand-held portable router.

Trim is the collective term for timber-moulded sections. They may be termed as either vertical or horizontal and are used to cover the joint between adjacent surfaces, such as wall and floor or wall and ceiling or the joint between plaster and timberwork. In addition, they can provide a decorative feature, and may also serve to protect the wall surface from knocks and scrapes, see Figure 8.2.

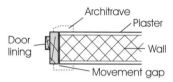
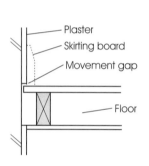

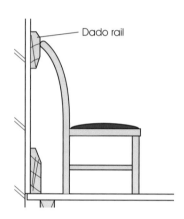

▼ **Figure 8.2** Covering of gaps and protecting plasterwork

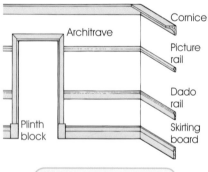

▲ **Figure 8.3** Types of trim

The common range and location of timber trim is shown in Figure 8.3.

- **Architrave.** The decorative trim that is placed internally around door and window openings to mask the joint between wall and timber and conceal any subsequent shrinkage and expansion.

- **Plinth block.** The block of timber traditionally fixed at the base of an architrave to take the knocks and abrasions at floor level. Also used to ease the fixing problems at this point that can occur when the skirting is thicker than the architrave.

- **Skirting.** The horizontal trim that is fixed around the base of a wall to mask the joint between the wall and floor. It also protects the plaster surface from knocks at floor level.

- **Dado rail.** A decorative moulding applied to the lower part of interior walls at about waist height approximately 1 m from the floor. It is also known as a chair rail as it coincides with a tall chair back height to protect the plaster. It also provides a line for the application of textured wallpaper down to the skirting, which was a feature in Victorian and Edwardian times.

- **Picture rail.** A moulding applied to the upper part of interior walls between 1.8 m and 2.1 m from the floor. Special clips may be hooked over the rail in order to suspend picture frames.

- **Cornice.** The moulding used internally at the wall and ceiling junction. It is normally formed from plaster, only rarely from timber. The area between the cornice and the picture rail is known as the *frieze*.

Heritage Link

Wooden moulding planes are still to be found in the tool chests of cabinet makers and specialist joiners undertaking restoration work.

Materials and profiles

Solid timber, either softwood or hardwood, is used for the majority of mouldings. However, MDF (medium-density fibreboard) and low-density foamed-core plastics are used to a limited extent for moulding production. They are normally ready to install, machined to a range of standard profiles (shapes) by the manufacturer or supplier, see Figure 8.4.

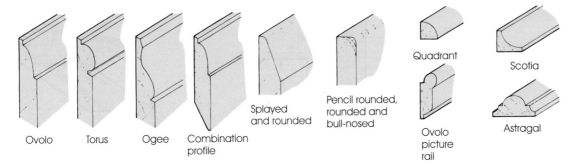

Ovolo Torus Ogee Combination profile Splayed and rounded Pencil rounded, rounded and bull-nosed Quadrant Ovolo picture rail Scotia Astragal

▲ **Figure 8.4** Standard trim sections

Skirting is often mass-produced using a combination profile, for example, with an ovolo mould on one face and edge and a splayed and rounded mould on the other. This enables it to be used for either purpose and also reduces the timber merchant's stock range. In addition, the moulding profile has the effect of an undercut edge enabling it to fit snugly to the floor and wall junction.

The following cutting and fixing details are generally suitable for all three of these materials. Consult manufacturer's instructions prior to starting to fix other proprietary mouldings or trim.

Installing architraves

Architraves are normally cut and fixed after hanging the doors. A set of architraves for one side of a room door consists of a horizontal head and two vertical jambs or legs, see Figure 8.5.

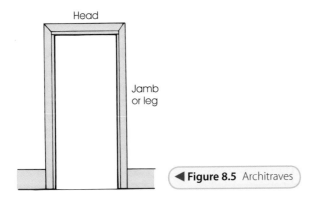

Head

Jamb or leg

◀ **Figure 8.5** Architraves

Jointing architraves

A 6 mm to 9 mm **margin** is normally left between the frame or lining edge and the architrave (see Figure 8.6). This margin provides a neat appearance to an opening; an unsightly joint line would result if architraves were to be kept flush with the edge of the opening.

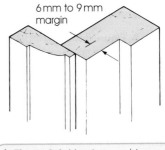

6 mm to 9 mm margin

▲ **Figure 8.6** Margin to architraves

Mitres

The return corners of a set of architraves are mitred, see Figure 8.7. For right-angled returns (90°) the mitre is 45° (half the total angle) and cut using either a mitre box/block or a mitre frame saw. A mitre chop saw speeds up work when a large number of mitres have to be cut.

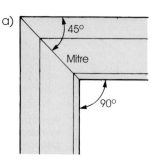

▶ **Figure 8.7** Mitring architraves

a)

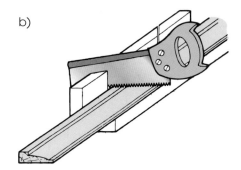

b)

Mitres for corners other than right angles should be half the angle of intersection, see Figure 8.8. They can be practically found by marking the outline of the intersecting trim on the frame/lining or wall, and joining the inside and outside corners to give the mitre line. Moulding can be marked directly from this. Alternatively an adjustable bevel can be used.

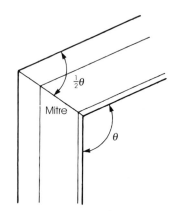

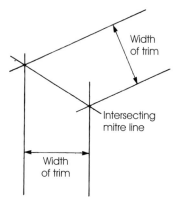

▶ **Figure 8.8** Determining mitre for corners other than right angles

The head is normally marked, cut and temporarily fixed in position first. The jambs can then be marked, cut, eased if required and subsequently fixed. Where the corner is not square or you have been less than accurate in cutting the mitre, it will require easing, either with a block plane or by running a tenon saw through the mitre.

Cutting and fixing architraves to door openings

Fixing is normally direct to the door frame or lining at between 200 mm and 300 mm centres using typically 38 mm or 50 mm long oval or lost-head nails. These should be positioned in the fillets or quirks (flat surface or groove in moulding, see Figure 8.9) and punched in. In addition, very wide architraves are often fixed back to the wall surface using either cut or masonry nails.

▶ **Figure 8.9** Fixing architraves

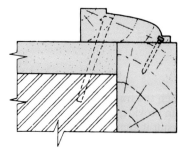

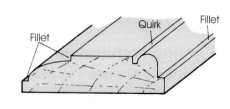

Mitres should be nailed through their top edge to reinforce the joint and to ensure both faces are kept flush, see Figure 8.10. 38 mm oval or lost-head nails are suitable for this purpose.

Do not forget to wear eye protection when driving masonry nails.

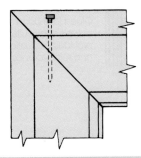

▲ **Figure 8.10** Nailing mitre joints at corners of architraves

A typical fixing procedure is as follows:

1. Mark a margin line around the door opening (see Figure 8.11). Check-lines the width of the architrave may also be marked on the wall at the top of the opening.
2. Starting with the headpiece, cut the mitre on one end, hold the piece in position and mark the second mitre.
3. Cut the second mitre and temporarily fix in position with two nails. Leave the nails protruding in case later adjustment is required.
4. Hold the first jamb in position and mark the mitre. Cut the mitre, fit with a block plane if required and temporarily nail in position.
5. Repeat the last stage with the second jamb.
6. Complete nailing at 200 mm to 300 mm intervals, including nailing across the mitres. Punch all nails below the surface in preparation for filling later by the painter.
7. Use a glass paper block to remove the arris around the frame, architraves and mitres.

Plinth blocks. Where used, these may be butt jointed to the architrave, see Figure 8.12, but traditionally they were pre-joined together using a barefaced tenon and screws.

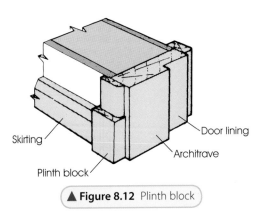

Skirting

Door lining

Architrave

Plinth block

▲ **Figure 8.12** Plinth block

Corner blocks. Sometimes used at the return corners of a set of architraves in place of mitres (see Figure 8.13). These may be simply pinned in place and the architrave butt jointed to them. Alternately, they may be prefixed to the architrave using dowels or a biscuit. To mask the effect of subsequent block shrinkage, the head and leg architraves may be housed into them.

Mark margin

Corner check line

Margin

1

Cut first mitre, mark second

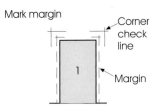

2

Cut second mitre and fix head

3

Mark mitre to first jamb

4

Cut mitre, fix first jamb

5

Mark mitre to second jamb

6

Cut mitre, fix second jamb

7

Punch in nails and remove arrises

8

▲ **Figure 8.11** Marking and fixing architraves

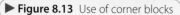
► **Figure 8.13** Use of corner blocks

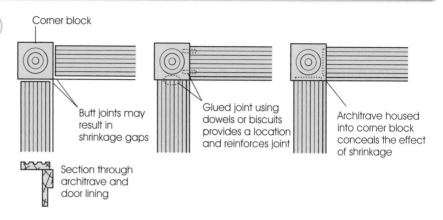

Cutting and fixing architraves to four-sided openings

In addition to the normally three-sided use of architraves for doors, they are also fixed around all four sides of wall serving hatches, ceiling hatches and even wall surfaces without an opening to create a decorative panel effect.

1. Mark a margin line all around the opening, see Figure 8.14. Mark the width of the architrave from the margin line on the wall to provide a check line.

► **Figure 8.14** Fixing architraves to a four-sided opening

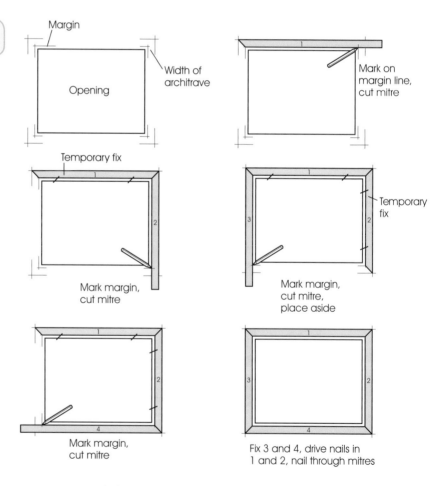

2. Starting with the top piece for wall hatches or one of the shorter sides of a ceiling hatch. Cut a mitre on one end of piece (1). Hold the piece in position and mark a second mitre.

3. Cut the second mitre and temporarily fix in position with two nails. Leave the nail heads protruding in case later adjustment is required.

4. Cut and fit, if required, a mitre on one end of piece (2). Hold the piece in position and mark a second mitre. Cut the mitre and temporarily fix in position.

5. Cut and fit, if required, a mitre on one end of the piece (3). Hold the piece in position and mark a second mitre. Cut the second mitre and place the piece aside.

6. Cut and fit, if required, a mitre on one end of the piece (4). Hold the piece in position and mark a second mitre. Cut the mitre.

7. Hold pieces (3) and (4) in position to check the second mitre. Fit if required. Fix both pieces in position.

8. Complete nailing at 200 mm to 300 mm centres, including nailing across the mitres. Punch all the nails below the surface, in preparation for filling later by the painter.

9. Use a glass paper block to remove arris around the frame, architrave and mitres.

Scribing architraves

Architraves should be scribed (one member cut to fit the contour of another) to fit the wall surface, where frames or linings abut a wall at right angles.

1. Temporarily fix the architrave jamb in position, keeping the overhang the same all the way down, see Figure 8.15.

2. Set a compass to the required margin plus the overhang or, alternatively, use a piece of timber this size as a gauge block.

3. Mark with the compass or gauge up the line to be cut. Saw along the marked line by slightly undercutting the edge (making it less than 90°). It will fit snugly to the wall contour, see Figure 8.16.

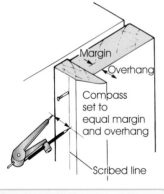

▲ **Figure 8.15** Scribing architraves

▲ **Figure 8.16** Scribing should fit snugly to wall surface

Quadrant and Scotia moulds

A quadrant mould or Scotia mould is often used to cover the joint to provide a neat finish to the reveal of external door frames (see Figure 8.17). These moulds may also be used in place of an architrave jamb where the frame/lining joins to a wall at right angles. The arris on the back of the mould is best pared off with a chisel to enable the mould to sit snugly into the plaster/timber intersection.

All nails used for fixing moulds should be punched below the surface on completion of the work. This is in preparation for subsequent filling by the painter.

▲ **Figure 8.17** Use of quadrants as an alternative to scribing

Installing skirting

Skirting is normally cut and fixed directly after cutting and fixing the architraves.

Jointing skirting

Internal corners

Corners that are 90° or less are best scribed, one piece being cut to fit over the other, see Figures 8.18 and 8.19.

▶ **Figure 8.18** Internal corners

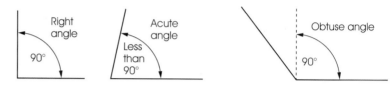

Scribes can be formed in one of two ways:

- **Mitre and scribe.** Fix one piece and cut an internal mitre on the other piece to bring out the profile (see Figure 8.20). Cut the profile square on mitre line to remove waste. Use a coping saw for the curve.

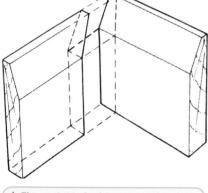

▲ **Figure 8.19** Scribing internal corners

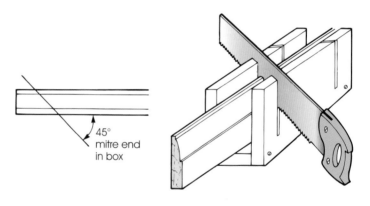

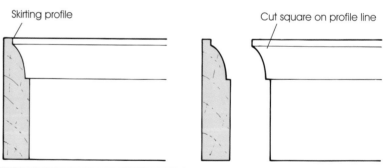

▲ **Figure 8.20** Cutting an integral scribe (mitre and scribe)

- **Compass scribe.** Fix one piece and place the other piece in position (see Figure 8.21). Scribe with a compass. Cut square on the scribed line to remove waste.

Scribing is the preferred method especially where the walls are slightly out of plumb. The mitred scribe would have a gap, but the compass scribe would fit the profile neatly.

Mitres are not normally used for internal corners of skirting, as wall corners are rarely perfectly square, making the fitting difficult. In addition, mitres open up as a result of shrinkage, forming a much larger gap than scribes.

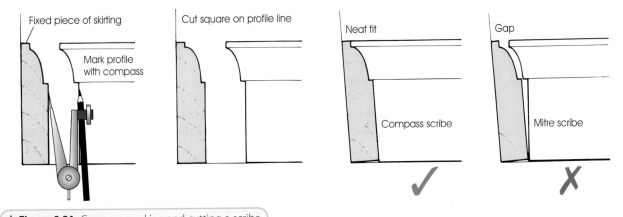

▲ Figure 8.21 Compass marking and cutting a scribe

However, internal corners on bull-nosed or pencil-rounded skirtings may be cut with a **false mitre** or partial mitre on the top rounded edge and the remaining flat surface scribed to fit (see Figure 8.22).

Internal corners over 90°. These are called obtuse angles (see Figure 8.18) and are best jointed with a mitre.

Mitring external corners

These return the moulding profile at a corner (see Figure 8.23), rather than a butt joint, which would otherwise show unsightly end grain. Mitres for 90° external corners may be cut in a mitre box or with the aid of a frame saw or mitre chop saw.

▲ Figure 8.22 False mitre and scribe

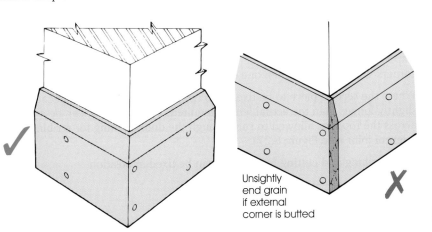

◀ Figure 8.23 Mitring external corners

Mitres for both internal and external corners over 90° can be marked out using the following method and then cut freehand:

1. Use a piece of skirting to mark line and width of skirting on floor, either side of the mitre (see Figure 8.24).
2. Place length of skirting in position.

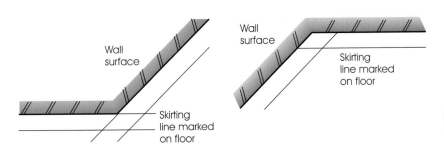

◀ Figure 8.24 Line of skirting for corners over 90°

3. Mark position of plaster arris on top edge of skirting. For internal corners this will be the actual back edge of the skirting (see Figure 8.25).
4. Mark outer section on front face of skirting.
5. Use try square to mark line across face and back surface of skirting.
6. Cut mitre freehand or with the aid of a frame saw or chop saw set to the required angle.
7. Repeat marking out and cutting procedure for other piece.
8. Fix skirting to wall; external corners should be nailed through the mitre.

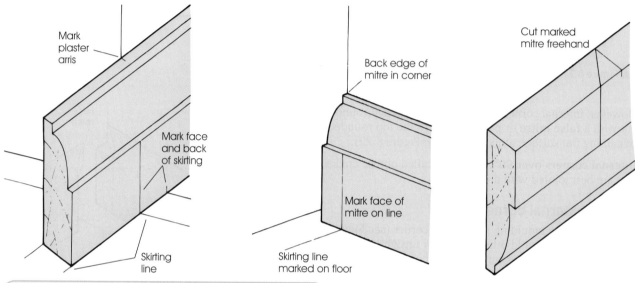

▲ **Figure 8.25** Marking out external and internal corners over 90°

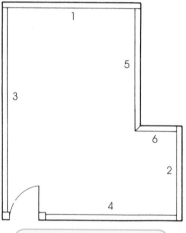

▲ **Figure 8.26** Order of fixing (trapped pieces first)

Long lengths are fixed first (see Figure 8.26), starting with those having two trapped ends (both ends between walls). Marking and jointing internal corners is much easier when one end is free.

Where the second piece to be fixed also has two trapped ends, a piece slightly longer than the actual length required, by say 50 mm, can be angled across the room or allowed to run through the door opening for scribing the internal joint (see Figure 8.27).

After scribing and cutting to length it can be fixed in position.

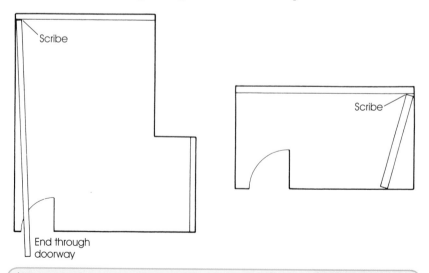

▲ **Figure 8.27** Extend through doorway to permit scribing of joint, or angle and scribe when both ends are trapped

Very short lengths of skirting returned around projections may be fixed before the main lengths. The two short returns are mitred at their external ends, cut square at their internal ends and fixed in position (see Figure 8.28).

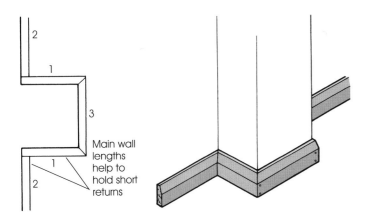

◀ **Figure 8.28** Small pieces may be fitted first

Main wall lengths are then scribed and fixed in position. These help to hold the short returns. Finally the front piece is cut and fixed in position by nailing through the mitres.

Heading joints can be used where sufficiently long lengths of skirting are not available. Mitres are preferred to butts, because the two surfaces are held flush together by nailing through the mitre (see Figure 8.29). In addition, mitres mask any gap appearing as a result of shrinkage.

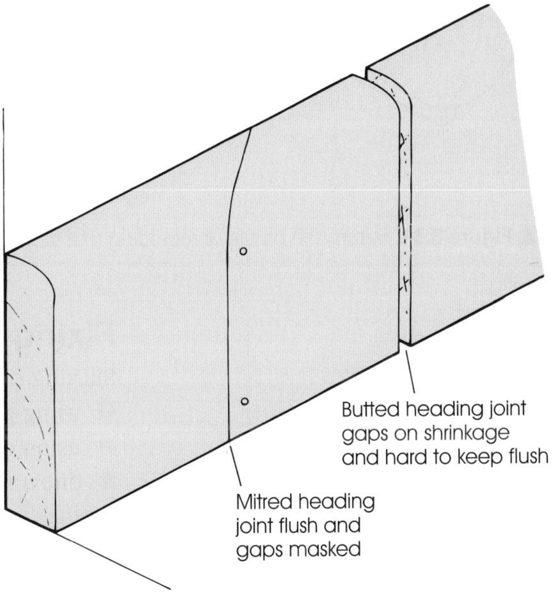

Installing skirting on a curved surface

Where skirtings and other trim are to be fixed around a curved surface, the back face will almost certainly require kerfing. This involves putting saw cuts in the back face at regular intervals to effectively reduce the thickness of the trim. The kerfs, which may be cut with a tenon saw, should be spaced between 25 mm and 50 mm apart. The tighter the curve, the closer together the kerfs should be. The depth of the kerfs must be kept the same. They should extend through the section to the maximum extent, but just keeping back from the face and top edge (see Figure 8.30).

Care is required when fixing the trim back to the wall to ensure that it is bent gradually and evenly. It is at this time that accuracy in cutting the kerfs evenly spaced to a constant depth is rewarded. Any over-cutting and the trim is likely to snap at that point.

Mitres at the ends of curved sections may be marked out and cut using the same methods as described for other obtuse angles.

▲ **Figure 8.29** Mitres are preferred for heading joints

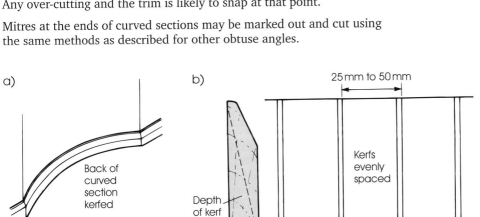

◀ **Figure 8.30** Fixing to a curved surface using kerfing

Returned stop mouldings

Skirtings and other mouldings may occasionally be required to stop part way along a wall, rather than finish into a corner or another moulding such as skirting to a landing. In these circumstances the profile should be returned to the wall or floor.

- **Returned to wall.** This can be achieved by mitring the end and inserting a short mitred return piece giving a true mitre for good-quality work. Alternatively, the return profile can be cut across the end grain of the main piece. However, end grain will remain exposed (see Figure 8.31).
- **Returned to floor.** This involves mitring the moulded section (see Figure 8.32), cutting out the waste and inserting a return moulded section (cut from an off-cut) down the end of the board.

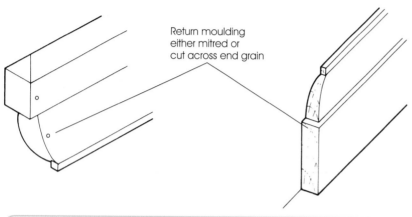

Return moulding either mitred or cut across end grain

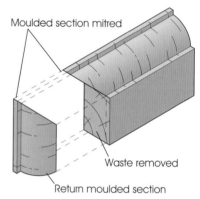

Moulded section mitred

Waste removed

Return moulded section

▲ **Figure 8.31** Return the profile of moulding that stops part way along a wall

▲ **Figure 8.32** Stopped moulding returned to floor

Fixing skirtings

Skirting can be fixed back to walls:

- using grounds
- using timber twisted plugs
- direct to the plaster or wall surface.

Grounds

Traditionally, timber battens or grounds were fixed around the base of a room just before plastering to provide a guide for the plasterer and also a means for the later fixing of the skirting. Today, grounds are rarely used in new work but may be encountered in restoration work. Where used, they are fixed just below the top of the skirting using cut nails into the mortar joint or masonry nails. One ground is required for skirtings up to 100 mm in depth. Deeper skirtings require either the addition of vertical soldier grounds at 400 mm to 600 mm centres or an extra horizontal ground (see Figure 8.33).

Packing pieces may be required behind the grounds to provide a true surface on which to fix the skirting. Check the line of ground with either a straightedge or string line. Skirtings can be fixed back to grounds using, typically, 38 mm or 50 mm oval or lost-head nails.

Twisted timber plugs

These are rarely used today. They are shaped to tighten when driven into the raked-out vertical brickwork joints at approximately 600 mm apart (see Figure 8.34) and are positioned before plastering. When all the plugs have been inserted, they should be marked with a straightedge or string line and cut off to provide a true line. An allowance should be made for the thickness of the plaster.

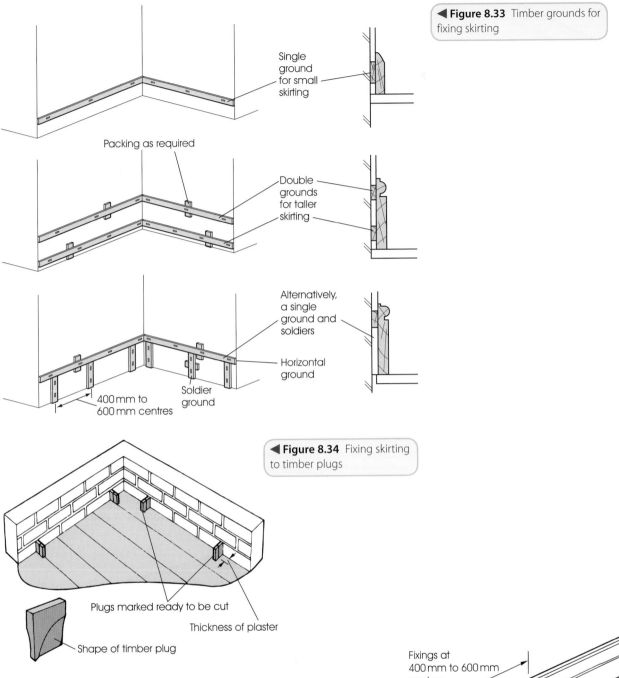

◄ **Figure 8.33** Timber grounds for fixing skirting

Single ground for small skirting

Packing as required

Double grounds for taller skirting

Alternatively, a single ground and soldiers

Horizontal ground

400 mm to 600 mm centres

Soldier ground

◄ **Figure 8.34** Fixing skirting to timber plugs

Plugs marked ready to be cut

Thickness of plaster

Shape of timber plug

Skirtings are fixed back into the end grain of the plugs using, typically, 50 mm cut nails. These hold better in the end grain than would oval or lost-head nails.

Direct to the wall

Skirtings are fixed back to the wall after plastering, using typically either 50 mm cut nails or 50 mm masonry nails depending on the hardness of the wall. Oval or lost-head nails may be used to fix skirtings to timber studwork partitions.

Fixings should be spaced at 400 mm to 600 mm centres (see Figure 8.35). These should be double nailed near the top and bottom edge of the skirting or, alternatively, they may be staggered between the top and bottom edge. Remember, all nails should be punched below the surface.

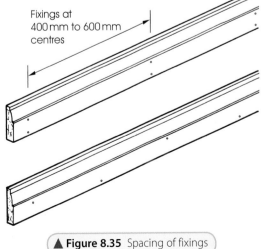

Fixings at 400 mm to 600 mm centres

▲ **Figure 8.35** Spacing of fixings

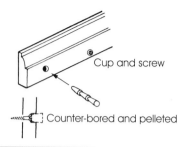

Cup and screw

Counter-bored and pelleted

▲ **Figure 8.36** Screws sometimes used to fix hardwood skirtings

Hardwood skirtings for very high-quality work may be screwed in position. These should be counter-bored and filled with cross-grained pellets on completion or brass screws and cups can be used (see Figure 8.36).

As a modern alternative to nails and screws, skirting and other trim fixed to wall surfaces may be bonded using a gun-applied, gap-filling, 'no nails' adhesive. One or two continuous 6 mm beads should be applied to the back of the trim before positioning and pressing in place. Strutting or temporary nails (see Figure 8.37) may be required to hold the trim in place while the adhesive cures. These should be left overnight before removal.

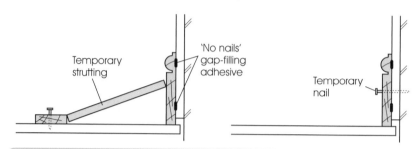

Temporary strutting

'No nails' gap-filling adhesive

Temporary nail

▲ **Figure 8.37** Use of gap-filling adhesive to fix skirting

Checking for hidden/buried services

Prior to fixing mouldings across any wall, a check should be made to see if any services are hidden below the wall surface. Fixing into electric cables and gas or water pipes is potentially dangerous and expensive to repair. Wires to power points normally run vertically up from the floor. Wires to light switches normally run vertically down from the ceiling. Therefore, keep clear of these areas when fixing. Buried pipes in walls are harder to spot.

Vertical pipes may just be seen at floor level; outlet points may also be visible (see Figure 8.38). Assume both of these run the full height of the wall, both up and down. Therefore, again, keep clear of these areas when fixing. When in doubt, an electronic device can be used to scan the wall surface prior to fixing. This gives off a loud noise when passing over buried pipes and electric cables.

▶ **Figure 8.38** Locating hidden services

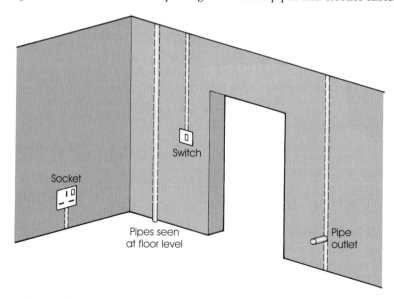

Socket

Switch

Pipes seen at floor level

Pipe outlet

▲ **Figure 8.39** Keeping skirting tight to floor surface when fixing

Holding down and scribing skirtings

When fixing narrow skirtings, say 75 mm to 100 mm in depth, they may be kept tight down against a fairly level floor surface with the aid of a kneeler (see Figure 8.39). This is a short piece of board placed on the top edge of the skirting and held firmly by kneeling on it.

Deeper skirtings and/or uneven floor surfaces may require scribing to close the gaps before fixing. This is carried out after jointing but prior to fixing:

1. Place cut length of skirting in position.
2. Use gauge slip or compass set to widest gap to mark on the skirting a line parallel to the uneven surface (see Figure 8.40).
3. Trim skirting to line using either a handsaw or plane.
4. Undercut the back edge to ensure the front edge snugly fits the floor contour.

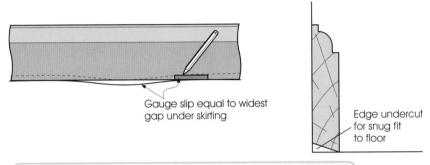

Gauge slip equal to widest gap under skirting

Edge undercut for snug fit to floor

▲ **Figure 8.40** Scribing and cutting skirting to an uneven floor surface

Installing dado and picture rails

Dado and picture rails are mainly horizontal mouldings. They may be cut and fixed using similar methods to those used for skirting.

Start by marking a level line in the required position around the walls (see Figure 8.41). Use a straightedge and spirit level or a water level and chalk line or, alternatively, a laser level. The required position may be related to a datum line where established.

When working single-handed, temporary nails can be used at intervals to provide support prior to fixing (see Figure 8.42).

Simple sections can be scribed at internal corners, as are skirtings; otherwise, use mitres. External corners should always be mitred.

> ### Related information
>
> Also see Book 1, Chapter 7, page 324, which covers the procedures for establishing datum lines and the use of water levels and laser levels.

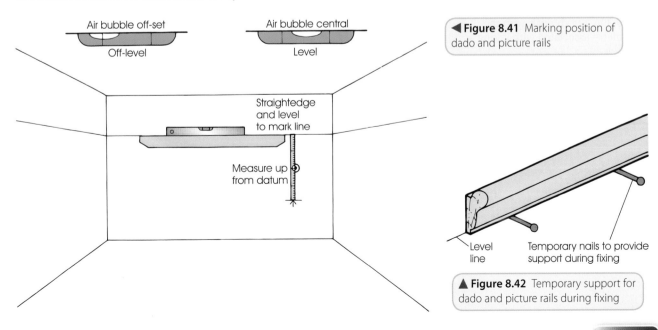

Air bubble off-set

Off-level

Air bubble central

Level

Straightedge and level to mark line

Measure up from datum

◄ **Figure 8.41** Marking position of dado and picture rails

Level line

Temporary nails to provide support during fixing

▲ **Figure 8.42** Temporary support for dado and picture rails during fixing

Fixings are normally direct to the wall surface at about 400 mm centres using typically either 50 mm cut nails or 50 mm masonry nails depending on the hardness of the wall. 50 mm oval or lost-head nails can be used when fixing mouldings to timber studwork partitions. Alternatively, a gun-applied, gap-filling, 'no nails' adhesive may be used to bond them to the wall surface.

Mitres around external corners should be secured with nails through their edge. Typically 38 mm oval or lost-head nails are used for this purpose. Remember, the carpenter should punch all nails below the surface ready for subsequent filling by the painter.

Raking moulds

These are mouldings that are not fixed horizontally, such as those that follow the pitch of the stairs. Where raking moulds join up with a level moulding at a mitred corner, the actual profile of one mould will need to change. Either a special mould is required or, more practically, the level mould can be returned around the corner a short distance and mitred to the raking mould of the same section, see Figure 8.43.

▶ **Figure 8.43** Raking moulds

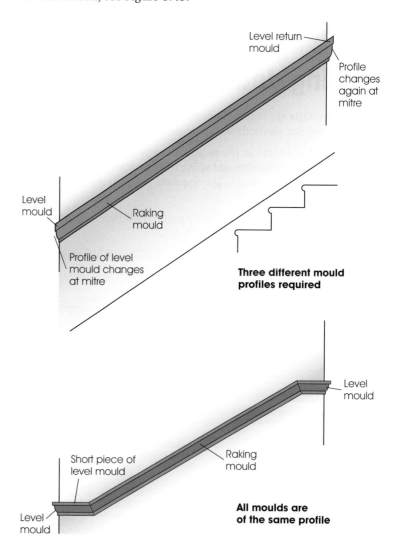

Level return mould

Profile changes again at mitre

Level mould

Raking mould

Profile of level mould changes at mitre

Three different mould profiles required

Level mould

Short piece of level mould

Raking mould

Level mould

All moulds are of the same profile

Related information

Also see Book 1, Chapter 9, page 396, which deals with the geometry associated with raking moulds.

Chapter Nine

Units, fitments, seating and miscellaneous items

This chapter covers the work of both the site carpenter and bench joiner. It is concerned with principles of constructing and installing units, fitments, seating and other items. It includes the following:

- Types, manufacture and installation of purpose-made and proprietary units.
- Types, manufacture and installation of built-in and free-standing fitments.
- Retail units and fitments.
- Types and manufacture of seating units.
- Miscellaneous joinery items.

Timber units and fitments here comprise cupboards, built in or fixed in, with worktops, shelves, drawers and doors, and special-purpose joinery items including seating, lecterns and tables. The details that follow can be used or adapted to suit a variety of units and fitments for various end uses, such as domestic, commercial, retail and leisure.

Timber units

Units fall into two distinct categories:

- **Purpose-made.** A unit made in a joiner's shop for a specific job. Most will be fully assembled prior to their arrival on site.
- **Proprietary.** A unit or range of units mass-produced to standard designs by a manufacturer. Budget-priced units are often sent in knockdown form (known as *flat packs*) ready for on-site assembly. Better-quality units are often ready-assembled in the factory (known as *rigid units*).

Terminology

- **Cupboard.** Historically derived from the name given to the simple open boards or shelves that were used to display cups, silver plate and other items ('cup board').
- **Carcass.** The main assembly or frames of a cupboard, excluding doors and drawers.
- **Pot board.** The lower shelf or base of a cupboard. Again, historically derived from the name given to a low board or shelf, raised just off the floor on which heavy cooking pots were stored.

Heritage Link

The photograph below shows a three-drawer Welsh dresser with an open pot board below.

- **Plinth.** The recessed base of a cupboard, which supports the pot board. It also provides a foot space for those standing in front of the cupboard and is often called a kickboard.
- **Standards.** The vertical end frames or panels and intermediate divisions of a cupboard. In some parts, end standards are known as 'gables' and intermediate standard as 'haffits'.

Methods of construction

The two main methods of carcass construction for both proprietary and purpose-made units are (see Figure 9.1):

- box construction
- framed construction.

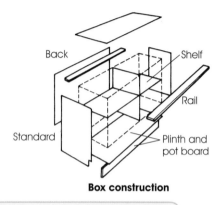
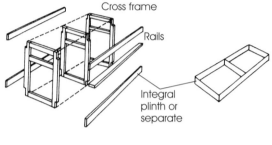

Box construction **Framed construction**

▲ **Figure 9.1** Types of unit construction

Box construction

This is also known as slab construction. It uses vertical standards and rails and horizontal shelves. A back holds the unit square and rigid. The plinth and potboard are often integral with the unit.

- **Proprietary units** are almost exclusively made from 15 mm to 19 mm thick melamine-faced chipboard (MFC) or medium-density fibre board (MDF).
- **Purpose-made units** may be constructed using MFC, MDF, blockboard, plywood or, more rarely, solid timber.
- **Flat packs** use **knock-down fittings** or screws to join the panels. Assembly is a simple process of following the manufacturer's instructions and drawings, coupled with the ability to use a screwdriver. See Figure 9.2.

Rigid and purpose-made units may be either doweled or housed and screwed together. Glue is used on assembly to form a rigid carcass. Typical carcass construction details are shown in Figure 9.3.

Framed construction

This is also known as **skeleton construction**. This uses frames either front and back joined by rails or standards, or cross frames joined by rails. The plinth and pot board are normally separate items.

The frames of proprietary units are normally dowelled, whereas purpose-made ones would be mortised and tenoned together. Typical carcass construction details are shown in Figure 9.4.

Adjustable shelves

In situations where the sizes of the items to be stored are not known or where they are subject to change, some form of shelf adjustment must be incorporated. Three of the many methods used are as follows (see Figure 9.5):

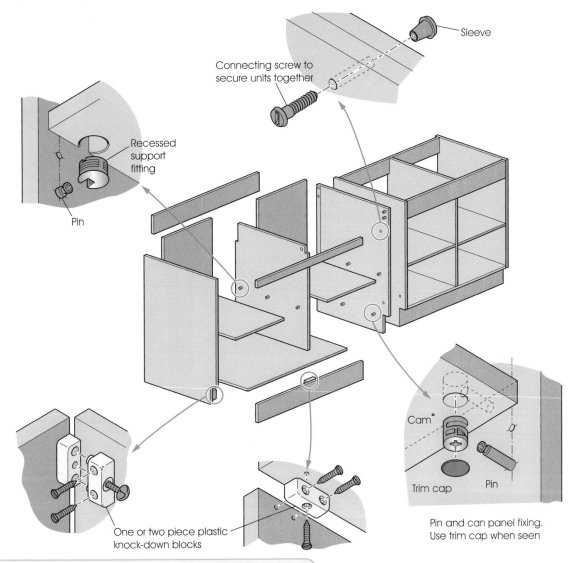

Sleeve

Connecting screw to
secure units together

Recessed
support
fitting

Pin

Cam°

Trim cap   Pin

Pin and can panel fixing.
Use trim cap when seen

One or two piece plastic
knock-down blocks

▲ **Figure 9.2** Box construction unit using typical knock-down fittings

- The traditional solution was the use of saw tooth supports and splayed-end push-in shelf bearers that can be fitted at any desired height.
- A very popular and efficient method is the use of tonks: bookcase strips and studs, which allow height adjustment in 25 mm units. The flush strip is designed to be recessed into the standards. A smaller, deeper groove must also be run to give clearance for inserting the tongues of the studs. Alternatively, a surface-fixed strip can be used. This overcomes the need for grooving out and weakening members, but results in an inferior finish.
- Sockets tapped into blind holes, which have been drilled at intervals down the standards, and used with push-in studs, are suitable for a lighter range of applications.

Drawers

Drawers may be incorporated into units in order to provide storage and security (when fitted with a lock) for smaller items. The size of a drawer will be related to the items it is intended to store, but in general will range between 100 mm and 200 mm in depth. When they are vertically stacked in one unit the deeper drawers should be located at the bottom.

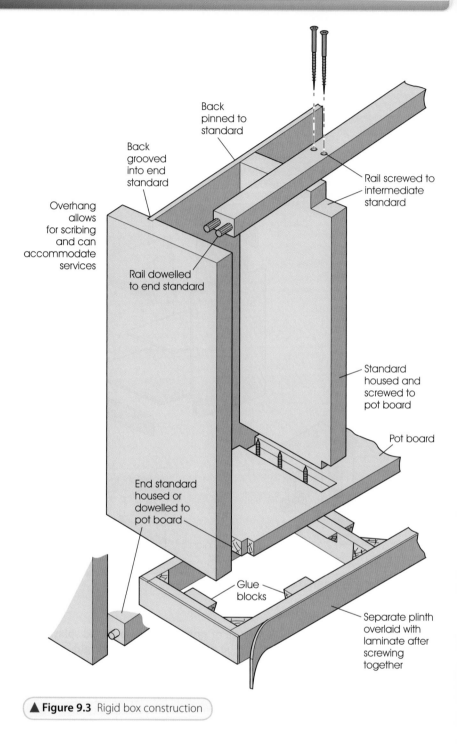

Back
pinned to
standard

Back
grooved
into end
standard

Overhang
allows
for scribing
and can
accommodate
services

Rail screwed to
intermediate
standard

Rail dowelled
to end standard

Standard
housed and
screwed to
pot board

Pot board

End standard
housed or
dowelled to
pot board

Glue
blocks

Separate plinth
overlaid with
laminate after
screwing
together

▲ **Figure 9.3** Rigid box construction

A traditional method of drawer construction uses through dovetails at the back and lapped dovetails at the front, see Figure 9.6. The plywood bottom is grooved into the front and sides and is pinned to the bottom of the drawer back. Small glue blocks are positioned under the bottom to provided additional rigidity and assist sliding. Rounded machine-made dovetails, produced on a router or spindle moulder, are a more economic alternative for better-quality work where repetitive production is required. Sky slopes formed using plastic laminate inserts, grooved into the rear of the drawer front, may be incorporated into better-quality drawers. These aid both the sliding out of the drawer contents and cleaning.

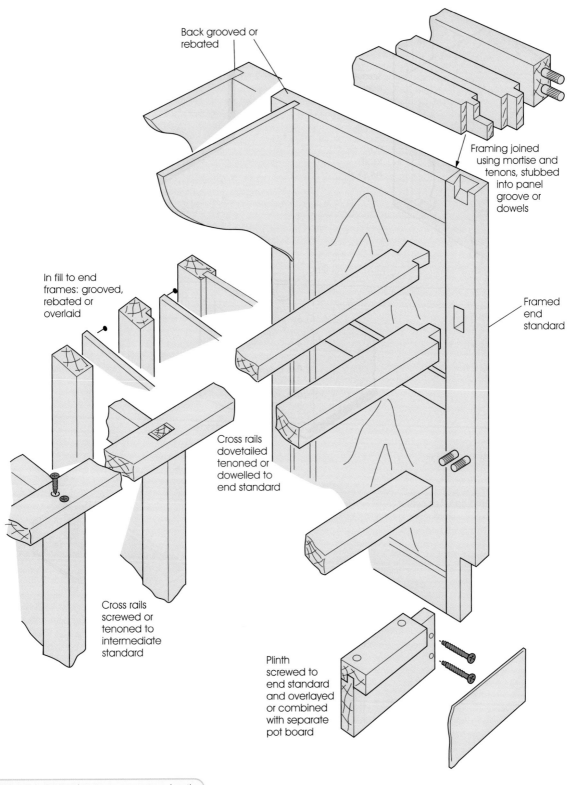

Back grooved or rebated

Framing joined using mortise and tenons, stubbed into panel groove or dowels

Framed end standard

In fill to end frames: grooved, rebated or overlaid

Cross rails dovetailed tenoned or dowelled to end standard

Cross rails screwed or tenoned to intermediate standard

Plinth screwed to end standard and overlayed or combined with separate pot board

▲ **Figure 9.4** Framed unit construction details

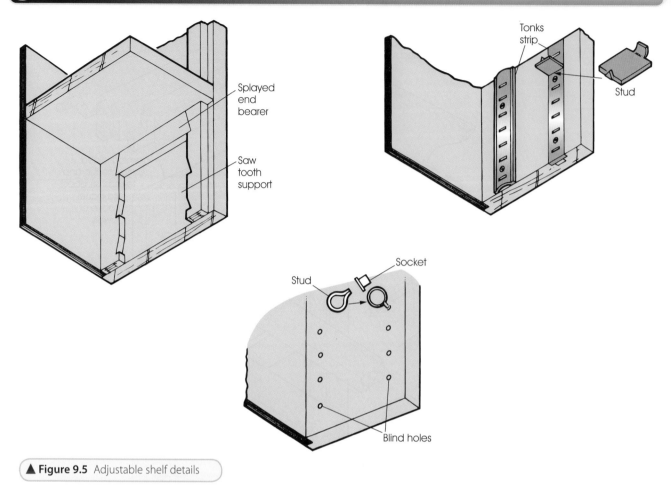

▲ Figure 9.5 Adjustable shelf details

▶ Figure 9.6 Traditional drawer construction

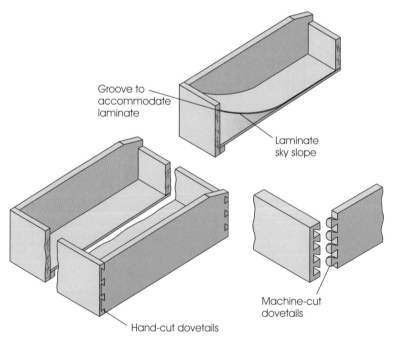

Modern mass-produced drawers are made in MFC or MDF. The corner joints can be housed and pinned, butted and screwed, dowelled or biscuited. The bottom of the drawer is typically either nailed or screwed directly to the underside of the drawer sides. A separate false slab front is secured to the drawer by screws from the inside, see Figure 9.7.

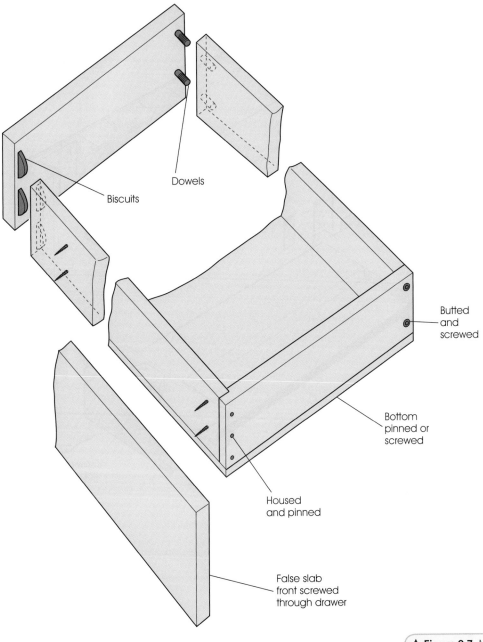

Dowels

Biscuits

Butted
and
screwed

Bottom
pinned or
screwed

Housed
and pinned

False slab
front screwed
through drawer

▲ **Figure 9.7** Modern drawer construction

Various methods can be used to suspend and slide drawers. Traditionally, rails may be incorporated into a unit for this purpose, see Figure 9.8. The dust board shown grooved into the rails is mainly used on better-quality work. Its purpose is to separate the drawer and cupboard spaces. A drawer kicker is fitted between the top rails to prevent the front of the drawer falling downward as it is pulled out. When closed, the drawer front should finish flush with the unit. This can be achieved by pinning small plywood drawer stops to the front of the drawer rail.

Another traditional method of suspending and sliding drawers is shown in Figure 9.9. This uses grooved drawer sides, preferably of a hard wood with good wearing qualities, which slide on hardwood runners glued and screwed to the unit's sides or standards. Where this method is used, the drawer is often fitted with a false slab front screwed from the inside of the drawer. This front has projecting ends to conceal the runner from view. Alternatively, fibre drawer slides may be used with one part fixed to the side of the drawer and the other to the unit.

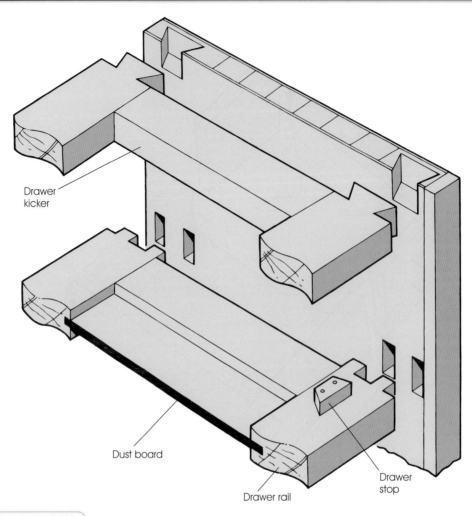

Drawer
kicker

Dust board

Drawer rail

Drawer
stop

▲ **Figure 9.8** Drawer rail details

▼ **Figure 9.9** Drawer slide details

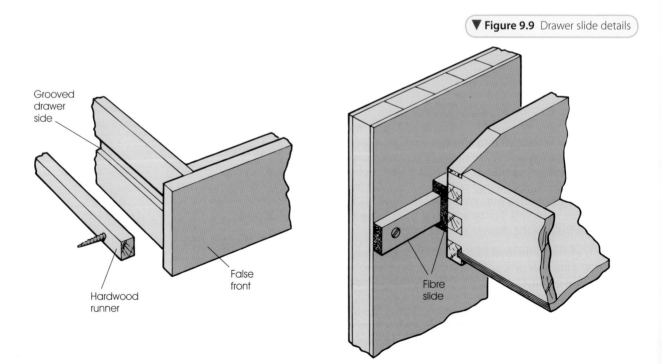

Grooved
drawer
side

False
front

Hardwood
runner

Fibre
slide

Modern methods of suspending and sliding drawers, see Figure 9.10, employ the use of metal side- or bottom-mounted runners. These incorporate a ball race or plastic roller to ensure ease of operation. Some also include a self-close or soft-close action.

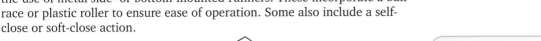 **Figure 9.10** Metal drawer runners

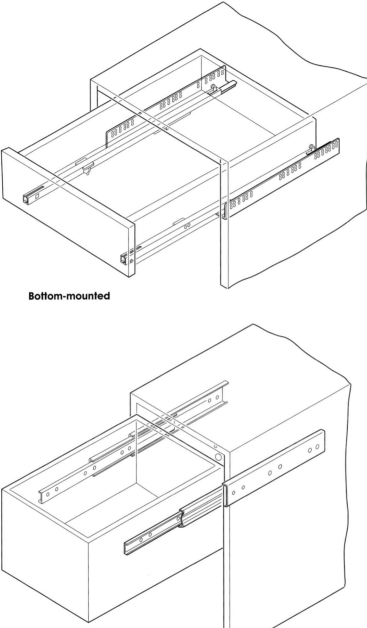

Bottom-mounted

Side-mounted

Doors

Doors can be incorporated to close the front of open units for reasons of tidiness, protection or security. They may be either side-hung or sliding doors. Side-hung doors allow maximum access but when they are open they project into the room, which can be restrictive and even hazardous in confined spaces. Figure 9.11 shows various methods of side-hanging cupboard doors:

- **Type A: flush hinge.** The door is set flush within the unit and hung on butt hinges. The use of flush hinges avoids the need for recessing and provides the necessary clearance joint.

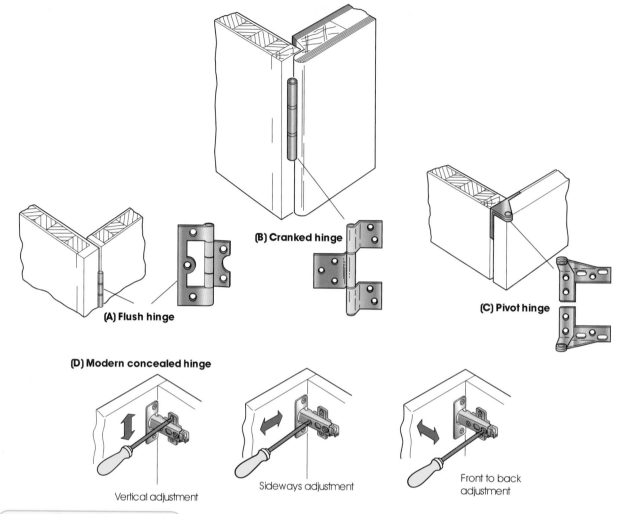

(B) Cranked hinge

(A) Flush hinge

(C) Pivot hinge

(D) Modern concealed hinge

Vertical adjustment

Sideways adjustment

Front to back adjustment

▲ **Figure 9.11** Side-hung cupboard doors

Type B: cranked hinge. Rebated doors hung on cranked hinges were at one time popular for mass-produced units. These doors do not require any individual fitting as the rebate, which laps over the face of the unit, conceals the very large clearance joint.

Type C: pivot hinge. Doors hung on the face of a unit are probably the simplest to make and fit. Although this arrangement is possible with standard butt hinges, the use of cranked or special extended pivot hinges permits the door to open within the width of the unit.

Type D: concealed hinge. Doors are again face hung. A concealed cabinet hinge is bored into the rear face of the door and its mounting plate fixed to the inside of the standard. This type, which is extensively used in modern kitchen units, allows the door to be adjusted in height, sideways, backwards and forward. They include a degree of self-closing in the final stage of swing and so do not require a catch.

Sliding doors

There are various methods available for making cupboard doors slide. Glass, thin plywood or MDF doors are often made to slide in nylon or fibre tracks (see Figure 9.12). The deeper channel track is used at the top, so the doors can be inserted and removed by pushing them up into the top track, clearing the bottom one. Figure 9.12 also shows how a cupboard door may be made to slide on fibre tracks grooved into the pot board. Two nylon sliders are recessed and screwed to the underside of each door. These run on the fibre track and provide a smooth sliding action, which wears well. The top edge of the door is usually rebated to engage in a groove run in the underside of the

(a)

(b)

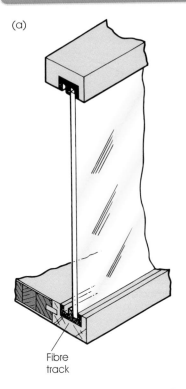

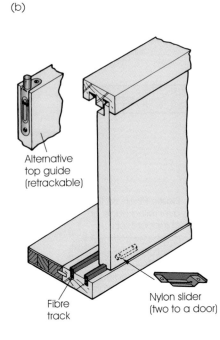

Alternative
top guide
(retrackable)

Fibre
track

Nylon slider
(two to a door)

Fibre
track

top rail. Sufficient clearance for insertion and removal must be made at this
point to allow the bottom of the door to clear the pot board when pushed
up into the top groove. Retractable top guides similar to flush bolts can be
used instead of rebating the top edges of the doors.

Heavyweight cupboard doors are best top hung to achieve a smooth running
action. Figure 9.13 shows one of the simplest types of top-hung cupboard
door sliding track. It consists of a surface-fixed aluminium top track and
bottom guide. The door is suspended by two nylon hangers/sliders fixed to
its top edge.

The handle position of bottom-sliding doors is best kept nearer the bottom
of the door for smooth sliding action. There will be a tendency for the door
to tip and judder if the handle is positioned higher than a distance equal to
the door's width. The best action is achieved with top-hung doors when the
handle position is kept as high as possible.

Worktops

The main types that are in common use are shown in Figure 9.14.

■ **Solid timber.** Made up from narrow-edge jointed boards. Provision
for moisture movement needs to be considered when fixing them to
a carcass. Mitred return ends may be specified; these serve to hold
the top flat and also cover the exposed end grain.

■ **Post-formed.** A chipboard or MDF base covered with a plastic
laminate, which has been formed over a rolled edge. The most
popular type of worktop for kitchen units, it is ready finished and
simply requires fixing in place.

■ **Edged sheet material.** A chipboard or MDF base covered with
either a melamine face, plastic laminate or wood veneer. Matching
edging is applied to the seen edges. The undersides of sheet material tops
should be the same as the top face or be sealed to prevent distortion. It is
good-practice to use a double-faced, melamine or wood veneer board. On
good-quality work a balancer laminate is applied to the underside of
laminate-faced tops to relieve the stresses that would result from facing
one side only. A cheaper alternative is to seal the underside with an
application of varnish or adhesive.

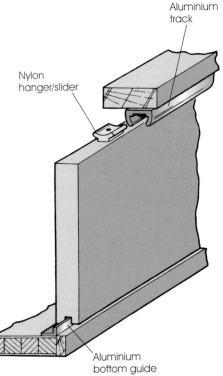

Aluminium
track

Nylon
hanger/slider

Aluminium
bottom guide

▲ **Figure 9.13** Top-hung sliding door

▶ **Figure 9.14** Worktop details

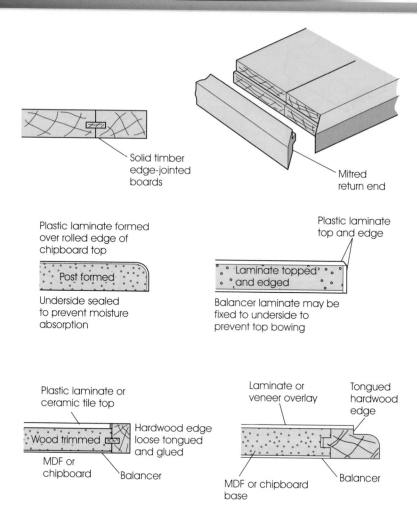

Solid timber
edge-jointed
boards

Mitred
return end

Plastic laminate formed
over rolled edge of
chipboard top

Post formed

Underside sealed
to prevent moisture
absorption

Plastic laminate
top and edge

'Laminate topped'
and edged

Balancer laminate may be
fixed to underside to
prevent top bowing

Plastic laminate or
ceramic tile top

Wood trimmed

MDF or
chipboard

Hardwood edge
loose tongued
and glued

Balancer

Laminate or
veneer overlay

Tongued
hardwood
edge

MDF or chipboard
base

Balancer

- ■ **Wood trimmed.** A chipboard or MDF base covered with either melamine face, plastic laminate or wood veneer. A hardwood lipping or edging is simply glued; rebated, glued and screwed; or tongued and glued to the seen edges, providing a neat finish. Return corners are better mitred than butted.
- ■ **Wood trimmed and over-laid.** A chipboard or MDF base edged in hardwood and overlaid with either laminate or wood veneer. The laminate is trimmed off on the spindle moulder or by use of a hand-held power router. Often a decorative edge feature is incorporated in the trimming process.

Built-in fitments

These normally use the wall, floor and/or ceiling as part of the construction.

Cupboards

Figure 9.15 shows details of a cupboard that is built into a reveal at the side of a fireplace. It has been framed up using 38 mm × 75 mm framing. The skirting is continued across the front of the frame to match in with the existing timber work. The doors, which have rebated meeting stiles, are also framed up and a 9 mm plywood is used for the panels. The top, base, shelf and front framework are fixed to 25 mm × 50 mm battens that have been plugged and screwed to the walls.

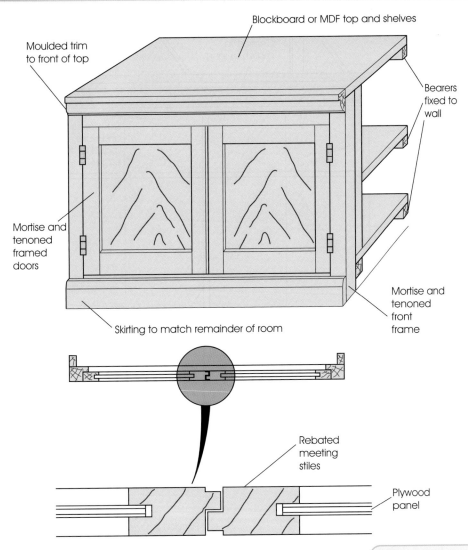

Blockboard or MDF top and shelves

Moulded trim to front of top

Bearers fixed to wall

Mortise and tenoned framed doors

Mortise and tenoned front frame

Skirting to match remainder of room

Rebated meeting stiles

Plywood panel

▲ **Figure 9.15** Framed front built-in fitment

Wardrobes

Shown in Figure 9.16 are details of a built-in wardrobe. This can be made up using 18 mm or 25 mm MFC or MDF for the base, partitions, shelves and doors. To provide a good finish, the exposed edges should be either lipped with 10 mm timber edging or taped with an iron-on edging, see Figure 9.17.

The outside doors are hung on 25 mm × 50 mm battens that have been fixed to the walls, while all the remaining doors are hung on the partitions. These can be hung inside the opening using flush hinges, or on the face using concealed cabinet hinges.

The base is made up on a plinth board to match the height of the existing skirting as shown. An in-fill piece is used at the top to drop the head of the wardrobe down, so that a cornice to match the existing one can be fixed along the ceiling line if required.

Assembly procedure for units and fitments

The assembly of framed items follows closely that of other framed joinery items. The following procedure is shown in Figure 9.18:

1. Dry assemble to check fit of joints, overall sizes, square and winding.
2. Clean up inside edges of all framing components and both faces of infill panels and so on.

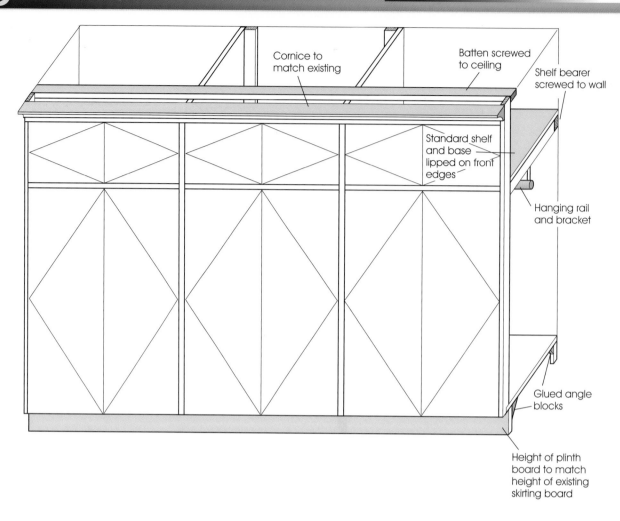

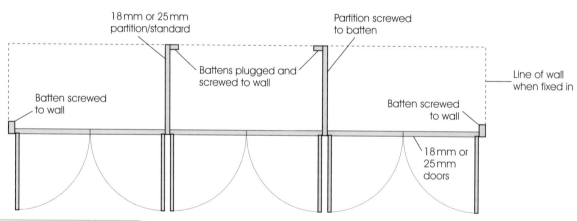

▲ **Figure 9.16** Built-in wardrobe

3. Glue, assemble, cramp up and wedge each individual frame. Re-check for square and winding.
4. Clean up internal faces of individual frames.
5. Clean up any rails, shelves, top, pot board and so on.
6. Glue, assemble and cramp up individual frames and other members to form the unit carcass. Check for square.
7. Clean up external surfaces and prepare for finishing.
8. Fix worktop if separate.
9. Install drawers, hang doors and fix any ironmongery.

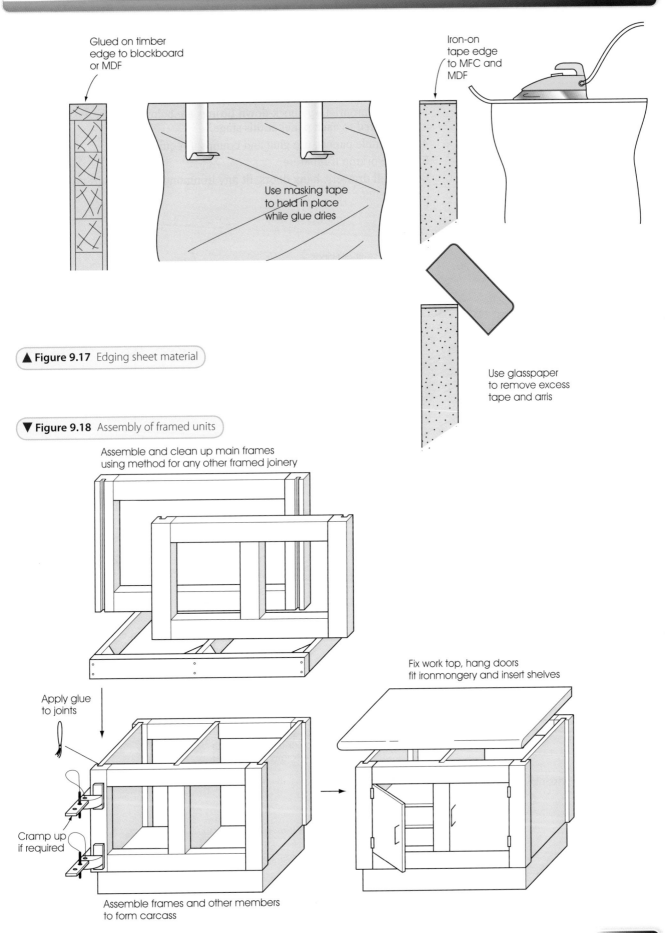

Glued on timber edge to blockboard or MDF

Use masking tape to hold in place while glue dries

Iron-on tape edge to MFC and MDF

Use glasspaper to remove excess tape and arris

▲ **Figure 9.17** Edging sheet material

▼ **Figure 9.18** Assembly of framed units

Assemble and clean up main frames using method for any other framed joinery

Apply glue to joints

Cramp up if required

Assemble frames and other members to form carcass

Fix work top, hang doors fit ironmongery and insert shelves

The assembly of box construction units can be carried out using the following procedure, see Figure 9.19:

1. Check measurements of panels.
2. Carry out any necessary handwork, such as squaring out corners, iron-on edging, fittings for shelves, drawers and so on.
3. Where proprietary knock-down fittings are being used, these should be pre-fitted to each panel at this stage.
4. Assemble panels, use glue and cramps if required.
5. Fix worktop if separate.
6. Install drawers, hang doors, fit any ironmongery and insert any shelving.

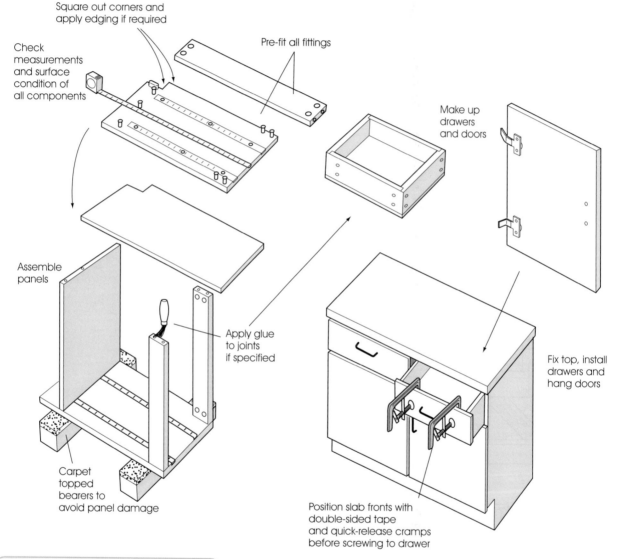

▲ **Figure 9.19** Assembly of box construction

Retail units and fitments

Although there are many variations to suit a wide range of retail outlets, hotels, bars, office receptions and so on, they may all be constructed using a similar carcass. The main differences are size, front treatment and worktop or counter top treatment.

Figure 9.20 is a pictorial impression of a bank or building society counter. Wide counter tops and glass screens are used as a security measure, in 'high cash' branches. However, the modern trend is to do away with the glass screens to create a more user-friendly environment.

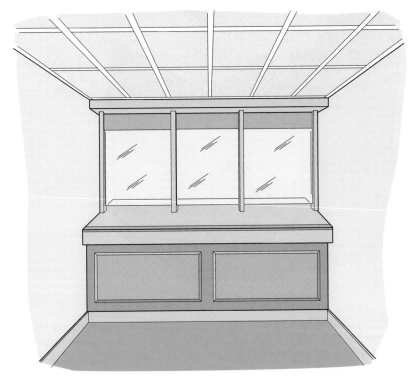

◀ **Figure 9.20** Pictorial view of bank or building society counter

A section through a typical bank or building society counter is shown in Figure 9.21. Veneered blockboard or MDF lipped with a matching hardwood on exposed edges and housed together has been used for the main carcass. This sits on a separately framed plinth. The counter top, which is also made from blockboard or MDF, has a boxed-out overhanging front edge, finished with a deep hardwood edge trim. Leather, PVC, lino or plastic laminate could be used for the actual top finish. The top rails are continuous over the intermediate standards, which should be cut out to receive them. The front of the counter has been given a decorative feature by fixing hessian- or PVC-covered panels to the carcass. Alternatively, a more conservative, traditional effect can be achieved by fixing framed dado panelling to the front of the counter carcass. The staff side of the counter is fitted with drawers and sliding doors.

▼ **Figure 9.21** Section through bank or building society counter

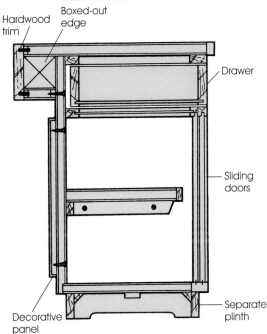

Bar counters and fitments

A typical bar counter and over-bar unit is shown in Figure 9.22. This incorporates a lifting counter flap and inward-opening door to permit through access. When closed, the flap and door should be continuous with the front and top surface so that they are hardly noticeable. Counter flap hinges are used to hang the flap.

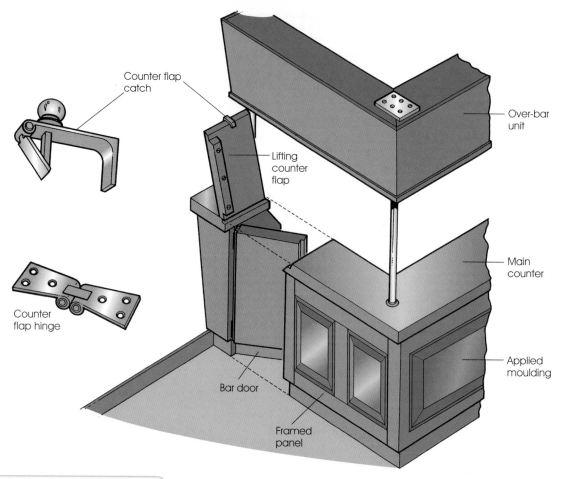

▲ **Figure 9.22** Bar counter and over-bar unit

These have no projection above the counter surface when the flap is closed and allow the flap to open through 180° to rest flat on the counter top when required. When the flap opens against a wall a counter flap catch should be fitted to hold it in the open position.

Alternative sections through the main carcass showing construction details is shown in Figure 9.23. The inside of this and other food-service counters require easily cleaned hygienic surfaces. In the examples shown, one uses a moisture-resistant, melamine-faced MDF (MR MF MDF) and the other uses plastic laminate-faced, moisture-resistant MDF (MR MDF). Internally, the counter will typically contain shelved sections for bottles, boxes, glasses, wash sink and so on; and open sections to accommodate equipment such as fridges and glass washers. Externally, the counter front may simply be an inclined or plumb plain flat panel; allied moulding may be planted on to give a panelled effect; alternatively, it could be formed from traditionally framed panelling.

The bar top itself has been formed from a solid piece of 32 mm hardwood thickened at its front edge to give a deeper impression. This would be fixed to the main carcass using shrinkage plates and finished with polyurethane varnish or other suitable treatment to give a heat-, water- and spirit-resistant surface.

Solid timber bar tops and other counter tops of any width will probably require a number of pieces to be edge-jointed together, normally by a well-glued joint incorporating a loose tongue for strength and to keep the faces flush during assembly. The ends can be cramped, see Figure 9.24. This mitred cramp, into which the top is tenoned, serves to keep the top flat and prevents any unsightly end grain showing.

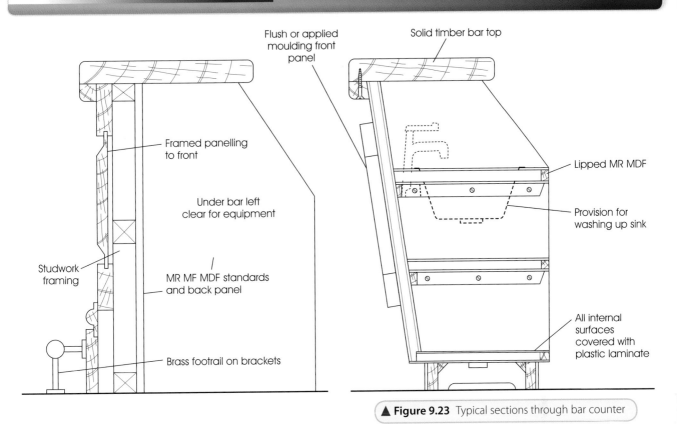

Flush or applied moulding front panel

Solid timber bar top

Framed panelling to front

Under bar left clear for equipment

Studwork framing

MR MF MDF standards and back panel

Brass footrail on brackets

Lipped MR MDF

Provision for washing up sink

All internal surfaces covered with plastic laminate

▲ **Figure 9.23** Typical sections through bar counter

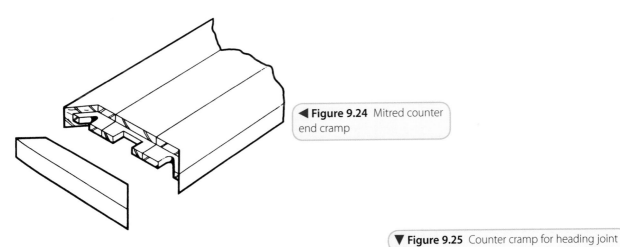

◀ **Figure 9.24** Mitred counter end cramp

▼ **Figure 9.25** Counter cramp for heading joint

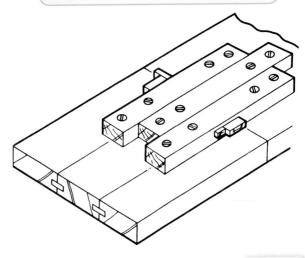

Heading joints in a worktop are sometimes required where long runs or a change of direction are necessary. These can be formed using a loose tongue or dowels to keep the top surface flush. Traditionally, the joint is pulled up tight with a counter cramp that remains permanently under the counter (see Figure 9.25). Today, worktop connection bolts are more common.

The cramp consists of three short pieces mortised to take wedges. They are screwed to the underside of the top across the joint initially at one end only; the two outside pieces to one half, and the middle piece with its mortise slightly off centre to the other half. After driving the wedges to pull up the joint, the remaining screws can be driven. Alternatively, small blocks could be fixed to the underside of the top and a sash cramp then used to pull up the joint.

Construction details of the over-bar unit are shown in Figure 9.26. This utilises lipped veneered MDF housed together and surfaced on the inside with plastic laminate. This unit is positioned above the bar and held typically using timber posts or brass support poles. The ends are fixed back to the structure in order to provide it with a degree of stability.

▶ **Figure 9.26** Over-bar unit

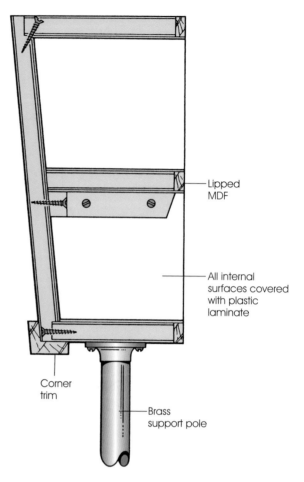

Lipped MDF

All internal surfaces covered with plastic laminate

Corner trim

Brass support pole

Curved counters

Curved counters may be formed using curved rails, shelves, pot board and plinth, see Figure 9.27. The ends and intermediate standards are positioned so that they radiate normal to the curve. For obvious reasons (short graining) both the rails and the plinth cannot be cut in the usual way from solid timber. Instead they may be made from plywood, MDF, or formed using glulam methods. The counter front could be finished in 6 mm bendy plywood or MDF bent around the curve. This would be glued and pinned to the main carcass. Alternatively, the facing may be built up using a double layer of 3 mm ply or MDF on tight curves.

Perimeter walls, shelving and gondolas

The walls of retail outlets are often lined in timber panels such as MFC, MF MDF and veneered MDF. These can be used to incorporate slat wall inserts or slotted strips to display goods or support shelves, see Figure 9.28. Separate plinths may be added at floor level to form a lower shelf.

Gondolas are free-standing shelf or rail units. These often take the form of a metal supporting framework cladded in timber panels. Again, they can incorporate slat wall inserts or slotted strips to display goods and support shelves and so on, as shown in Figure 9.29.

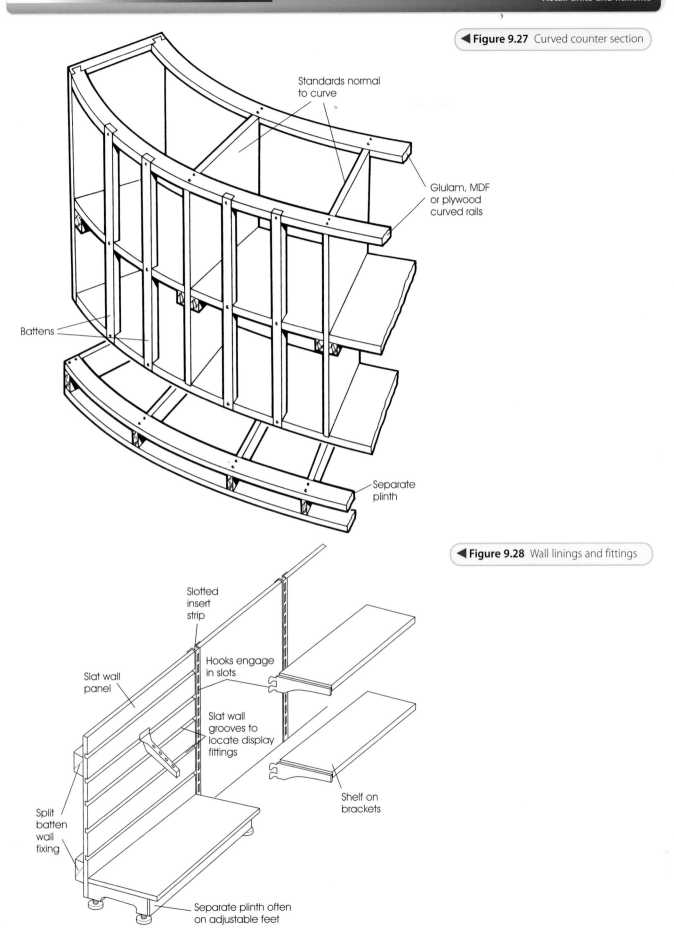

◀ **Figure 9.27** Curved counter section

Standards normal to curve

Glulam, MDF or plywood curved rails

Battens

Separate plinth

◀ **Figure 9.28** Wall linings and fittings

Slotted insert strip

Hooks engage in slots

Slat wall panel

Slat wall grooves to locate display fittings

Shelf on brackets

Split batten wall fixing

Separate plinth often on adjustable feet

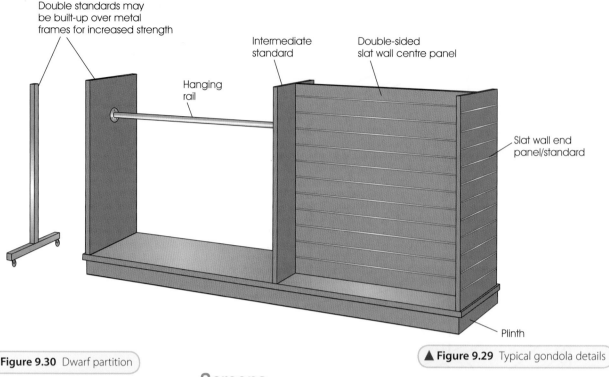

Double standards may be built-up over metal frames for increased strength

Hanging rail

Intermediate standard

Double-sided slat wall centre panel

Slat wall end panel/standard

Plinth

▲ Figure 9.29 Typical gondola details

▼ Figure 9.30 Dwarf partition

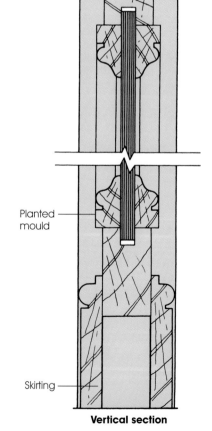

Capping

Planted mould

Skirting

Vertical section

Screens

Dwarf screens or partitions, less than the full height of a room, are used in a variety of buildings to provide visual separation between adjacent areas. Figure 9.30 shows an elevation and sectional detail of a fixed dwarf partition about a metre in height. It consists of a framed panelled section fixed between two built-up end posts. A grooved capping for additional stiffening tops this. A skirting to match that of the room is fixed on both sides.

Where the partition is continuous between two walls, a door or **wicket gate** may be provided for access. The hanging and closing edges would be rebated to prevent a through joint and also act as an effective stop for the closed position. A radiused rule joint, see Figure 9.31, will have to be used on the hanging edge of the overhanging cap to prevent binding on opening. Alternatively, parliament hinges could be used, as these would extend the pivot point outside the capping.

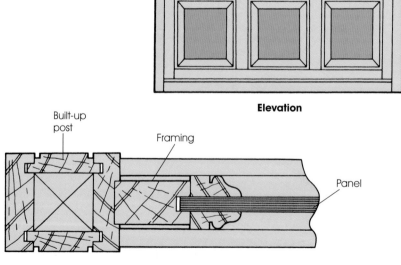

Elevation

Built-up post

Framing

Panel

Part horizontal section

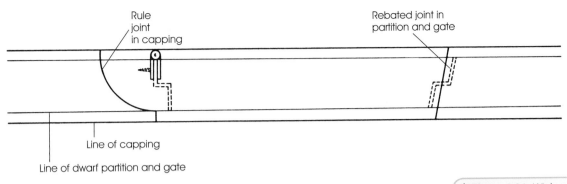

▲ **Figure 9.31** Wicket gate

Figure 9.32 shows details of a typical free-standing **movable screen** often used to divide up and give a degree of privacy to open-plan offices. The framework, which incorporates the feet, is of polished hardwood, while the in-fill panels fixed to softwood battens may be of veneered plywood (as shown). MDF or chipboard, either veneered or melamine-faced, may be used. Alternatively, the in-fill panels can be made of softboard or other acoustic material with the hollow core between them filled with fibreglass or rock wool. This would give them excellent sound absorption qualities, making them ideal for screening noisy machinery or other activities. In addition, this type of screen is suitable as a pinboard for poster or information display purposes.

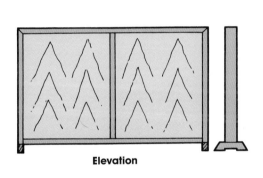

Elevation

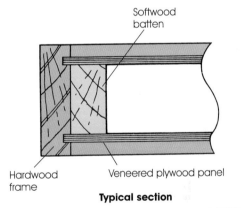

Typical section

▲ **Figure 9.32** Free-standing screen

Leisure or hotel casegoods

Much of the built-in guestroom furniture or casegoods in budget and mid-range hotels is constructed using the box construction method with knock-down fittings. At the higher end, individual items of furniture tend to be used.

In budget hotels the carcasses will be in a low cost MFC or MF MDF and edged PVC, ABS or melamine lippings. These are often left open fronted, with no moving parts such as doors or drawers, in order to save on costs in both manufacturing and ongoing cleaning and maintenance, see Figure 9.33. These are normally partially assembled in the joinery works with the final assembly undertaken when they are installed.

As the grade of hotel moves upwards, carcasses will be made of veneered MDF with either veneer or solid hardwood lippings and details; tops may be overlaid in laminate or glass for increased durability; doors and drawers are added to close off fronts and provide easily accessible but concealed storage, see Figure 9.34. These are normally fully assembled in the joinery works, however, where access is limited the final assembly may be undertaken when they are installed.

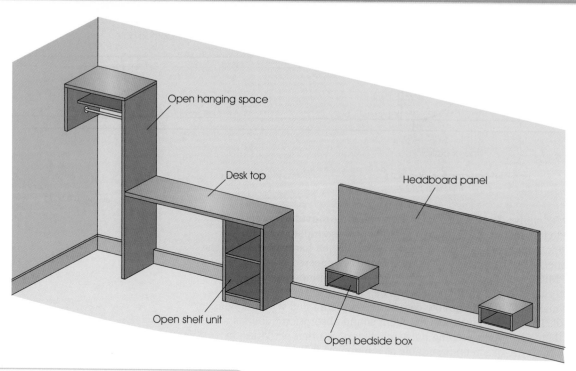

▲ **Figure 9.33** Typical open-fronted budget casegoods

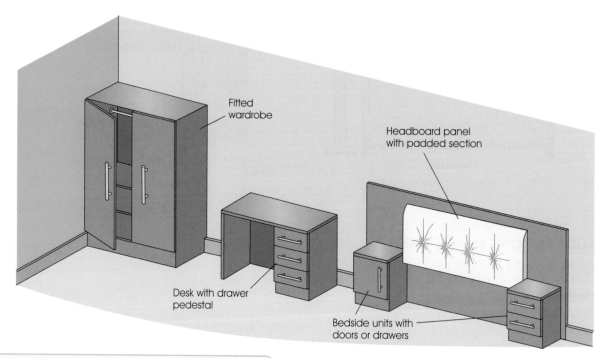

▲ **Figure 9.34** Typical close-fronted mid-range casegoods

The modern trend in hotels is towards more a minimal uncluttered look using wall-hung panels with cantilevered shelves and tops, see Figure 9.35. Panels can be in MFC, MF MDF or veneered MDF and edged with melamine, PVC, ABS, veneer or solid hardwood lippings according to the hotel grade and the clients' budget. Split battens are used to simply hang the panels on the wall, shelves normally include some form of concealed metalwork for support and tops may be supported with a timber standard or metal leg.

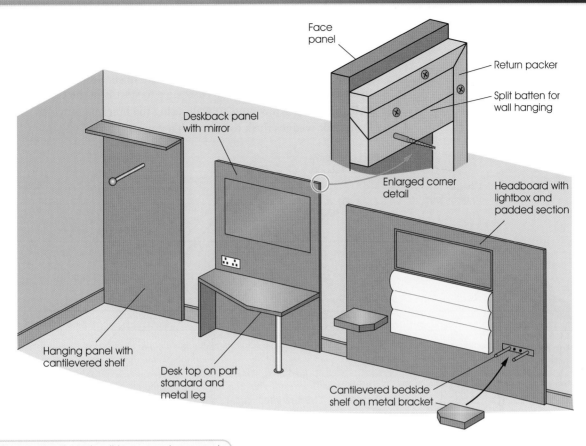

▲ **Figure 9.35** Typical wall-hung panel casegoods

Fixing joinery items

This category includes screens, partitions, cupboards and so on. These will have been made in a joiner's shop, and most are fully assembled prior to their arrival on site. However, budget-priced units (kitchen cupboards, wardrobes, etc.) are often sent in flat-pack form for simple on-site assembly, normally involving either screws and dowels or proprietary joint devices.

Flat-pack unit assembly and installation

The method of assembling and installing flat-pack units will vary from manufacturer to manufacturer. However, each unit is supplied with its own instructions. It is most important to take the time to read through these prior to commencing work. In general this is a three-stage process:

1. **Assembly.** Put carcasses together. Unpack and assemble units one at a time and check contents. Open more than one and you risk confusing the parts!
2. **Installation.** Fix base units to the wall starting with a corner base unit and working outwards from either side. Finally, install wall units, again working from the corner outwards.
3. **Finishing.** Fit worktops, drawers and doors. This should not be done until all units are firmly fixed to the wall and connected together.

Fixing base units

See Figure 9.36:

1. Position base units level and plumb using wedges provided if required. Ensure all top edges and fronts of units are flush. Secure units together using connecting screws. The best place to start is in the corner.
2. Drill, plug and screw units to wall, using the brackets provided.
3. Trim floor wedges flush to unit with a knife.

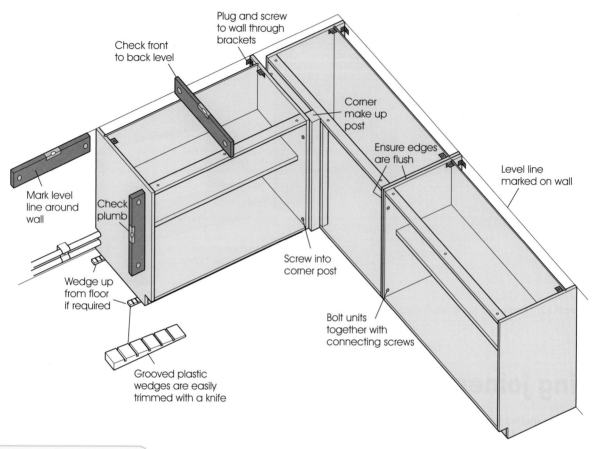

Plug and screw to wall through brackets

Check front to back level

Corner make up post

Ensure edges are flush

Level line marked on wall

Mark level line around wall

Check plumb

Screw into corner post

Wedge up from floor if required

Bolt units together with connecting screws

Grooved plastic wedges are easily trimmed with a knife

▲ **Figure 9.36** Fixing base units

Fixing wall units

See Figure 9.37:

1. Draw a level horizontal line from the top of the tall unit if being used. It is normal practice to keep tall units and wall units level with each other. Draw another line the depth of the wall units below this line. This marks the position of the underside of the wall units. Where tall units are not being used, wall units are normally fixed with a gap of 450 mm between their underside and the work surface.
2. Temporarily fix a batten on the marked level line, to act as a support while marking fixing holes and screwing.
3. Rest the unit on the batten. Ensure the tops of wall units and tall units are flush.
4. Mark the wall through the fixing holes.
5. Drill, plug and screw the unit to the wall.
6. Packing behind a fixing may be required on an uneven wall surface to ensure the units are plumb.
7. Position the remaining wall units in place one at a time.
8. Ensure the top edges and fronts are flush and secure together using connecting screws.

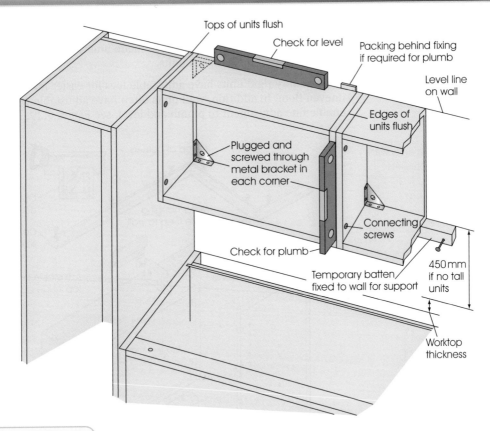

▲ Figure 9.37 Fixing wall units

Fixing worktops

See Figure 9.38 for post-formed worktops. See 'Worktops', in the previous section for other types.

1. Measure and cut the worktop to the required size.
2. Metal filler/joint strips or routed butt and mitre joints are used to connect worktops in corners. These joints can be pulled up tight using worktop connecting bolts.
3. Position worktop and screw in place through the fixing brackets.
4. Sawn bare ends of the top can be covered with a metal trim or a plastic pre-glued edge banding. Iron the edge banding into place, using a sheet of paper in between banding and iron for protection.

Fixing drawers

Insert drawers, made up previously, onto drawer runners.

Fixing doors

1. Lay the doors face down on a flat clean surface.
2. Locate the hinges over the previously fixed hinge plate and secure hinges with a mounting screw.
3. Adjust the hinges if required to ensure accurate door alignment.

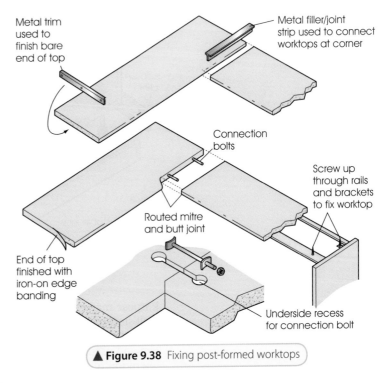

▲ Figure 9.38 Fixing post-formed worktops

Rigid and purpose-made unit installation

The installation of rigid and purpose-made units follows the general procedures used for flat-pack units.

Good-quality rigid units have adjustable legs for easier levelling on an uneven floor. In addition, wall units often have adjustable wall brackets that enable fine adjustment to plumb and level, see Figure 9.39.

▶ **Figure 9.39** Fixing units with adjustable legs and brackets

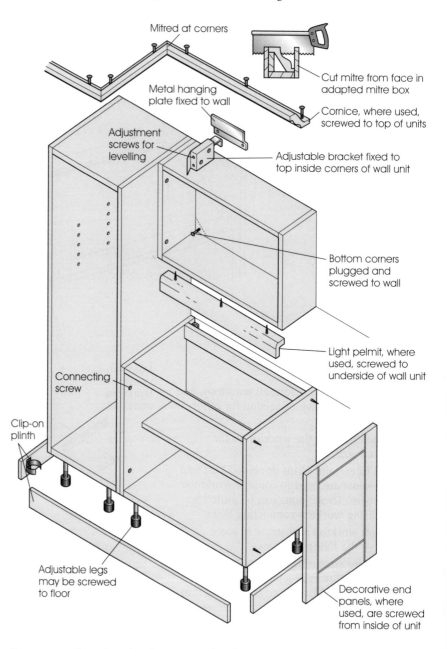

Mitred at corners

Metal hanging plate fixed to wall

Cut mitre from face in adapted mitre box

Cornice, where used, screwed to top of units

Adjustment screws for levelling

Adjustable bracket fixed to top inside corners of wall unit

Bottom corners plugged and screwed to wall

Light pelmit, where used, screwed to underside of wall unit

Connecting screw

Clip-on plinth

Adjustable legs may be screwed to floor

Decorative end panels, where used, are screwed from inside of unit

Purpose-made units often have provision for scribing to uneven floor and wall surfaces (see Figure 9.40). The units should be wedged up off the ground until they are plumb and level. A compass can be set to the widest gap and used to mark a line parallel with the floor. This is then cut and the operation repeated to scribe to the wall. The unit can then be screwed to battens fixed to the wall and floor. Alternatively, cover moulds can be used in place of scribing to mask any gap between the unit and an out-of-plumb wall. The cover moulds may themselves require scribing. Plastic laminate is often cut and glued to the plinth after fixing to provide a neat, easily cleaned finish.

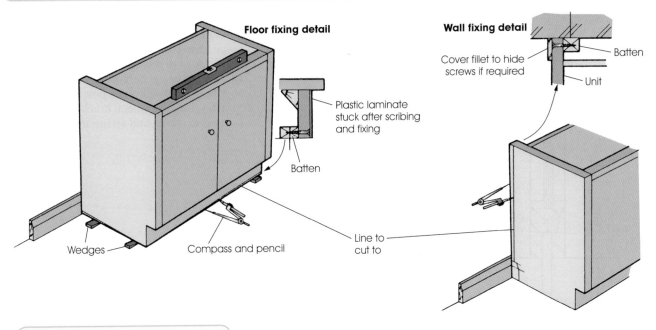

Floor fixing detail

Plastic laminate stuck after scribing and fixing

Batten

Wedges

Compass and pencil

Line to cut to

Wall fixing detail

Cover fillet to hide screws if required

Batten

Unit

▲ **Figure 9.40** Scribing to uneven surfaces

Installation of partitions and screens

Partitions and screens are normally made slightly smaller than the opening in which they are to fit. They can then be packed or wedged out and suitable cover strips or linings applied to conceal the joint (see Figure 9.41).

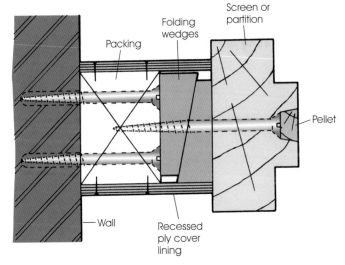

Packing

Folding wedges

Screen or partition

Pellet

Wall

Recessed ply cover lining

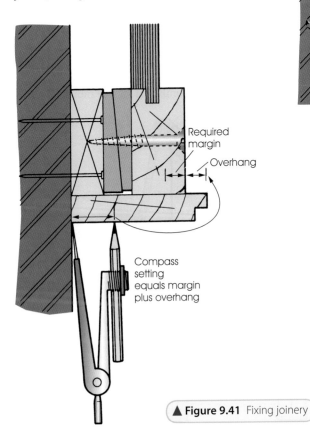

Required margin

Overhang

Compass setting equals margin plus overhang

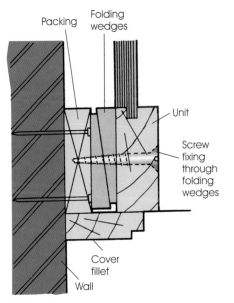

Packing

Folding wedges

Unit

Screw fixing through folding wedges

Cover fillet

Wall

▲ **Figure 9.41** Fixing joinery

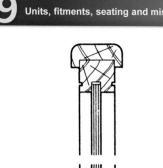

Figure 9.42 shows how a framed dwarf screen may be fixed to the floor by inserting steel dowels into the bottom rail and grouting these into holes in the floor. Often the free end of this type of screen terminates with a post. This may be fixed to the floor in a similar way to a newel post, either by inserting a steel dowel partly into the bottom of the post and grouting it into the concrete or, in the case of a timber floor, continuing the post through the floor and bolting it to a joist. Alternatively, L-shaped metal brackets screwed to both the framing and floor can be used to stiffen and secure the screen at an open end or door position.

Steel dowel grouted into floor

◀ **Figure 9.42** Fixing a dwarf screen

Seating

Normally the manufacture of seating is more of a requirement for specialist furniture or upholstery firms than the joiner, although purpose-made joinery works may be involved with the timber framework for upholstered seats and other forms of continuous public seating.

Banquet seating

A typical upholstered seat suitable for use in bars, hotels, office reception areas and restaurants is shown in Figure 9.43. This type of seating can be used on its own free standing, backing against a wall or in pairs back to back. The basic framework to take the sprung or foam upholstery work is normally made in beech. It consists of a number of simple mortised and

▼ **Figure 9.43** Bench seat upholstery and framework

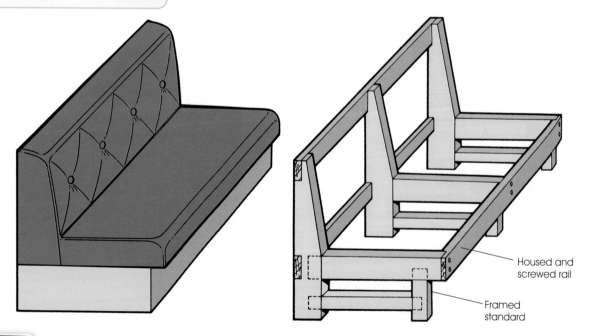

Housed and screwed rail

Framed standard

tenoned framed standards spaced at about 500 mm centres (the amount of space allowed for each person). The rails that tie the whole unit together are housed and screwed to these standards.

Figure 9.44 illustrates another upholstered bench seat. It can be clearly seen from the section shown that this is a more basic form of construction, where the seat and supports are framed from 45 mm × 45 mm softwood.

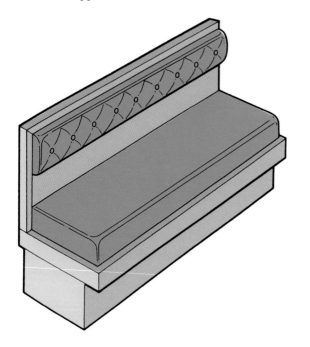

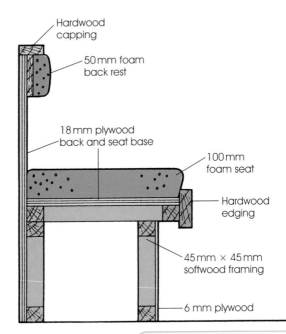

Figure 9.44 Alternative bench seat construction

The seat is formed using a removable upholstered foam cushion, which is supported by 18 mm plywood. The foam back rest is upholstered around 12 mm plywood and is fixed through the veneered plywood back. A matching hardwood edging and capping are used to finish both the seat and back rest. These are screwed and pelleted in place. The capping may also require scribing to an irregular wall surface. The space under the seat is closed in by pinning 6 mm veneered plywood in place.

This type of seating is suitable for fixing against a wall. Alternatively, two units can be fixed back to back forming island seating as shown in Figure 9.45, in which case a wider capping would be used to cover the joint between them.

Church seating

Pews are invariably un-upholstered and little attention is paid to making them comfortable, although the use of sloping backs and seats are an improvement to straight backs and flat seats. Traditionally, they were made from oak or pitch pine and the pew ends often decorated with ornate carvings.

A traditional style pew is shown in Figure 9.46. The pew ends are built up from four 45 mm boards, cross-tongued together. They are housed to receive the seat back and bookshelf. The back is framed up using 45 mm × 95 mm timber and 19 mm tongued and grooved (T&G) boarding, with muntins incorporated at 1 m centres to coincide with the intermediate standards. The seat, also made from 45 mm cross-tongued material, is housed into the ends, screwed and pelleted to the intermediate standards and tongued into the back panel for support. The bookshelf is also tongued into the back framing. Small brackets may be included for intermediate support at muntin positions. Both the pew ends and intermediate standards are tenoned and pinned to plinth blocks.

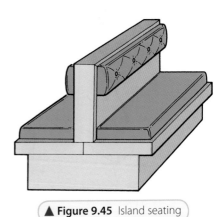

Figure 9.45 Island seating

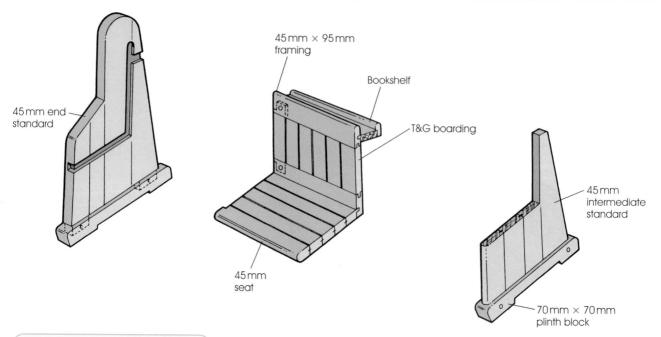

45mm end standard

45mm × 95mm framing

Bookshelf

T&G boarding

45mm seat

45mm intermediate standard

70mm × 70mm plinth block

▲ **Figure 9.46** Traditional church pew

A similar pew, more suited to a church of modern design, is shown in Figure 9.47. Glulam construction has been used to form all of the components, which are housed and screwed and pelleted together.

▼ **Figure 9.47** Modern church pew

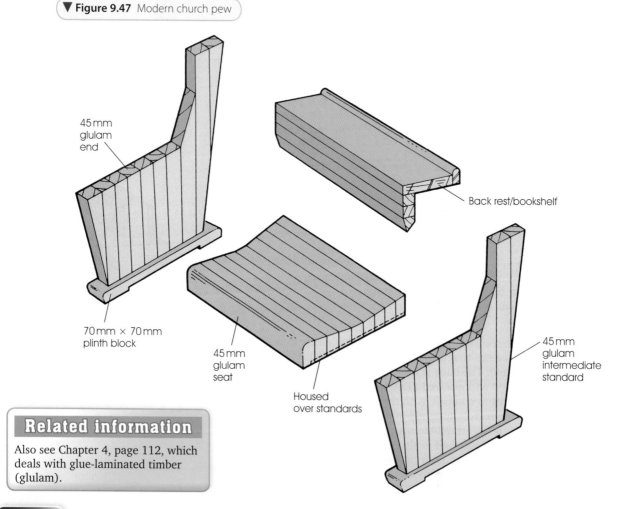

45mm glulam end

Back rest/bookshelf

70mm × 70mm plinth block

45mm glulam seat

Housed over standards

45mm glulam intermediate standard

Related information

Also see Chapter 4, page 112, which deals with glue-laminated timber (glulam).

External seating

A traditional external **park bench** is shown in Figure 9.48. Typically framed up in teak, it uses pinned mortise and tenon joints. The seat slats are fixed to the end frame and intermediate cross rails with brass recessed cups and screws. The back legs have been shaped from a 75 mm × 225 mm section, which gives a slope to the back rest and reduces the likelihood of the seat toppling over.

Related information

Also see Chapter 11, page 335, which covers the formwork moulds required for pre-cast concrete items.

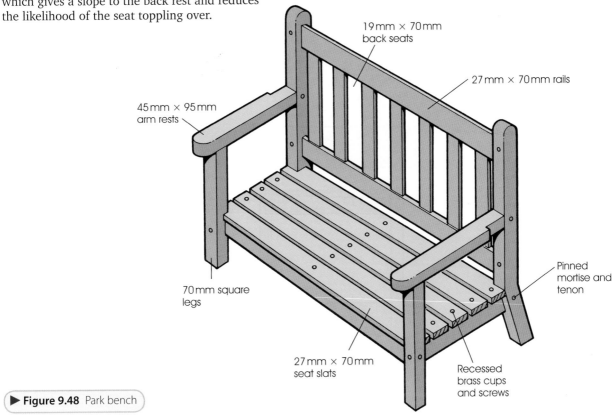

19 mm × 70 mm back seats

27 mm × 70 mm rails

45 mm × 95 mm arm rests

Pinned mortise and tenon

70 mm square legs

27 mm × 70 mm seat slats

Recessed brass cups and screws

▶ **Figure 9.48** Park bench

A more basic external bench is shown in Figure 9.49. This uses hardwood seat boards and back rest supported by pre-cast concrete standards spaced about 1.8 m apart. Intermediate standards can be introduced where longer seats are required. The hardwood boards forming the seat and back rest are coach bolted to the standards.

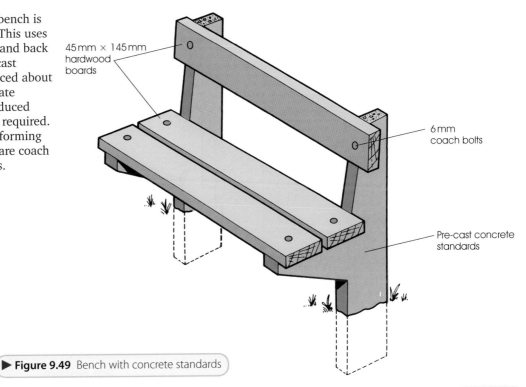

45 mm × 145 mm hardwood boards

6 mm coach bolts

Pre-cast concrete standards

▶ **Figure 9.49** Bench with concrete standards

Lecterns and tables

Lectern

This is a reading desk or bookstand, normally used in churches and lecture halls by readers in a standing position.

Figure 9.50 shows a modern type of lectern standing up to 1.2 m in height to the lower edge of the sloping top, which is angled at about 25°. The side support frames and cross rails are mortised and tenoned together. The sloping top, which is fixed by screwing up through the sloping bearers, has a raised lipping on the edge nearest the reader to prevent books and papers slipping off.

► **Figure 9.50** Lectern

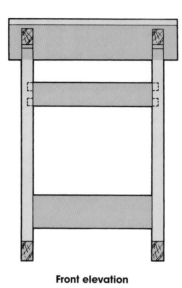
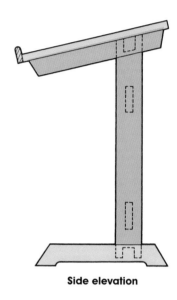

Front elevation **Side elevation**

Litany desk

This is similar to a lectern but lower in height, as it is intended for use either on a tabletop or on the floor of a church with the reader in a kneeling position. Figure 9.51 illustrates a litany desk consisting of two framed sides joined by a framed front panel and a sloping book top. Often the front panel

▼ **Figure 9.51** Litany desk

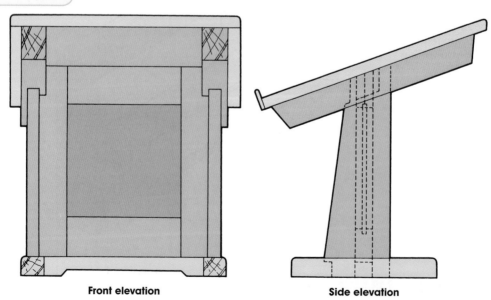

Front elevation **Side elevation**

is decorated with carved tracery or linen fold pattern, as shown in Figure 9.52. Linen fold panels, also known as drapery panels, are intended to imitate folded or draped curtains.

Table construction

Details of a typical table suitable for use in most situations are shown in Figure 9.53. The table consists of four legs that are joined by four rails. The joint shown in Figure 9.54 between the rails and legs is a table haunched mortise and tenon.

The tenon is barefaced and is mitred on its end to allow for the tenon of the other rail. Alternatively, dowels, biscuits or proprietary brackets may be used for the leg-to-rail joints. Also shown are different methods that can be used for fixing the table top to the framework. Pocket screwing and plastic blocks are suitable for sheet material, whereas the other methods allow for moisture movement when solid timber tops are used.

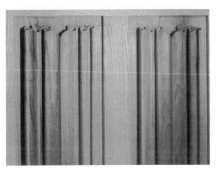

Carved tracery panel

Liner fold panel

▲ **Figure 9.52** Decorative panel details

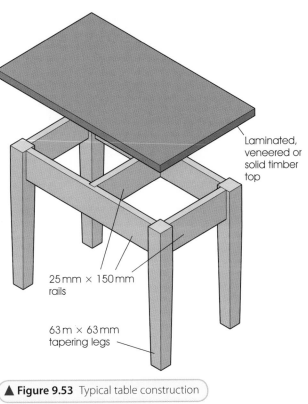

Laminated, veneered or solid timber top

25 mm × 150 mm rails

63 m × 63 mm tapering legs

▲ **Figure 9.53** Typical table construction

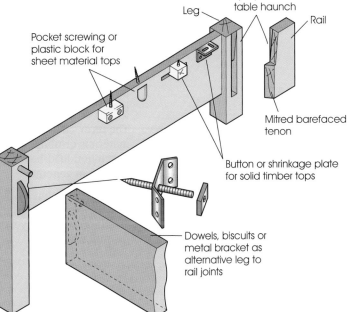

Leg

Splayed table haunch

Rail

Pocket screwing or plastic block for sheet material tops

Mitred barefaced tenon

Button or shrinkage plate for solid timber tops

Dowels, biscuits or metal bracket as alternative leg to rail joints

▶ **Figure 9.54** Table jointing details

10

Chapter Ten

Casings, false ceilings, panelling and cladding

This chapter covers the work of both the site carpenter and bench joiner. It is concerned with principles of constructing and installing casings, false ceilings, panelling and cladding and includes the following:

- Terminology.
- Types and installation of casings.
- Types and installation of false ceilings.
- Types, manufacture and installation of panelling.
- Types and installation of cladding.

Terminology

The term 'casing' or 'boxing' refers to the framework, cladding and trim used to form an enclosure in which service pipes are housed.

The term 'false ceiling' refers to a suspended ceiling below the structural soffit used to lower the ceiling height or conceal services.

The term 'panelling' refers to framing consisting of stiles, rails and muntins, mortised and tenoned together, with panels infilled between framing members. It is used to line walls and sometimes ceilings. The term is also loosely applied to wall linings made up of sheet material.

The term 'cladding' refers normally to the application of boarding, sheet materials or tiling for external protection. However, cladding can also refer to the non-load-bearing skin or protective covering of casing walls and ceilings in general.

Much preparation work can be done off site by the joiner and incorporated into the building by the carpenter at the second fixing stage. External cladding is normally fixed by the carpenter as part of the carcassing operations.

Casings

Service pipework is cased or boxed in to conceal the pipes so providing a neat, tidy appearance, which blends with the main room decoration when decorated. In addition, they must also provide access to stop valves (stop cocks), drain down valves and cleaning or rodding points or for electrical connections.

Guidelines for encasing services

In many situations, encasing services is a simple process of forming an L- or U-shaped box from timber battens covered with a plywood or hardboard/MDF facing (see Figures 10.1 to 10.3). Consider the following simple rules when planning and fixing casings.

- Use standard sections of timber where possible.
- Where casing is to be tiled, ensure the dimensions are simple widths of whole or half tiles.
- Stop valves or other fittings may require access. Fit a separate length of facing board over this section.
- Use WBP (weather and boil proof) plywood or an oil-tempered hardboard/MR MDF for casings in wet areas, for example, kitchens, bathrooms and laundries.

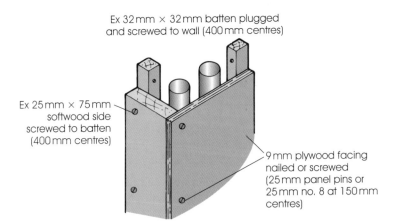

Ex 32 mm × 32 mm batten plugged and screwed to wall (400 mm centres)

Ex 25 mm × 75 mm softwood side screwed to batten (400 mm centres)

9 mm plywood facing nailed or screwed (25 mm panel pins or 25 mm no. 8 at 150 mm centres)

◀ **Figure 10.1** Pipe casings

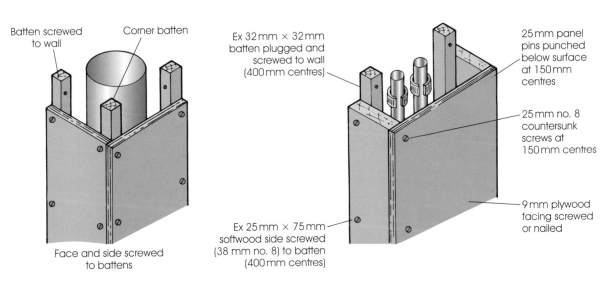

Batten screwed to wall

Corner batten

Face and side screwed to battens

Ex 32 mm × 32 mm batten plugged and screwed to wall (400 mm centres)

Ex 25 mm × 75 mm softwood side screwed (38 mm no. 8) to batten (400 mm centres)

25 mm panel pins punched below surface at 150 mm centres

25 mm no. 8 countersunk screws at 150 mm centres

9 mm plywood facing screwed or nailed

Battens, typically 32 mm × 32 mm, may be fixed to the wall using plugs and screws or nails, cut or masonry, depending on the wall hardness.

When using 6 mm facings, timber **ladder frames** (see Figure 10.2) are required for support. These are typically 25 mm × 50 mm softwood half lapped together. For casing sides direct to battens without a supporting framework, 9 mm and 12 mm facings can be used. Facing can be nailed or screwed to battens and frames.

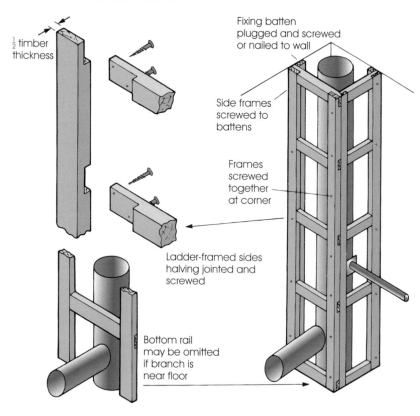

▶ **Figure 10.2** Use of ladder frame for pipe casing

Access panels can be screwed in position using brass cups and screws, typically 25 mm no. 8s. Alternatively, they may be hinged as a small door.

Casings have to be scribed to wall surfaces and finished more neatly if they are to be painted or papered rather than tiled. All nails and screws should be punched or sunk below the surface ready for subsequent filling by the painter. The sharp corner arris needs to be removed with glass paper or, as an alternative, may be covered with a timber, metal or plastic trim, see Figure 10.3.

Where casings are located in living rooms or bedrooms they can be packed out with fibreglass, mineral wool or polystyrene in order to quieten the noise of water passing through.

L-shaped casings

These are used for pipes in a corner.

1. Mark plumb lines on walls. Use a spirit level and straightedge, see Figure 10.4.
2. Fix battens to the marked lines.
3. Fix the side to the batten.
4. Fix the front facing to the side and batten.

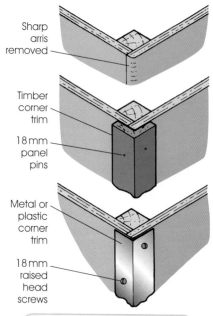

▲ **Figure 10.3** Alternative corner treatments

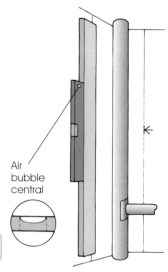

▶ **Figure 10.4** Marking plumb line on wall

Alternatively, make up and fix a ladder frame and fix the facings.

Where pipes branch off the main one, the side will require notching or scribing over them. For small pipes simply mark the side, drill a hole and saw the side to form a notch, see Figure 10.5. Larger branch pipes are best scribed around with the face split on the pipe's centre line as shown.

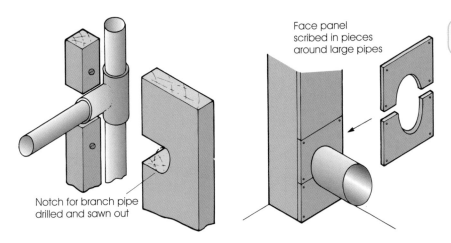

Face panel scribed in pieces around large pipes

▶ **Figure 10.5** Cutting around branch pipes

Notch for branch pipe drilled and sawn out

U-shaped casings

These are used for pipes in the middle of a wall.

1. Mark plumb lines on the wall.
2. Fix battens to the marked lines.
3. Fix the sides to the battens.
4. Fix the front to the sides.

Uneven wall surfaces

Sides and faces that fit to an uneven wall surface will require scribing, see Figure 10.6.

1. Position the side against the wall. Place a pencil on the wall surface and move it down to mark a parallel line on the side. Plane to the line, slightly undercutting to ensure a tight fit.
2. Alternatively, cut the facing oversize (say, 15 mm over required width).
3. Position the facing against the wall.

▼ **Figure 10.6** Scribing and cutting to fit an uneven wall

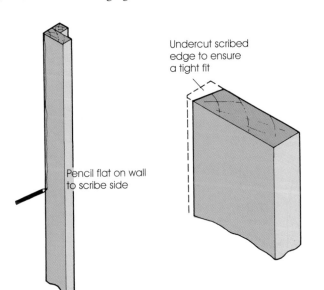

Pencil flat on wall to scribe side

Undercut scribed edge to ensure a tight fit

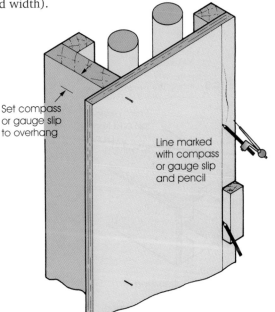

Set compass or gauge slip to overhang

Line marked with compass or gauge slip and pencil

4. Temporarily nail the facing in position, keeping the overhang on the side the same distance from top to bottom.

5. Set a compass or gauge slip to the width of overhang. Mark a parallel line on the facing.

6. Plane or saw to this line, slightly undercutting to ensure a tight fit.

Access panels

The edges of access panels are often chamfered, see Figure 10.7. This breaks the straight joint and permits a better paint finish. In addition, removal of the panel is eased without risk of damage. Simply remove screws, run a trimming knife along chamfered joints to cut paint film and lift off panel. Alternatively, access panels may be hinged or screwed into a lining and finished with an architrave trim as shown.

▶ **Figure 10.7** Access panels

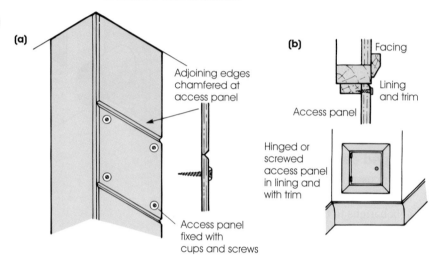

(a)

Adjoining edges chamfered at access panel

Access panel fixed with cups and screws

(b) Facing

Lining and trim

Access panel

Hinged or screwed access panel in lining and with trim

Horizontal pipe casings

These are usually at skirting level. They can be formed using the same construction as vertical casings (see Figure 10.8). Alternatively, they can be formed using a skirting board fixed to a timber top as shown.

Taller horizontal casings may have their top extended over the front facing in order to form a useful shelf.

1. Mark a level line on the wall using a spirit level and straightedge.

2. Mark a straight line on the floor.

3. Fix battens to the marked lines.

4. Fix the top and front facing.

▼ **Figure 10.8** Horizontal pipe casings

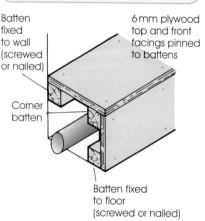

Batten fixed to wall (screwed or nailed)

6 mm plywood top and front facings pinned to battens

Corner batten

Batten fixed to floor (screwed or nailed)

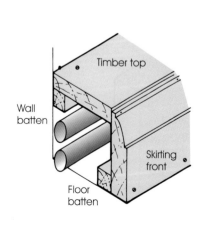

Timber top

Wall batten

Skirting front

Floor batten

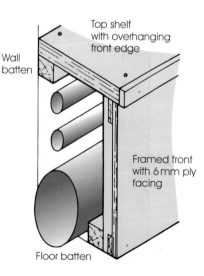

Top shelf with overhanging front edge

Wall batten

Framed front with 6 mm ply facing

Floor batten

Bath front casings

These are also termed bath panels. They can be either a standard (normally plastic) set or purpose made.

Standard bath panels

Normally simply fit up under the bath rim and fix along their bottom edge to a floor batten, see Figure 10.9.

Read the specific instructions supplied with the panel for details.

Purpose-made panels

These have to be fixed (nailed or screwed) to a batten framework, see Figure 10.10. This is typically made from ex 25 mm × 50 mm PAR (planed all round) softwood, halved and screwed together.

The panels can be formed from a variety of materials. For example:

- 9 mm plywood or MDF covered with tiles.
- 9 mm plywood or MDF covered with carpet.
- 9 mm veneered plywood or MDF with applied mouldings to create a traditional panelled effect.
- Matchboarding, TGV (tongued, grooved and vee jointed).
- Melamine-faced hardboard or MF MDF.

Sheet materials should be of a waterproof or moisture-resistant quality.

Access panels

An access panel is often formed at the tap end of the bath for maintenance purposes, rather than removing the whole panel. The panel may be screwed in position or hinged on, see Figure 10.11.

1. Ensure the batten framework is set back far enough under the bath rim to allow for the panel thickness.
2. Mark plumb lines on the walls.
3. Mark a straight line on the floor.
4. Fix wall and floor battens to the marked lines.
5. Make and fix the battened framework.
6. Fix the panel to the framework.

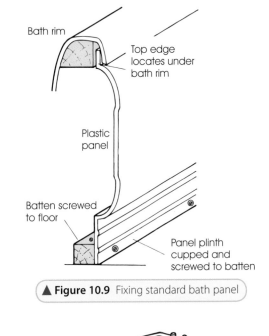

▲ **Figure 10.9** Fixing standard bath panel

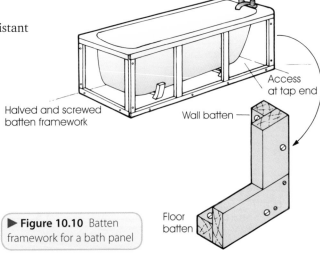

▶ **Figure 10.10** Batten framework for a bath panel

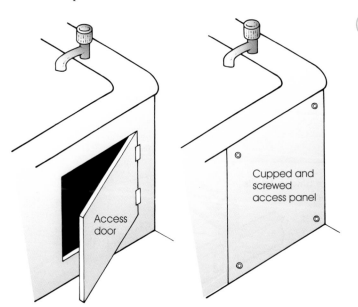

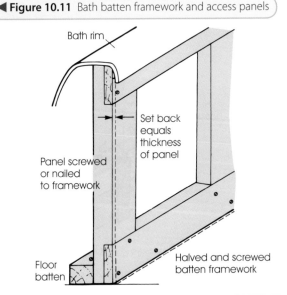

◀ **Figure 10.11** Bath batten framework and access panels

False ceilings

These are also known as suspended ceilings. They are mainly used in concrete or steel-frame buildings, such as offices, shops, factories and schools, to provide a decorative finished ceiling.

The main reasons for the installation of suspended ceilings are as follows:

- To improve the thermal insulation and so reduce heat loss.
- For sound insulation purposes.
- For structural fire protection.
- To conceal structural beams.
- To route and conceal services, heating ventilation ducts, sprinkler systems and so on.
- To create integrated lighting.

In addition, false ceilings may be used in houses, particularly older ones with high ceilings. This is usually to reduce the height of the ceiling, so creating a better-proportioned room, but also to conceal an existing ceiling that may be in a bad state of repair.

Ceilings may be suspended from the underside of roof or floor structures using either a timber framework or a proprietary system.

Timber-framed suspended ceilings

A timber and plasterboard suspended ceiling is shown in Figure 10.12. This consists of ceiling joists spanning the shortest dimension of the area and skew nailed between the wall pieces, which have been previously fixed at the

> ▼ **Figure 10.12** Timber and plasterboard suspended ceiling and binder details

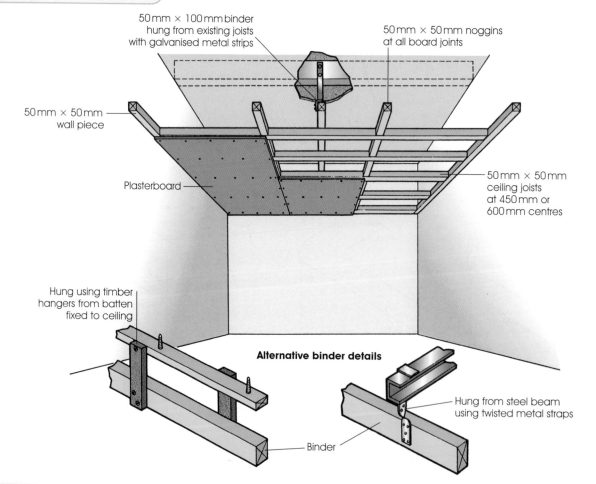

50mm × 100mm binder hung from existing joists with galvanised metal strips

50mm × 50mm noggins at all board joints

50mm × 50mm wall piece

Plasterboard

50mm × 50mm ceiling joists at 450mm or 600mm centres

Hung using timber hangers from batten fixed to ceiling

Alternative binder details

Hung from steel beam using twisted metal straps

Binder

required height. Binders are hung at intervals from the main structure to provide intermediate support to the ceiling joists and prevent them sagging. The method used to hang the binders will depend on the type of main structure and the distances between it and the suspended ceiling.

The actual ceiling finish will normally be tongued-and-grooved (T&G) matchboarding, acoustic tiles or plasterboard. Where plasterboard is used, the boards should be staggered and noggins fixed between the ceiling joists to support the joints.

Counter-battened suspended ceiling

This method, which is also known as **brandering** or **underboarding**, is used when the existing ceiling base surface is not level enough or is otherwise unsuitable for the application of the required ceiling finish. The counter-battens are fixed to the ceiling base surface, lined in and packed down as required to provide a level surface, see Figure 10.13.

The ceiling finish itself can again be either T&G matchboarding, acoustic tiles or plasterboard. Figure 10.13 shows an acoustic tile finish. These tiles have a stepped T&G joint to enable a secret nailed fixing, although they may also be stuck in position with an impact or 'no nails' adhesive. The joint between the wall and ceiling in this example has been masked with a plasterboard cove.

▼ **Figure 10.13** Counter-batten suspended ceiling with acoustic tiles

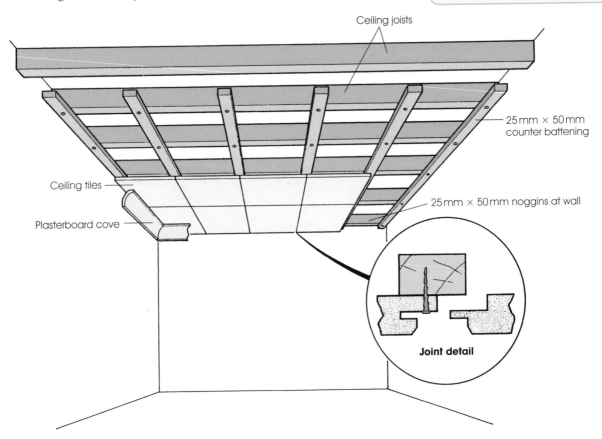

Ceiling joists

25 mm × 50 mm counter battening

Ceiling tiles

25 mm × 50 mm noggins at wall

Plasterboard cove

Joint detail

Framed panelled suspended ceilings

These may be formed in the same way as wall panelling and fixed to a supporting framework similar to that shown in the previous examples, although a more straightforward method, giving a similar appearance, is shown in Figure 10.14.

This is a *timber-framed drop-in panel suspended ceiling*. It consists of polished hardwood ceiling joists spanning the shortest dimension of the area cut over hardwood wall pieces, which have been previously fixed to the top of the wall

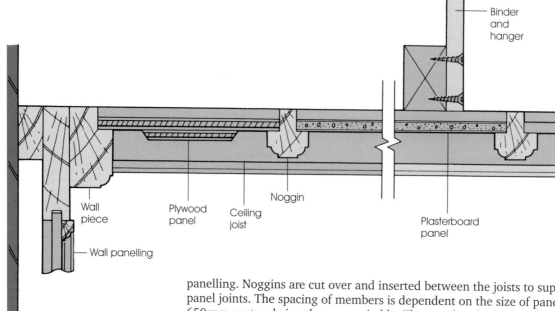

▲ Figure 10.14 Panelled ceiling

panelling. Noggins are cut over and inserted between the joists to support the panel joints. The spacing of members is dependent on the size of panels, 650 mm centres being the most suitable. The panels, which lay in the rebate run around the framing members, could be of veneered plywood, MDF or plasterboard covered with an embossed paper or acoustic tiles. Again, binders are hung at intervals from the main structure to prevent sagging.

The advantage of this method is that it is simply pocket screwed together on site (see Figure 10.15). Also, each panel may be pushed up to provide access to the services that are often run in the ceiling void. As much of the preparation work as possible should be carried out in the workshop, leaving only the wall pieces and one end of the ceiling joists to be cut on site.

▶ Figure 10.15 Panelled ceiling assembly details

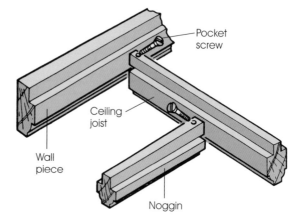

Proprietary systems

There are two main metal suspended grid ceiling systems in use: the exposed grid system and the concealed Z-bar system.

Exposed grid system

The exposed grid system is used to support either 600 mm × 1200 mm or 600 mm square plasterboard, or acoustic or other decorative infill tiles, see Figure 10.16. The grid consists of galvanised steel main T-section runners spaced at 1200 mm centres and hung on wires from the structural soffit or on straps fixed to timber joists. 1200 mm cross-tees interlock into the main runners at 600 mm centres. Where 600 mm square tiles are to be used, 600 mm cross-tees are interlocked into the centre of the 1200 mm cross-tees.

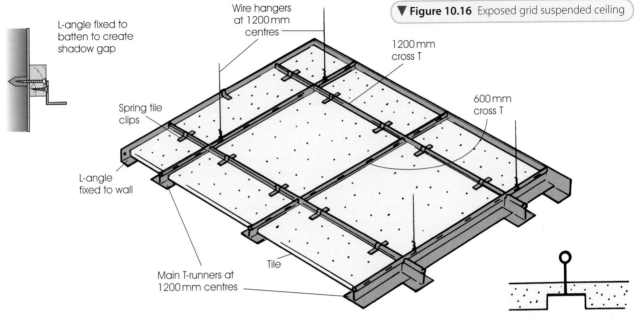

L-angle fixed to batten to create shadow gap

Wire hangers at 1200 mm centres

1200 mm cross T

▼ **Figure 10.16** Exposed grid suspended ceiling

600 mm cross T

Spring tile clips

L-angle fixed to wall

Main T-runners at 1200 mm centres

Tile

Alternative rebated edge tile to give three-dimensional effect

An L-sectioned angle is fixed around the perimeter of the room to support and finish the edge. This may be fixed directly to the wall or a batten can be used to form a shadow gap.

The exposed parts of the grid are stove enamelled or capped in either an aluminium foil, plastic foil or a timber veneer, making a wide variety of finishes.

The tiles are simply dropped in place between the T-sections and secured using spring clips. A number of tiles are normally left unclipped in order to provide a convenient access to the space above. The ceiling may be given a three-dimensional effect by using rebated edge tiles.

Concealed Z-bar system

This is a concealed grid system for supporting various types of grooved-edge tiles, see Figure 10.17. The system consists of galvanised main channels that support the Z-sections. The Z-sections are clipped to the main channels at either 300 mm or 600 mm centres (depending on the tile size).

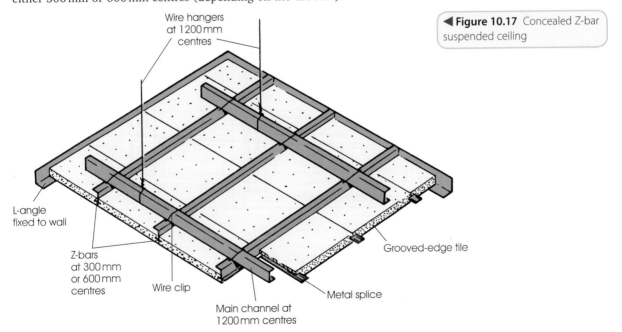

Wire hangers at 1200 mm centres

◀ **Figure 10.17** Concealed Z-bar suspended ceiling

L-angle fixed to wall

Z-bars at 300 mm or 600 mm centres

Wire clip

Grooved-edge tile

Main channel at 1200 mm centres

Metal splice

Related information

Also see Chapter 7 in Book 1, page 328, which covers the use of laser levels.

The bottom flanges of the Z-section provide support for the two opposite sides of the tiles, the other sides of the tiles being joined by a metal splice or noggin tee. L-section angles are used to support the tiles and finish the ceiling where it abuts the walls. Again, these may be fixed directly to the wall or to a batten.

Before fixing, the layout of the components must be planned so that the tiles are centralised and any cut tiles around the perimeter are the same width on opposite sides of the room.

Levelling

The levelling of both timber framed and proprietary suspended ceilings is extremely important, as any distortion or misalignment will be evident in its finished appearance.

To achieve a level ceiling in each case, the first operation is to establish the position and fix the wall pieces. All of the other components must then be lined into these. The position of the wall pieces may be marked out by measuring up a set distance from a datum line. This will have been previously established using either a water level or an optical levelling instrument. However, specialist erectors working on large office or industrial contracts generally use electronic and laser levelling instruments to establish datums or to project the fixing lines directly on the wall.

Panelling

Wall panelling is the general term given to the covering of the internal wall surface with timber or other materials to create a decorative finish. All panelling may be classified in one of two ways, either by its method of construction or by its height.

Dado-height panelling

This is panelling that extends about 1 m from the floor up the walls to the windowsill level or chair-back height and is known as dado panelling, see Figure 10.18.

▶ **Figure 10.18** Dado-height panelling

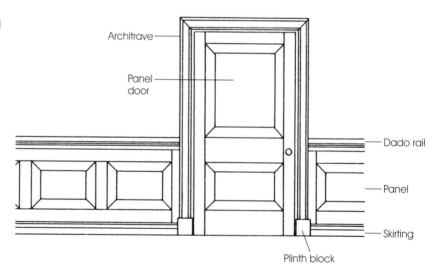

Architrave

Panel door

Dado rail

Panel

Skirting

Plinth block

Three-quarter-height panelling

This type of panelling is also known as **frieze-height panelling**. It extends about 2 m from the floor up the walls to the top of the door. Traditionally a plate shelf was incorporated on top of this type of panelling to display plates and other frieze ornaments, see Figure 10.19.

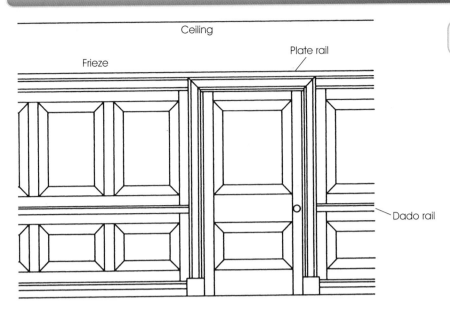

Full-height panelling

This, as its name suggests, covers the whole of the wall from floor to ceiling, see Figure 10.20.

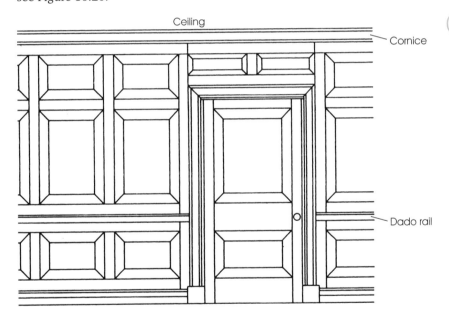

◄ **Figure 10.20** Full-height panelling

Methods of construction

Traditionally all panelling was made up from solid timbers. It consisted of framing, mortised and tenoned together to receive decorative, flat or raised panels and other solid mouldings. The modern practice is to utilise sheet materials as far as possible, either in a traditional framed surround or on its own without framing to give a flush appearance.

This flush appearance can be given a traditional look by tacking planted mouldings to its surface to form mock panels. In addition to framed or sheet panelling, strip panelling (internal cladding) made from narrow strips of boarding (TGV) is also used.

Vertical and horizontal sectional details of three-quarter traditional framed panelling are shown in Figure 10.21. Different panel and mould details have been included to show the various treatments possible. For example, panels are sometimes fitted into rebates and beaded rather than using plough grooves. This method enables the polishers to work on the panels and framing before they are finally put together. Also shown are finishing details at door and window openings.

In this type of work it was common for the large architraves to be halved and mitred together at their top joint and barefaced tenoned into the back of the plinth block at the bottom (see Figure 10.22).

▼ **Figure 10.21** Framed panelling details

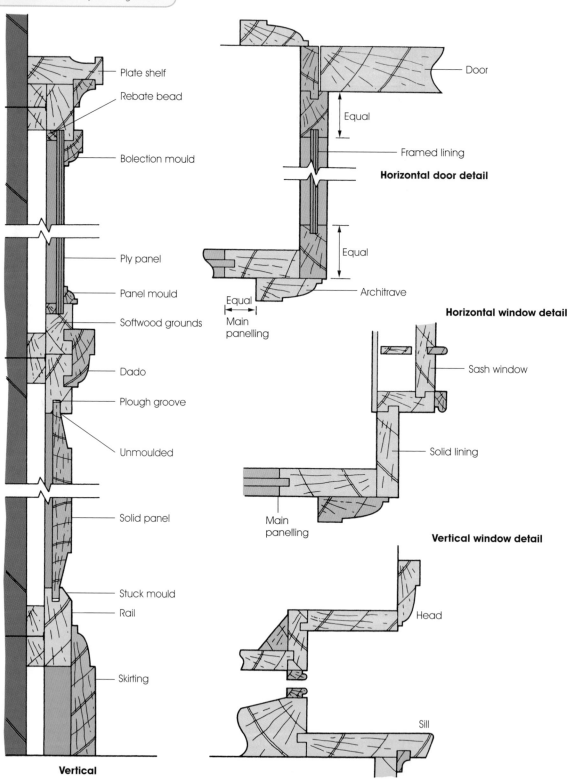

Plate shelf
Rebate bead
Bolection mould
Ply panel
Panel mould
Softwood grounds
Dado
Plough groove
Unmoulded
Solid panel
Stuck mould
Rail
Skirting

Vertical

Door
Equal
Framed lining

Horizontal door detail

Equal
Architrave

Equal
Main panelling

Horizontal window detail

Sash window
Solid lining

Main panelling

Vertical window detail

Head
Sill

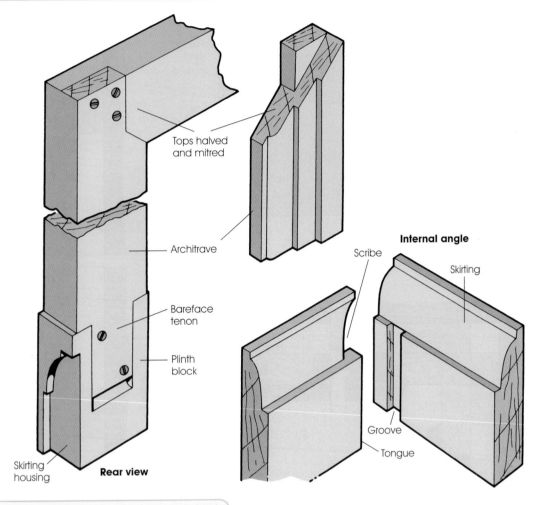

Tops halved
and mitred

Architrave

Bareface
tenon

Plinth
block

Skirting
housing

Rear view

Internal angle

Scribe

Skirting

Groove

Tongue

▲ **Figure 10.22** Architrave, skirting and plinth block details

The whole set of architraves and plinth blocks are glued and screwed together prior to fixing. The purpose of the plinth blocks is to take the knocks and abrasions at floor level that would easily damage the large moulded section architraves if they were continued down.

In addition, they also ease the fixing problems and provide a neat finish when the skirting is thicker than the architrave. Skirtings are mitred in the normal way at external angles, but often scribed and tongued at internal angles and housed into the plinth block to conceal any movement.

The elevation and sectional details of a more modern design of three-quarter height panelling is shown in Figure 10.23. This is formed of sheet panelling that is fixed on the surface of a framed surround. The panels could be made of a variety of material, for example, insulation board covered with hessian or other material for good acoustics, ply or MDF with a veneer, plastic laminate, cork, lino or plastic applied finish. Other sheet materials that are suitable have either a chipboard, MDF, hardboard or ply base and a factory-finished decorative surface including melamine, simulated stonework, brickwork and vee-jointed matchboarding.

The simple framework, the bottom rail of which forms the skirting, may be painted or polished to contrast with the panel finish. Also shown is a method of forming a vertical joint for use where the total length of panelling might cause problems with handling or access.

Whichever method of construction is used, most of the work will have been carried out off site in a joiner's shop. The panels should be delivered to site at the latest stage possible in easy-to-handle sections of up to 3.6 m in length.

Related information

Also see Chapter 8, page 253, which covers the cutting and fixing of skirting and architrave.

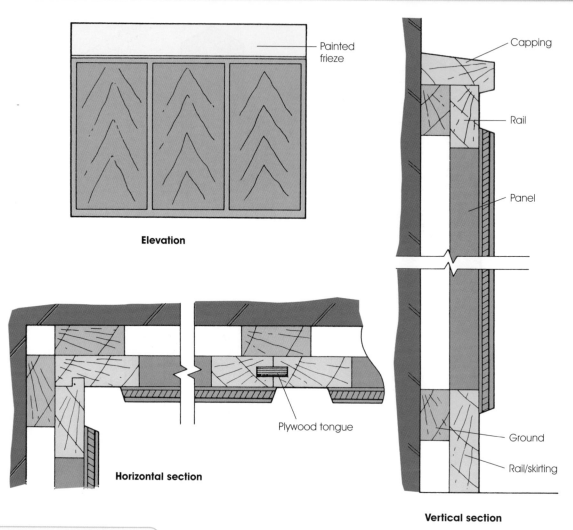

Elevation

Painted frieze

Capping

Rail

Panel

Horizontal section

Plywood tongue

Ground

Rail/skirting

Vertical section

▲ **Figure 10.23** Sheet panelling details

Grounds

Probably the most important requirement of panelling is a straight and level surface on which to fix it. Any distortion will be exaggerated in the finished panelling and will mar its appearance. Battening out the walls with grounds can provide a straight and level surface. Grounds are normally preservative-treated softwood. They may have been framed up using halvings or mortise-and-tenon joints or, alternatively, supplied in lengths for use as separate grounds or counter-battening, see Figure 10.24.

The method used to fix the grounds will vary depending on the material they are being fixed to, for example, brick, block, concrete or steel, but will be either by plugging and screwing, nailing into twisted timber plugs, nailing direct to the wall with cut or hardened steel nails, or by using a cartridge fixing tool.

Uneven wall surfaces will require packings behind the grounds to achieve a flat surface. A level and straightedge can be used to test the wall in order to find any high spots. The grounds can then be fixed, packed out and lined in to this level, see Figure 10.25. As framed grounds are made up with the panelling in the joiner's shop, they will have been planned to suit each other, but when separate grounds or counter-battens are used it is essential that these are correctly positioned in order to provide the desired fixings for the panelling. These positions will be shown on the full-size setting-out rods of the panelling supplied by the joiner's shop.

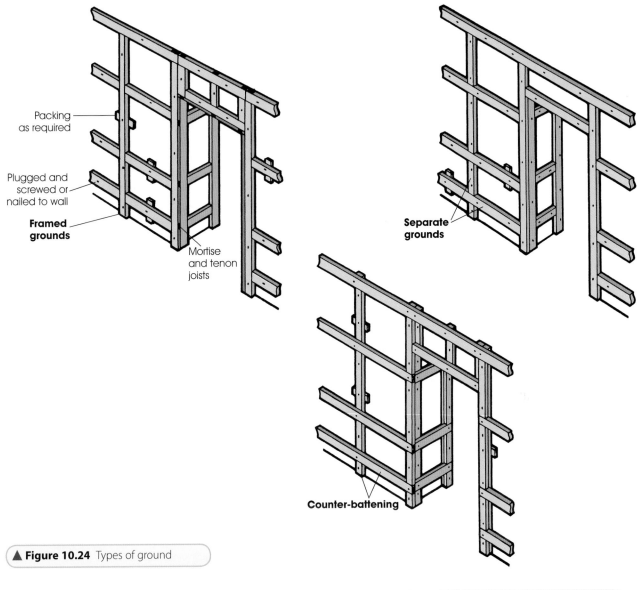

Packing as required

Plugged and screwed or nailed to wall

Framed grounds

Mortise and tenon joists

Separate grounds

Counter-battening

▲ **Figure 10.24** Types of ground

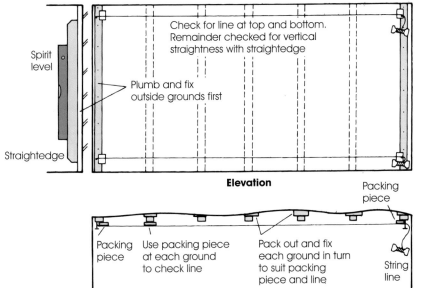

Spirit level

Check for line at top and bottom. Remainder checked for vertical straightness with straightedge

Plumb and fix outside grounds first

Straightedge

Elevation

◄ **Figure 10.25** Plumbing and lining grounds

Packing piece

Packing piece

Use packing piece at each ground to check line

Pack out and fix each ground in turn to suit packing piece and line

String line

Plan

Fixing panelling

The fixing of panelling to the grounds should be concealed as far as possible. There are various methods of achieving this. Figure 10.26 shows the following methods of concealed or 'secret' fixings:

- **Interlocking grounds.** Splayed or rebated grounds are fixed to the back of the panelling and the wall. As the panelling is lowered it is hooked in position on the grounds.
- **Slot screwing.** Keyhole-shaped slots are prepared in the back of the panelling and corresponding countersunk-head screws are driven into the grounds. The slots are then located over the projecting screws and the panelling tapped down so that the head is driven along the slot to provide a secure completely secret fixing. This method was also widely used in traditional panelling for fixing skirtings, architraves and plinth blocks.
- **Slotted and interlocking metal plates.** The use of keyhole slotted plates is similar to slot screwing, in that the plates are recessed in the back of the panelling. Interlocking plates are similar in principle to interlocking grounds, in that the cranked plates are fitted in corresponding positions

▼ **Figure 10.26** Concealed fixings

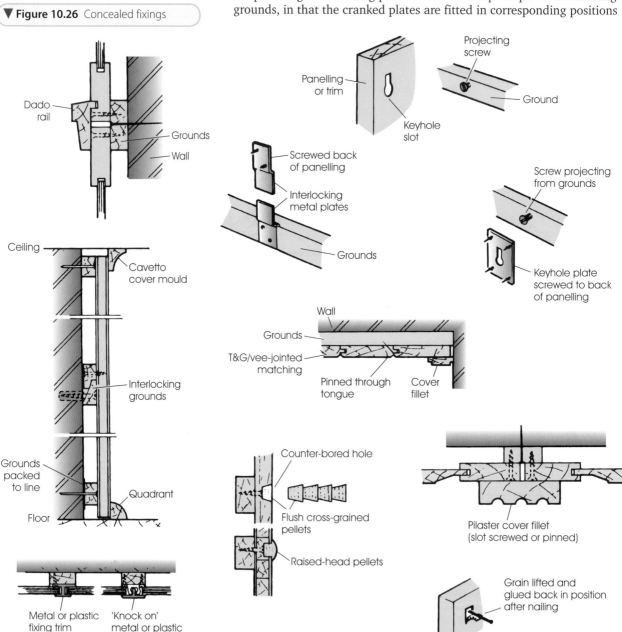

in the back of the panelling and the face of the grounds. An alternative method is to fit the plates so that the panels can slide in sideways.

- **Pellets.** Screw-holes are counter-bored and cross-grained pellets, made from the same material as the panelling, are glued and inserted in the hole. Care must be taken in matching the grain and cleaning the pellet back flush with the surface. Dome-headed pellets are also available to provide a featured fixing.

- **Cover fillets.** Panelling can be surface screwed to the grounds and then covered with a fillet pinned in place. This may be moulded to form a feature or may in fact be the skirting, cornice, frieze rail, dado rail or pilaster.

- **Nailing.** Strip panelling such as TGV matchboarding may be secret fixed by pinning through the tongue. In other sections the nails may be partly concealed by pinning through a quirk in the moulding and filling with a matching stopping or hard wax, or secret fixed through the surface by lifting a thin sliver of grain with a chisel, nailing behind this and gluing the grain back in position. (Not suitable if pre-finished.)

- **Metal or plastic trim.** These are mainly used to fix sheet material panelling and in most circumstances provide a raised feature joint.

Corner details. The method of forming internal and external angles will depend on the type of panelling, but in any case they should be adequately supported by grounds fixed behind. Figure 10.27 shows various details. T&G joints, loose tongues, rebates or cover fillets and trims have been used to locate the panelling members and at the same time conceal the effects of moisture movement.

▼ **Figure 10.27** Corner details

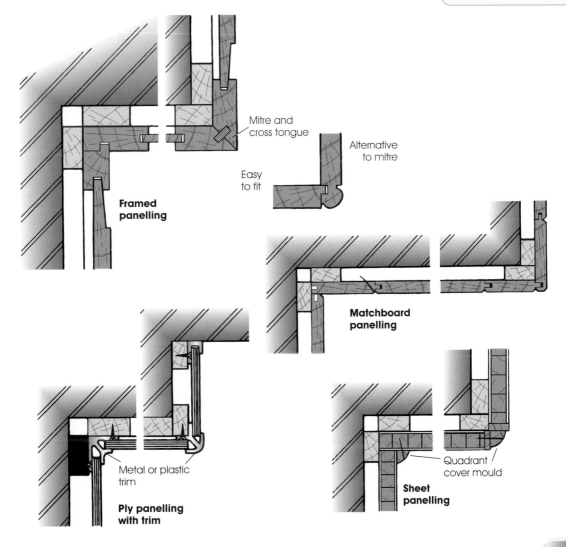

Mitre and cross tongue

Alternative to mitre

Easy to fit

Framed panelling

Matchboard panelling

Metal or plastic trim

Ply panelling with trim

Quadrant cover mould

Sheet panelling

Panel moulds. Where bolection mouldings are used, these should be slot screwed through the panel from behind. The length of the slot must be at right angles to the grain in order to allow panel movement without any possibility of splitting (see Figure 10.28). Also shown is a method used to fix planted panel moulds. They should be pinned to the framing only and not the panel. This again allows a certain amount of panel movement without splitting.

▶ **Figure 10.28** Bolection moulds and planted panel moulds

Related information

Also see Chapter 6, page 155, which details other decorative panel sections.

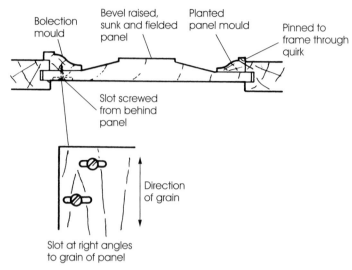

Matchboarding. Where matchboarding or similar timber strips are used for panelling, they are fixed back to the grounds using secret nails through the tongue. The boards at either end of a wall should be of equal width and may be surface nailed (see Figure 10.29).

▶ **Figure 10.29** Layout of matchboard panelling

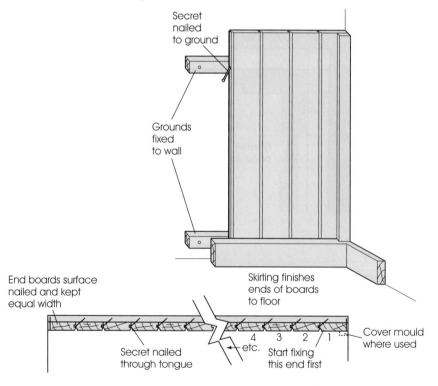

General requirements of panelling

- Before any panelling commences it is essential that the wall construction has dried sufficiently.
- All timber should be of the moisture content required for the respective situation, that is, the equilibrium moisture content (MC).

- The backs of the panelling sections should be sealed prior to fixing, so preventing moisture absorption.
- Timber for grounds should be preservative treated.
- A ventilated air space is desirable between the panelling and the wall.
- Provision must be made for a slight amount of moisture movement in both the panelling sections and the trim.
- The positioning of the grounds must be planned to suit the panelling.
- The fixing of the panelling to the grounds should be so designed that it is concealed as far as possible.

Cladding

Cladding is the non-load-bearing skin or covering of external walls for weathering purposes, for example, timber or plastic boarding, sheet material and tile hanging, see Figure 10.30. Most profiled boards are fixed horizontally, with the exception of matchboarding, which is fixed vertically or diagonally. Square-edged boards used for cladding are normally fixed vertically. In all situations it is essential that the cladding is positioned to shed rainwater quickly off the surface and prevent moisture penetration at joints and overlapping edges.

The Building Regulations place restrictions on the amount and location of combustible materials used to clad external walls. These restrictions are due to the risk of fire spreading from one building to another and are more severe the closer the cladding is to the boundary. The amount of unprotected area, which includes door and window openings as well as combustible cladding, must be calculated. For example, the total amount permitted for small residential buildings is around $5\,m^2$ at $1\,m$ from the boundary, progressively increasing to no limit at $6\,m+$ from the boundary. Cladding that has been treated with an approved exterior flame retardant may be permitted to exceed these limits, but approval from the Building Inspector will be required.

▼ **Figure 10.30** Cladding

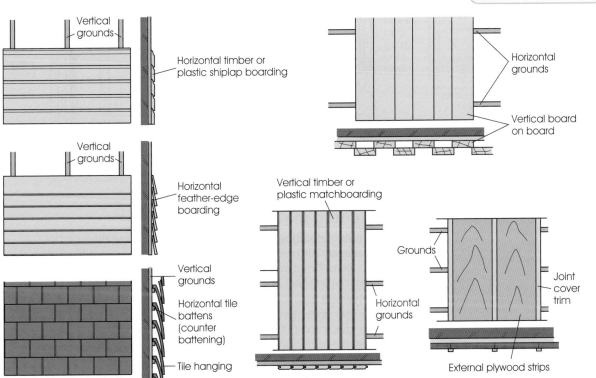

Related information

Also see Chapter 7, setting out, working and levelling, page 315 in Book 1, which covers the use of a range of levelling equipment for establishing internal datums.

Grounds

Cladding is normally fixed to timber battens or grounds spaced at a maximum of 600 mm centres. A moisture barrier is fixed below the cladding to battens in timber-framed buildings to provide a second line of protection to any wind-driven rain that might penetrate the cladding. This is often termed a breather paper, as it must allow the warm air vapour to pass or breathe through it from inside the building and not get trapped in the wall. The moisture barrier is often omitted for claddings over brick or blockwork.

The battens are typically a minimum of 38 mm wide and at least 1½ times the thickness of the cladding boards. They are fixed to the studs of the timber frame or direct to the brick or block surface. The battens will run in the opposite direction to the cladding, unless counter battens are used where the first layer is fixed parallel with the cladding. They are lined and levelled to provide a flat surface in the same way as internal grounds.

Venting the cavity

Venting the space behind the cladding is important in order to reduce the risk of timber decay. Venting allows any moisture vapour passing through the wall from the inside or any moisture absorbed by the cladding from the air to be safely dispersed, see Figure 10.31.

Venting of the cavity can be considered as an increased fire hazard and cavity barriers may require fitting in order to prevent the passage of smoke and flames. Again, approval from the Building Inspector will be required.

▼ **Figure 10.31** Venting the cavity

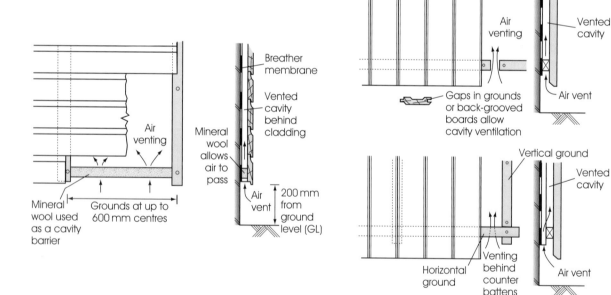

Cladding materials

- **Timber cladding** is normally specified as 16 mm thick and varies between 100 mm and 150 mm in width. Feather-edged boards will taper to about 6 mm at their thin edge. Boards are often back grooved to relieve internal stresses. This also aids ventilation of the space behind.

- **Plywood** used for cladding must be WBP rated. This means that its veneer layers have been glued using a weather- and boil-proof adhesive to give it a very high resistance to all weather conditions.

- **MDF** must be an external moisture-resistant grade and *hardboard* must be of the oil-tempered kind for weather resistance. Before fixing, hardboard sheets will require conditioning. This involves brushing up to 1 l of water into the back face (rough surface) of each sheet 24 to 48 hours before fixing. The purpose of conditioning is to expand the sheet, which ensures that it dries out and shrinks on its fixings and remains flat. If this were not done it could expand after fixing resulting in a bowing or buckling of the surface.

Related information

Also see Chapter 4, which deals with cladding to timber-framed buildings.

Preservative treatment

Most softwood claddings are not naturally durable and must be preservative treated and finished using a permeable microporous paint or stain, which allows the wood to breath. Traditionally, natural durable timber cladding, such as Western Red Cedar, was used without preservative treatment or any subsequent finish. However, as the life expectancy is greatly reduced by adverse climates it is recommended that even this is treated.

It is recommended that all timber used for cladding, the grounds as well as the boarding, is preservative treated before use. There is little point in treating the face of cladding after it has been fixed, leaving the joints or overlapping areas, back faces and grounds untreated. Any preservative-treated timber cut to size on site will require re-treatment on the freshly cut ends and edges. Applying two brush flood coats of preservative can carry this out.

Cladding details

Careful detailing is required where claddings finish around door and window openings or where different claddings join, see Figure 10.32. This is to allow for differing movement and prevent rain penetration.

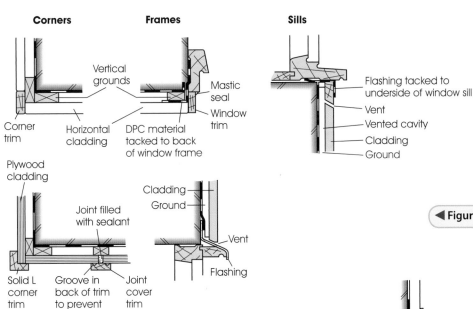

Figure 10.32 Typical cladding details

Cladding fixings are normally nails at least 2½ times in length of the cladding's thickness. Ferrous metal (metal that will rust) nails should be galvanised or sherardised to resist corrosion. Copper or aluminium nails must be used with Western Red Cedar as it accelerates rusting in ferrous metals and causes unsightly timber staining. Boards should be single nailed to each batten and care must be taken to ensure the nails do not go through the board below where they overlap, as splitting will occur when the cladding is subjected to moisture movement, see Figure 10.33.

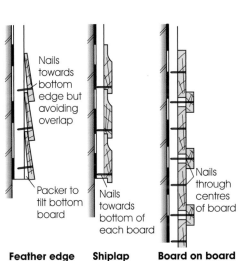

▶ Figure 10.33 Typical cladding fixings

11

Chapter Eleven

Temporary construction and site works

This chapter covers the work of the site carpenter, much of which is of a temporary nature and includes the following:

- Centring to support arched brickwork under construction.
- Shoring to support brickwork that is either failing or being extensively altered.
- Formwork to shape and support wet concrete until it cures sufficiently to become self-supporting.
- Formwork inspection and safety.
- Timbering to support the sides of trench excavations while construction work is being carried out.
- Site hoardings and footpaths.
- Boundary fences and gates.

Centring

To enable bricklayers to build an arch they require a temporary framework, which is normally made of timber and is known as centring. There are two main types of supporting framework.

Turning piece

This is the simplest form of centre and is cut from a solid piece of timber. It is only suitable for flat or segmental arches, which are of low rise, approximately 150 mm from the springing line (see Figure 11.1).

Built-up centres

These are used where the rise exceeds 150 mm. Centres for low rises up to approximately 200 mm can have solid ribs, see Figure 11.2. For rises in excess of this, made-up or laminated rib frames are used. Figure 11.3 shows a built-up centre with rib frames made from a double layer of separate ribs held together at the bottom with tie. This is suitable for rises of approximately 500 mm.

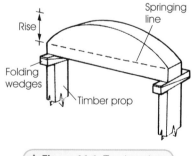

▲ **Figure 11.1** Turning piece

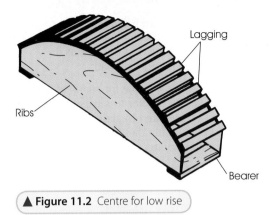

▲ **Figure 11.2** Centre for low rise

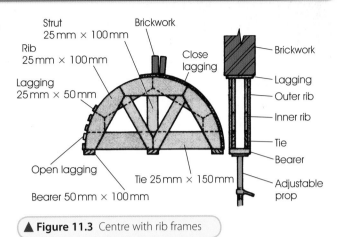

▲ **Figure 11.3** Centre with rib frames

Where the width of the brickwork exceeds half a brick, cross-braces should be used to stiffen the centre. The two sets of ribs are spanned around the centre by lagging and at the bottom by bearers. The laggings and bearers should be 20 mm shorter than the depth of the arch, so ensuring that the centre will not obstruct the bricklayer in levelling and plumbing the brickwork. Timber battens or plywood may be used for the lagging, depending on the type of work in hand. In general plywood or closed timber lagging is used for brickwork and open timber laggings for stonework.

Gothic arch centre

Figure 11.4 shows a centre for an equilateral or Gothic arch. Each rib frame is made from two separate ribs, held together at the top and bottom with a tie and a strut in the centre for additional support. Cross-braces are fixed to the struts to stiffen the centre.

Solid rib centre

Figure 11.5 shows a centre made up using two solid ribs cut from a sheet of 18 mm plywood. Laggings join the ribs around the curve and the bearers at the bottom. The vertical and the horizontal timber stiffeners have been included to provide rigidity. These take the place of the cross-braces used in more traditional centres.

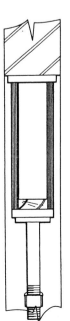

▲ **Figure 11.4** Centre for Gothic arch

◀ **Figure 11.5** Centre with two solid plywood ribs

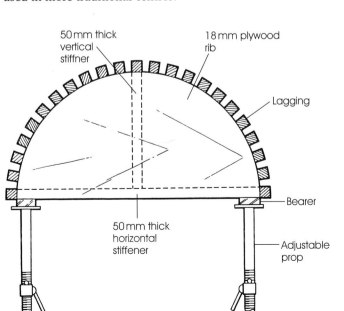

Large spans

The previous turning pieces and centres were suitable for arches with spans of up to approximately 1 m. Centres for larger spans can be made using the same form of construction, the only difference being that the size of the timber used should be increased to support the additional load.

Figure 11.6 shows a trussed arch centre, which is suitable for spans of up to 3 m. The method of making up this centre is very similar to a trussed rafter. The components are butt-jointed together and secured by nailing plywood gussets or nail plates on either side. Before assembly the timber used should be prepared and individual ribs pre-cut to a template.

▶ **Figure 11.6** Gusseted arch centre

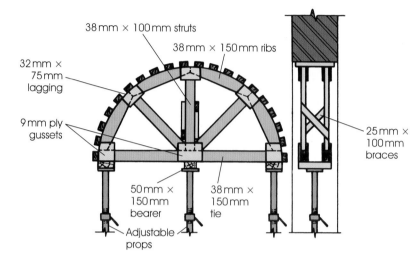

Arches with spans in excess of approximately 3 m are rarely encountered in normal building work. They are normally restricted to civil engineering work, such as bridge or tunnel construction, where proprietry steel centres are used.

▼ **Figure 11.7** Semi-elliptical gusseted arch centre

Semi-elliptical arch centre

Figure 11.7 shows a semi-elliptical gusseted arch centre made up using the trussed gusset plate method. This method can be used for making up any arch centre irrespective of its outline shape.

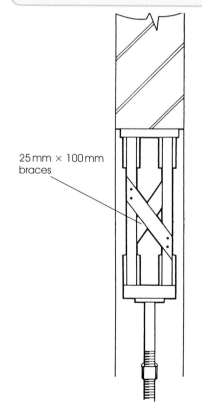

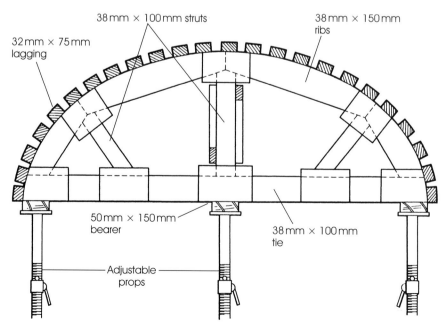

Setting out and assembly

Segmental centres and turning pieces are set out using either the triangular frame method or a radius rod. Where the radius rod is used, the required radius will have to be calculated or the centre point found by bisection. Other arched centres are normally set out using a radius rod or trammel.

When making the outline of an arch an allowance must be made for the thickness of the laggings.

The rib frames for built-up centres should be assembled over a full-size outline of the centre that has been marked either on a sheet of ply or the workshop floor. This ensures the ribs are made up to the required shape. Figure 11.8 shows one rib frame for a small semi-circular centre. The inner ribs, outer ribs, struts and tie are clench nailed together.

The outline of the arch has been marked and the rib frame is ready to be cut on the band saw. Once the rib frames have been assembled the lagging bearers and braces can be nailed in position.

On site, where a band saw may not be available, a portable power jigsaw can be used. The ribs for larger centres are normally pre-cut to a template before the rib frames are assembled.

Support of centres

Turning pieces and centres can be supported, levelled and adjusted by means of either adjustable steel props or timber posts and folding wedges. The props or posts transfer the load imposed on the centre to the ground or other support: the adjustment given by the props or wedges enables the centre to be levelled and, in addition, facilitates the easing and striking (slight lowering and removal) of the centre without damage to the finished brickwork. Large-span centres require additional intermediate props in order to support their imposed load. Once the arch has been built and the mortar joints have set sufficiently, the centre can be removed; but before this, props or wedges should be gradually eased until the arch takes up the load of the structure.

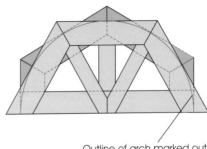

Outline of arch marked out on ribs ready for cutting

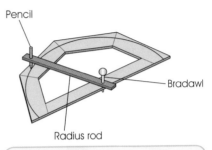

Pencil

Bradawl

Radius rod

▲ **Figure 11.8** Marking out semi-circular arch

> ### Related information
>
> Also see Chapter 9 in Book 1, which deals with the geometry required to set out a range of arch centres.

Shoring

When a building is to be extensively altered or is considered to be structurally unsound, possibly owing to ground subsidence, it may be necessary to give temporary support to the walls of the building. This temporary support is called shoring. The three main types used are shown in Figure 11.9, these being:

- vertical shores
- raking shores
- horizontal shores.

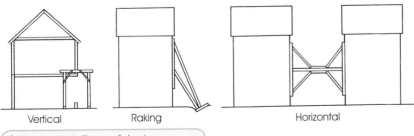

Vertical   Raking   Horizontal

▲ **Figure 11.9** Types of shoring

Vertical shores give vertical support, horizontal shores provide horizontal support, whereas raking shores fall somewhere between the two, providing a certain amount of both vertical and horizontal support. Often combinations of two or more types are required.

As an alternative to timber, tubular scaffolding or proprietary steel systems are often used. The overview in the following section is restricted to timber.

Vertical shores

The purpose of vertical shores is to support the superimposed or dead loads of a section of a building during alteration or repair work. A typical situation where vertical shores are used would be where an opening is to be made in a load-bearing wall. The wall above the opening must be supported while a lintel is inserted and the wall made good around the lintel. Figure 11.10 shows a typical detail for vertical shores.

▼ **Figure 11.10** Detail for vertical shores

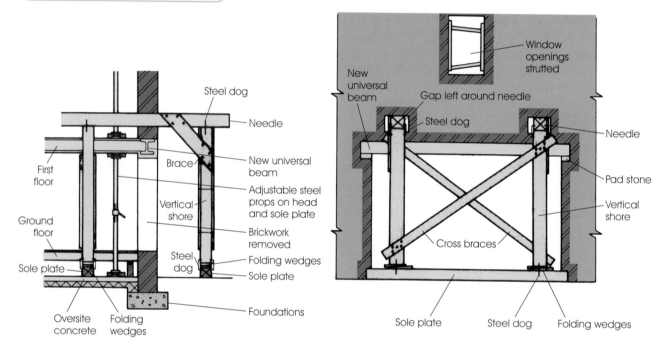

Sequence of operations

1. Strut all window openings. This is done by placing lengths of 50 mm × 75 mm timber in the window reveals and driving struts tightly between them.
2. Prop floors using 75 mm × 225 mm head and sole plates and adjustable steel props. On hollow ground floors, some boards should be removed so that the prop can extend down to the oversite concrete.
3. Cut holes in the wall for the needles. These should be spaced at 1.2 m centres under solid brickwork and not placed under windows and so on.
4. Place the sole plates on firm ground both inside and outside the building, as these are going to spread the load. It is a good idea to place the lintel in position on the floor before the shoring is erected, as access may be restricted afterwards.
5. Insert the needles through the wall and support them on dead shores. Drive hardwood folding wedges between the vertical shores and the sole plates to bring the needles tight up under the brickwork. Packing may be needed between the needles and the brickwork. Where cement mortar is used for the packing it must be allowed to harden before any load is placed upon it.
6. Securely fix all joints between the sole plates, the vertical shores and the needles with steel dogs.

7. Cross-brace the vertical shores to prevent any possibility of them overturning.
8. Check all of the shoring for tightness (props, wedges, struts, bracing and so on) and adjust if required.
9. Remove wall and build piers and padstones to support the lintel.
10. Insert the lintel. This is usually either a universal steel beam or a concrete lintel.
11. Make good the brickwork around the lintel and allow at least seven days for this to thoroughly set.
12. Ease and remove the shoring and make good the holes in the wall through which the needles passed.

The easing entails partially slackening the wedges. After about 1 hour these should be checked. If they are still loose then all of the shoring can be removed using the reverse order to that in which it was erected. However, if the wedges have tightened, some form of settlement has occurred. The new brickwork around the lintel should be checked for any signs of cracking. If this is satisfactory then only some initial settlement has taken place and the easing procedure can be repeated.

Timber sizes

The size of timbers used for vertical shores will depend on the building to be supported, but an approximate method of determining a suitable section of timber for a two-storey building is to divide the shore's height by 20. For example, vertical shores that are 3 m high require a section of 3000 ÷ 20 = 150 mm. Therefore the section required is 150 mm × 150 mm.

Where the building exceeds two storeys, 50 mm should be added to the section for each additional storey. For example, vertical shores which are 3 m high for a three-storey building require a section of 150 mm + 50 mm. Therefore, the section required is 200 mm × 200 mm.

In order to save on costs, second-hand timber can be used providing it is sound and free from defects. Where large-section timber is not available, this can be built up or laminated from smaller sections; for example, three lengths of 50 mm × 150 mm bolted together will give a 150 mm × 150 mm section.

Raking shores

These are shores that are inclined to the face of a building. Their purpose is to provide support to a building that is in danger of collapsing, or they may be used in conjunction with vertical shores for support during alteration or repair work.

In order to provide maximum support to the floors and to prevent 'pushing in' defective brickwork, the tops of the rakers must be positioned in relation to the floors (see Figure 11.11). When the joists are at right angles to the wall the centre line of the raker should, when extended, intersect with the centre of the joist's bearing. For joists that are parallel to the wall, the centre line of the floor, wall and raker should intersect.

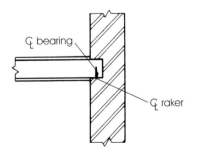
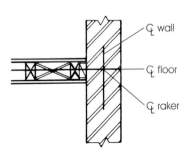

◀ **Figure 11.11** Position of raking shores

Single-raking shore

Figure 11.12 shows the front and side elevation of a single-raking shore. These should be positioned at each end of the building and over intermediate cross-walls at between 2.5 m and 5 m apart, depending on the building's condition.

▶ **Figure 11.12** Raking shore elevation

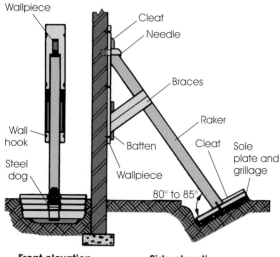

Front elevation   **Side elevation**

The term 'single-raking shore' does not refer to the number of shores used but to the number of rakers in each shore, single, double, triple and so on.

Figure 11.13 shows the jointing arrangement at the top of the raker. The raker is cut to fit the wall piece and notched around the needle. This locates it and prevents lateral movement. A straight-grained hardwood must be used for the needle, which is mortised through the wall piece and allowed to project into the brickwork by about 100 mm. The function of the wall piece is to spread the thrust of the raker evenly over a large area of the wall. In order to do this effectively it must be securely fixed back to the wall using wall hooks. The cleat, which is housed into the wall piece, stiffens the needle.

Figure 11.14 shows the detail at ground level. This consists of the raker bearing on the sole plate at an angle of about 80° to 85°. This angle must always be less than 90° to enable the raker to be tightened when it is levered forward.

Where the ground is weak, a grid or grillage of sleepers can be placed under the sole plate to spread the load over a larger area.

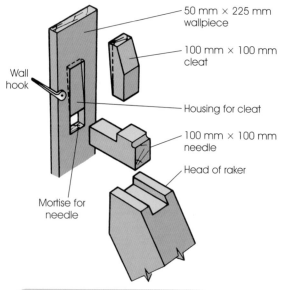

▲ **Figure 11.13** Jointing arrangement at top of raker

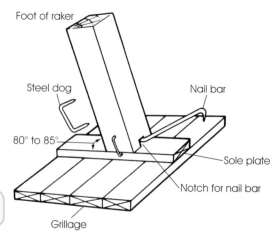

▶ **Figure 11.14** Raker detail at ground level

Double-raking shores

Figure 11.15 shows a double-raking shore to a two-storey building. The head of each raker must be positioned in relation to the floor as before. In any shore system the internal angles between the outer raker and the horizontal should be between 60° and 75°.

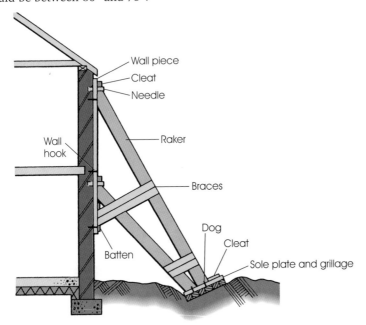

◀ **Figure 11.15** Double-raking shore to two-storey building

Labels: Wall piece, Cleat, Needle, Wall hook, Raker, Braces, Dog, Cleat, Batten, Sole plate and grillage

Multiple-raking shores

Figure 11.16 shows a multiple-raking shore to a four-storey building. Where a sufficient length of timber is not available for the outer raker it may be made up using two pieces. The longest available length is used for the top rider shore. The remaining length is made up with a backshore that lays against the adjacent raker. Adjustment is made by tightening the folding wedges between the rider and backshore.

Sequence of operations

1. Strut windows as before.
2. Cut out half-bricks in the wall to accommodate the needles.
3. Fix the needle and the cleat into the wall piece.
4. Place the prepared wall piece up against the wall and fix.
5. Prepare the sole plate and grillage.
6. Place the raker in position and tighten using a nail bar to lever it forward on the sole plate.

Note that over-tightening can be dangerous. The shore is there for support only. No attempt should be made to push back bulging brickwork.

7. Fix the foot of the raker with a cleat and dogs.
8. Brace the raker back to the wall piece. Diagonal cross-braces and ties should be used between adjacent shores.
9. Check the shoring for tightness periodically and adjust as required.

Timber sizes again depend on the buildings being supported, but as a guide 150 mm × 150 mm rakers are suitable for single-shore systems, 175 mm × 175 mm for double and 225 mm × 225 mm for triple. As before, these sections can be laminated by bolting a number of smaller sections together.

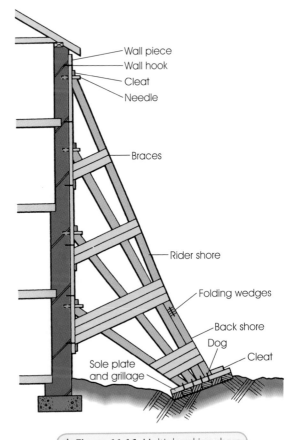

Labels: Wall piece, Wall hook, Cleat, Needle, Braces, Rider shore, Folding wedges, Back shore, Dog, Cleat, Sole plate and grillage

▲ **Figure 11.16** Multiple-raking shore

Horizontal shores

These are also known as flying shores and are used to provide horizontal support between the walls of buildings that run parallel to each other. They are mainly used for support when an intermediate building of a terrace has to be demolished prior to rebuilding, or across a narrow street or alley way, to support a weak wall off a sound one. In these situations the use of raking shores may be impracticable because of the limited space available or because of the obstruction they would cause at ground level.

Buildings of two storeys in height require only a single horizontal shore suitably strutted as shown in Figure 11.17. For taller buildings multiple horizontal shores may be required. Figure 11.18 shows the arrangement of a double horizontal shore.

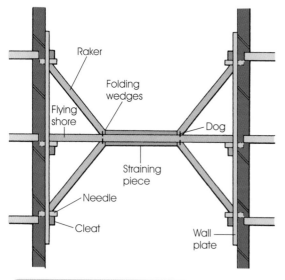

▲ **Figure 11.17** Single horizontal shore

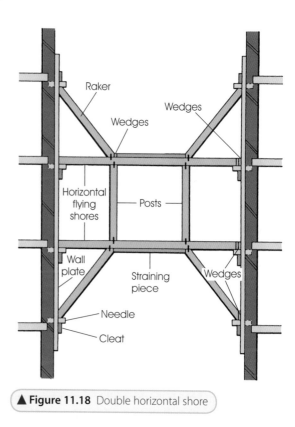

▲ **Figure 11.18** Double horizontal shore

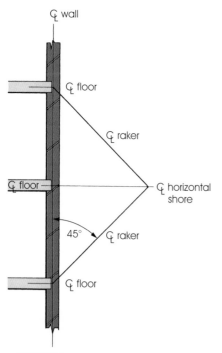

▲ **Figure 11.19** Centre-line intersection for horizontal shores

The arrangement and position of the shores, wall pieces, needles and cleats are similar to those used for raking shores.

Figure 11.19 shows the centre-line intersections of floor, wall and shoring timbers. The positions of needles and cleats are inverted to support the horizontal shore and take the downward thrust from the lower rakers. Hardwood folding wedges are used to tighten the shore. Enlarged details are shown in Figure 11.20.

Often more than one horizontal shore is required to provide sufficient support, in which case a system of shores can be used with individual sets of shores positioned near the ends of the opposite walls and at 2.5 m to 5 m intervals.

Figure 11.21 shows the use of a horizontal shore where one building is higher than the other. The floor levels of the two opposing buildings do not always coincide. In these cases the floor levels of the weaker building should be used to determine the positions of the shores. A thicker wall piece is used on the sound building to transfer the thrust to the floors and reduce the risk of 'pushing in' the wall.

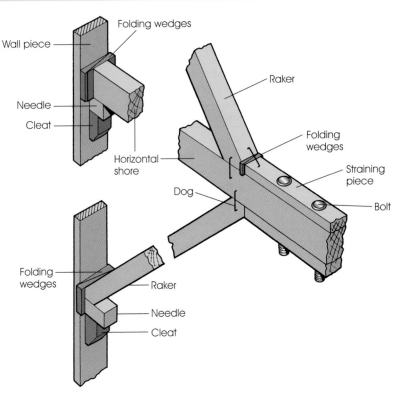

◀ **Figure 11.20** Enlarged details of intersections

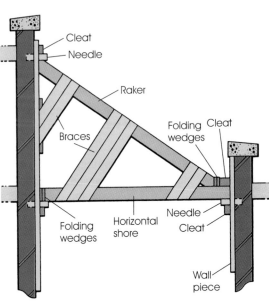

◀ **Figure 11.21** Horizontal shore for different-height buildings

Sequence of operations

1. Strut window opening (if any).
2. Cut out half-bricks in the wall to accommodate the needles.
3. Fix the needles and cleats into the wall pieces.
4. Place the prepared wall pieces up against the walls and fix.
5. Lay horizontal shores and posts in position and wedge up.
6. Place rakers and straining pieces in position and wedge up.
7. Nail all wedges and dog joints on completion to prevent lateral movement.
8. Fix diagonal cross-braces between adjacent sets of shores.
9. Check the shoring for tightness periodically and adjust by re-driving wedges if required.

Timber sizes in this case depend on the span of the horizontal shore and not on the height of the building. Typical sizes for members are given in Table 11.1. These can also be laminated by bolting a number of smaller sections together as before.

Table 11.1 Sizes for horizontal shore members

Span	Horizontal shore	Raker posts	Straining piece	Wall piece
Up to 4 m	100 mm × 150 mm	100 mm × 100 mm	50 mm × 100 mm	50 mm × 225 mm
4 m to 6 m	150 mm × 150 mm	100 mm × 100 mm	50 mm × 100 mm	50 mm × 225 mm
6 m to 9 m	150 mm × 225 mm	100 mm × 150 mm	50 mm × 100 mm	75 mm × 275 mm
9 m to 12 m	225 mm × 225 mm	150 mm × 150 mm	50 mm × 150 mm	75 mm × 275 mm

Shoring safety

Specialist firms normally carry out shoring. However, in all cases the work must be designed and closely supervised during erection by an experienced person with extensive knowledge of shoring operations.

Inspections

Prior to shoring erection a detailed inspection of the building should be made to determine its structural condition. Contact the local authority and service providers to determine the locations of underground sewers, water, gas, electricity mains and telephone cables. These must be avoided when positioning the sole plates of vertical and raking shores.

Irrespective of the type of shoring to be used, all materials and component parts must be thoroughly inspected before erection. Anything that is found to be defective must be clearly marked that it is unsafe for use. Timber should be discarded. Metal parts can often be returned to the manufacturer for straightening or repair.

After erection, every shoring system must be inspected at seven-day intervals and after adverse weather conditions for signs of distortion and movement.

Points to check:

- Distortion of members caused by overloading.
- Tightness of members. Slackness can be caused by timber drying out, or by soil movement due to repetitive freezing and thawing or soaking and drying of the ground.
- Displacement of timbers including wedges, possibly due to traffic vibration, moving plant or settlement of building.
- If further movement, cracking or bulging of the building is noticed, the designer of the shoring should be contacted immediately for advice, as such movement could possibly end in total collapse.

Removal of shoring

Horizontal or raking shores supporting a structurally unsound building cannot be removed until the defect has been remedied (walls rebuilt, foundations underpinned and so on) or the building has been completely demolished.

The procedure for striking is normally the reverse of that used for erection. This should be done in easy stages over several days so that the structure gradually takes up its full load. Look out for signs of building movement, such as cracks in the brickwork or the opening up of mortar joints. If such movement is found, consult a structural specialist immediately.

Raking or horizontal shores should be the first to be erected and the last to be removed when used in conjunction with vertical shores.

Formwork

Formwork can be defined as a temporary structure that is designed to shape and support wet concrete until it cures sufficiently to become self-supporting, and so includes not only the actual materials in contact with the concrete but all of the associated supporting structure.

Glossary of formwork terms and basic principles

See Figures 11.22 to 11.46.

- **Adjustable prop.** A proprietary steel prop with provision for adjustment so that its length may be varied (see Figure 11.22).
- **Beam box.** The formwork for a beam, which includes the sides and the soffit.
- **Beam cramp.** A proprietary cramp or timber yoke that maintains a constant beam width during casting (see Figure 11.23).

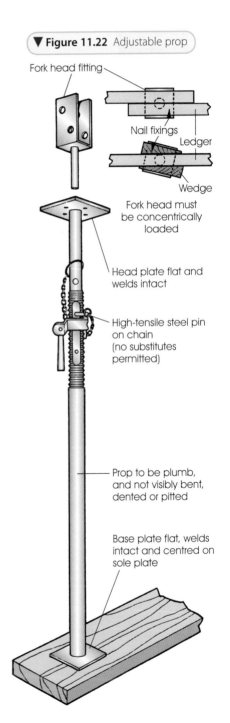

▼ **Figure 11.22** Adjustable prop

Fork head fitting

Nail fixings

Ledger

Wedge

Fork head must be concentrically loaded

Head plate flat and welds intact

High-tensile steel pin on chain (no substitutes permitted)

Prop to be plumb, and not visibly bent, dented or pitted

Base plate flat, welds intact and centred on sole plate

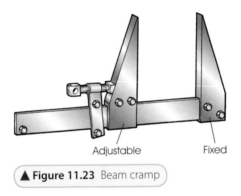

Adjustable   Fixed

▲ **Figure 11.23** Beam cramp

- **Blow hole.** A small hole or cavity in the concrete face caused by an air pocket, normally not more than 15 mm in diameter.
- **Box-out.** The formwork for a pocket or opening in the concrete item, for example bolt hole, service duct opening or door opening (see Figure 11.24).
- **Column box.** The formwork for a column.
- **Column cramp.** A proprietary steel cramp or timber yoke that holds the column box tightly together during casting (see Figure 11.25).
- **Cover.** The measurement between the concrete surface and the reinforcement. This protects the reinforcement from corrosion and increases the resistance to fire (see Figure 11.26).
- **Cube test.** A number of standard-size concrete cubes are cast on site from specified mixes. After curing, one is placed in a press and subjected to an increasing load to determine its crushing or compressive strength. After curing for 24 hours in the metal moulds, the cubes are stored in water until they are sent to the laboratory for testing, normally at seven and 28 days after casting.
- **Decking.** The sheeting or soffit of slab (floor and roof) formwork.
- **Distance piece.** A short member used to ensure the constant spacing of parallel wall and beam formwork.
- **Double-headed nail.** A wire nail with two heads used for formwork. The first head is driven home for strength, the second is left projecting, enabling easy withdrawal.

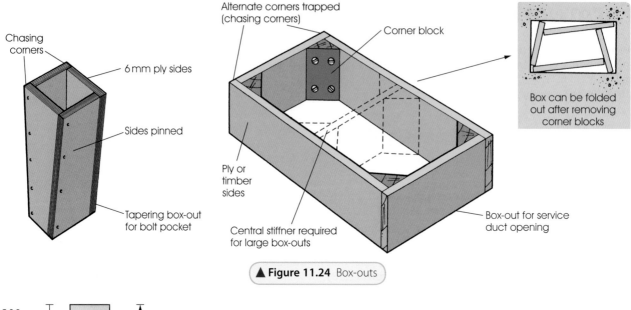

Chasing corners

6 mm ply sides

Sides pinned

Tapering box-out for bolt pocket

Alternate corners trapped (chasing corners)

Corner block

Box can be folded out after removing corner blocks

Ply or timber sides

Central stiffner required for large box-outs

Box-out for service duct opening

▲ **Figure 11.24** Box-outs

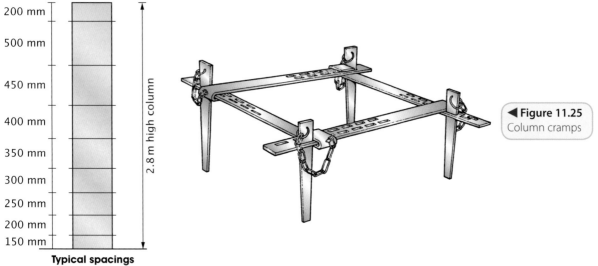

200 mm
500 mm
450 mm
400 mm
350 mm
300 mm
250 mm
200 mm
150 mm

2.8 m high column

Typical spacings

◀ **Figure 11.25** Column cramps

Stirrups or links

Main bar

Cover

Cover

Insufficient cover results in steel expansion and spalling concrete

▶ **Figure 11.26** Concrete cover to reinforcement

- **Draw.** The lead or taper of a fillet to allow its ease of removal from the cast concrete. Large features may require a false lead or draw (see Figure 11.27).
- **Edge form.** The vertical formwork to the edge of a slab, path or road and so on (see Figure 11.28).
- **Expanded metal.** An open-mesh sheet metal that is useful for forming box-outs and permanent stop ends. It is fairly easy to cut using shears. Reinforcing bars can be pushed through without difficulty. It has the advantage, when used as a stop end, that it provides a key for the subsequent pour.

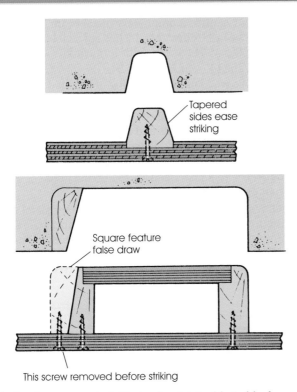

◄ Figure 11.27 Draw

Tapered sides ease striking

Square feature false draw

This screw removed before striking

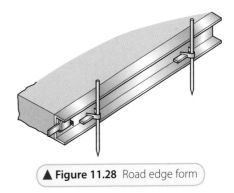

▲ Figure 11.28 Road edge form

- **Expanded polystyrene.** A rigid plastic, obtainable in blocks or sheets. Its main use is for forming box-outs and other small pockets for holding down bolts, metal handrails and balusters. After use, the polystyrene can be removed by breaking it up, burning it out or softening it out with petrol.

- **Fair face.** A high-quality plain concrete finish achieved straight from the form, without any subsequent making good or touching up.

- **False work.** The temporary support structure that supports the forms. Normally only used to describe the support structure for bridges and other major civil engineering works.

- **Fillet.** a small section of material fixed to the form face in order to produce a chamfer, bullnose or groove (see Figure 11.29).

▼ Figure 11.29 Fillet

Chamfer

Bullnose   Groove   Housed in to prevent grout getting underneath

✔

Not like this as feather edges cause this

✘

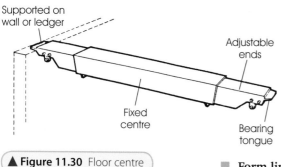

Figure 11.30 Floor centre

- **Floor centre.** A proprietary metal beam, normally of a lattice or box design, which is adjustable in length and used in place of ledgers in soffit formwork (see Figure 11.30).
- **Forkhead.** A fitting located at the top of adjustable steel props when used for decking, to provide a secure bearing for ledgers.
- **Form filler.** Filler used to repair surface damage to sheathing materials and fill small gaps, nail and screw holes for fair-faced work. Repairs are normally overfilled and sanded down flush once the filler has hardened.
- **Form lining.** The actual material used to line the inside of a form in order to produce the desired finish.
- **Form panel.** A standard panel either of a proprietary metal, or of a framed design, or specifically made up using a timber frame and ply sheathing, normally based on a standard sheet size. Prefabricated form panels can be used for a wide range of work including wall, slab construction and as baseboards for pre-cast work (see Figure 11.31).

▶ **Figure 11.31** Form panels

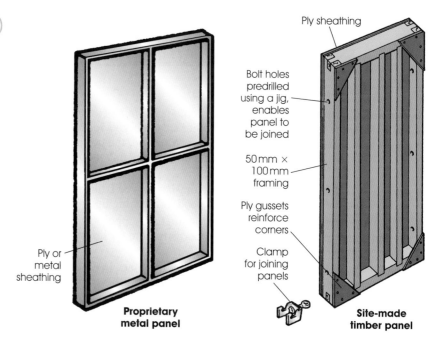

- **Form tape.** A waterproof adhesive tape used to seal joints in the form face, so preventing grout leakage (see Figure 11.32).
- **Form ties.** A bolt or tie rod, often called a wall tie, used to hold the retaining sides of formwork in the correct location. The main types in current use are coil tie, snap tie, she-bolt and through tie (see Figure 11.33).

▶ **Figure 11.32** Use of form tape

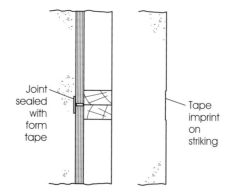

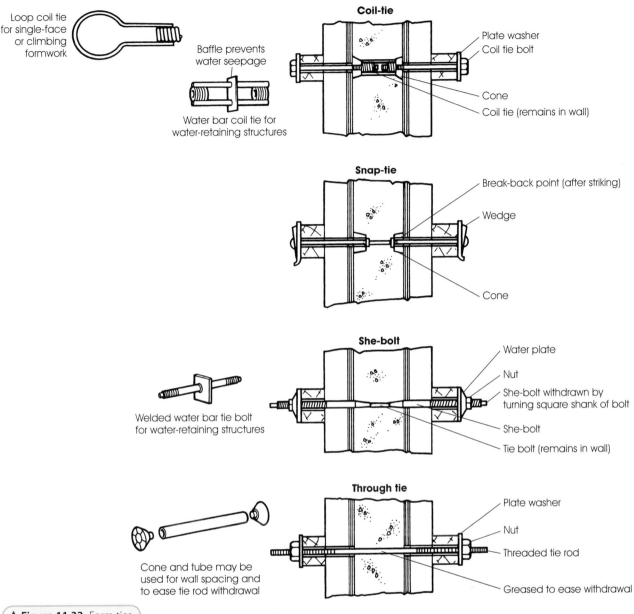

Loop coil tie for single-face or climbing formwork

Baffle prevents water seepage

Water bar coil tie for water-retaining structures

Coil-tie

Plate washer
Coil tie bolt

Cone

Coil tie (remains in wall)

Snap-tie

Break-back point (after striking)

Wedge

Cone

She-bolt

Welded water bar tie bolt for water-retaining structures

Water plate

Nut

She-bolt withdrawn by turning square shank of bolt

She-bolt

Tie bolt (remains in wall)

Through tie

Cone and tube may be used for wall spacing and to ease tie rod withdrawal

Plate washer

Nut

Threaded tie rod

Greased to ease withdrawal

▲ **Figure 11.33** Form ties

- **Gang mould.** A pre-cast mould to cast more than one unit in the same casting (see Figure 11.34).
- **Grout loss.** The leakage at form joints of a mixture of cement, sand and water.
- **Hydrostatic pressure.** The pressure exerted on the formwork by the liquid concrete.
- **Insert.** A dovetail timber block or other proprietary item that is cast in the concrete to provide a subsequent fixing (see Figure 11.35).
- *In situ* **cast concrete.** Where the product is cast in the actual location or position required.
- **Joist.** Normally 75 mm × 100 mm timbers used to support the decking material for concrete floor slabs. Proprietary aluminium joists are also available (see Figure 11.36).
- **Key.** A small indent or exposure of the concrete surface to provide a key for the next casting.

▼ **Figure 11.34** Gang mould

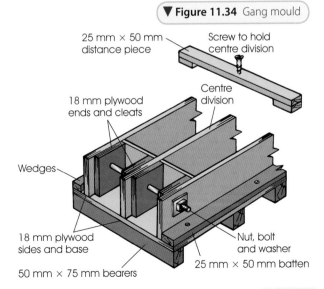

25 mm × 50 mm distance piece

Screw to hold centre division

18 mm plywood ends and cleats

Centre division

Wedges

18 mm plywood sides and base

Nut, bolt and washer

25 mm × 50 mm batten

50 mm × 75 mm bearers

339

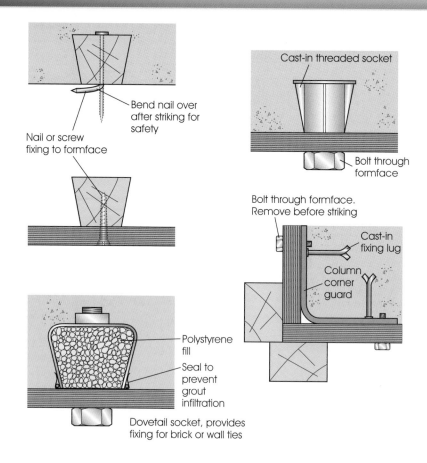

► **Figure 11.35** Formwork inserts

Bend nail over after striking for safety

Nail or screw fixing to formface

Cast-in threaded socket

Bolt through formface

Bolt through formface. Remove before striking

Cast-in fixing lug

Column corner guard

Polystyrene fill

Seal to prevent grout infiltration

Dovetail socket, provides fixing for brick or wall ties

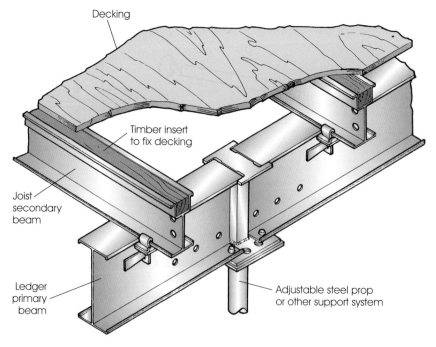

► **Figure 11.36** Proprietary aluminium joists and ledgers

Decking

Timber insert to fix decking

Joist secondary beam

Ledger primary beam

Adjustable steel prop or other support system

■ **Kicker.** A small upstand of concrete, the same plan size as the finished wall or column but only up to 150 mm in height. Its purpose is to locate the bottom of the form and prevent grout loss (see Figure 11.37).

■ **Lacing.** Horizontal members used to space and tie together vertical supports. Their use converts adjustable steel props into a support system, so greatly increasing their safe working load (see Figure 11.38).

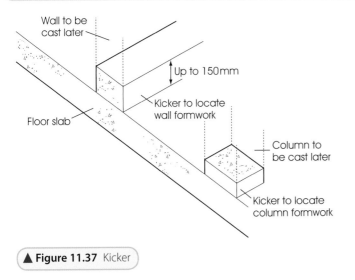

▲ **Figure 11.37** Kicker

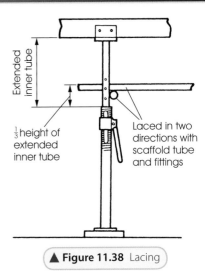

▲ **Figure 11.38** Lacing

- **Ledger.** Normally 75 mm × 150 mm timbers, supported by adjustable steel props, which in turn support the joists for concrete floor slabs. Also known as a runner. Proprietary aluminium ledgers are also available (see Figure 11.39).
- **Lift.** The height of concrete that is poured in one operation.
- **Live load.** The temporary load imposed on the formwork by working people and concreting equipment.

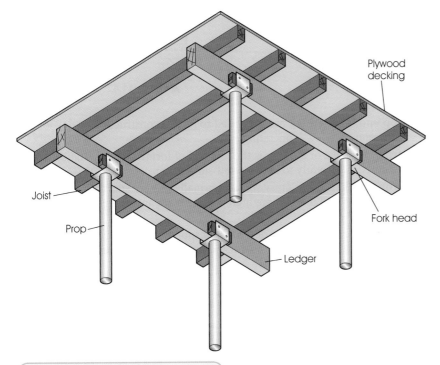

▲ **Figure 11.39** Timber ledger and joists

- **Mould.** The formwork for a pre-cast item.
- **Permanent form.** Formwork or the form face left permanently in position after casting.
- **Pre-cast concrete.** This is where the product is cast out of location, on site or in a factory.
- **Release agent.** A substance applied to the formwork to prevent adhesion between the form face and the concrete surface.

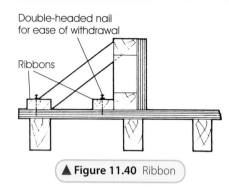

Double-headed nail for ease of withdrawal

Ribbons

▲ Figure 11.40 Ribbon

- **Ribbon.** A horizontal member fixed to a soffit, its purpose being to prevent beam sides and slab edges spreading. May be used to strut off (see Figure 11.40).
- **Sheathing.** The face of vertical forms.
- **Shuttering.** A term sometimes used when referring to formwork for *in situ* concrete structures.
- **Sill.** A horizontal member, also known as a sole plate, used under vertical supports to evenly spread the load.
- **Slump test.** An on-site test to measure the workability of concrete and compare its consistency between successive mixes of concrete. A metal slump cone is filled with concrete. After compaction the cone is removed and the slump measured. The workability of the mix increases with the slump.
- **Soffit.** The underside of a floor slab or beam; soffit formwork is known as decking (see Figure 11.41).
- **Soldier.** A vertical member to stiffen forms. Normally used in pairs in conjunction with form ties (see Figure 11.42).

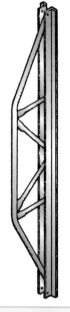

▲ Figure 11.43 Strong back

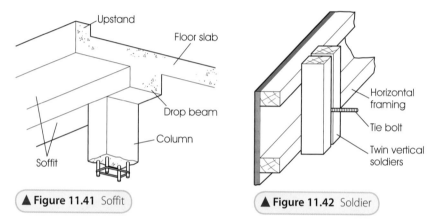

▲ Figure 11.41 Soffit

▲ Figure 11.42 Soldier

- **Stop end.** A sheathing member used to form a construction or day joint at the end of a day's concreting.
- **Strike.** To strike or striking is the removal of formwork after the concrete has cured sufficiently to become self-supporting.
- **Strong back.** A proprietary soldier normally of a substantial metal construction and used where high pressures are anticipated (see Figure 11.43).
- **Table form.** A proprietary decking and support system for soffits in the form of a table which can be struck, crane handled and repositioned in one piece (see Figure 11.44).

▼ Figure 11.44 Table form

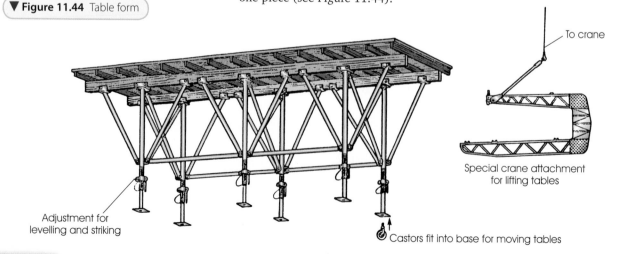

- **Trough form.** A hollow form used to produce trough-shaped recesses in the concrete soffit. Often made of glass-reinforced plastic.
- **Vibrator.** A mechanical tool that imparts vibrations into the concrete to assist in its compaction. Types include an immersion vibrator, known as a poker, or an external vibrator attached to the outside of the formwork.
- **Waffle form.** A domed hollow form, square on plan, used to make recesses in the concrete soffit. Normally made of glass-reinforced plastic (see Figure 11.45).
- **Waling.** A horizontal member to stiffen forms and used in conjunction with soldiers or strong backs and wall ties (see Figure 11.46).
- **Yoke.** The arrangement of members to encircle and secure a form, so preventing movement. For column formwork these have now been superseded by proprietary metal column cramps.

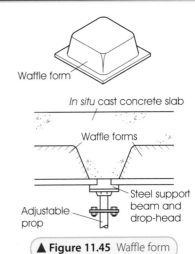

▲ **Figure 11.45** Waffle form

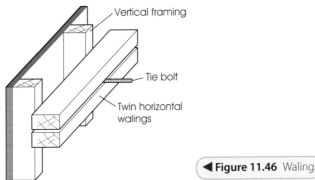

◀ **Figure 11.46** Waling

Materials for formwork

When choosing the materials for formwork the following points should be born in mind:

- Strength of materials.
- Economic use of materials.
- Ease of handling, making and erecting.
- Facilities for adjustment, levelling, easing and striking.
- The quality of the finish required.

Timber

Softwood is the most commonly used material. The main reasons for this are:

- availability
- economy
- structural properties
- ease in working and handling.

V-grade European redwood, European whitewood and Western hemlock are the species most widely used. For heavily loaded members the extra cost of using strength-graded timber can often be justified.

It is common practice to use PAR (planed all round) timber, not only because of its uniform cross-section in any one parcel but also because it is better to handle and easier to clean off after any concrete spillage or seepage.

The use of solid timber in formwork, for both quality of finish and economic reasons, is almost exclusively restricted to the support and framing members and rarely the actual soffit or sheathing. The exception is where decorative board-marked feature finishes are required, in which case the formwork is lined with either prepared or sawn softwood boards.

Related information

Also see Chapter 1 in Book 1, which covers timber and sheet material in detail.

A prior soaking in water or a sandblasting treatment may enhance the natural grain pattern that is transferred to the finished concrete.

The reuse potential of timber-boarded formwork is between one and 10 times, depending on the quality of finish required and the care taken in striking and handling.

Plywood

The two main types of plywood in common use are Douglas fir and birch. The grain pattern of Douglas fir plywood will be transferred to the concrete and can be used as a decorative feature, whereas birch plywood leaves a much smoother finish to the concrete.

Plywood for concrete formwork must be exterior quality (WBP – weather- and boil-proof). A/C or GIS grades are normally specified if a smooth finish to the concrete is required. B/C, BB/C or select sheathing is used where minor surface irregularities in the concrete finish are permissible. C/C or sheathing may be specified for underground or covered work where the concrete finish is not important.

Surfaced plywoods are used almost exclusively for high-quality repetitive work. This gives a greater number of reuses and a smoother concrete finish, and also reduces the effects of moisture and abrasion. These factory-applied surface treatments include:

■ barrier paints and varnishes
■ cellulose film surface, impregnated with a phenolic resin
■ overlay of glass fibre impregnated with polyester or phenolic resin – GRP (glass fibre reinforced plastic) overlay.

On site, any cut edges or holes must be sealed with a barrier paint or other waterproofing agent to prevent moisture absorption into the exposed layers, which would result in swelling and subsequent delamination.

Care in the handling of all types of plywood is essential, but more especially with surfaced plywoods as damaged surfaces are not so easily repaired.

The reuse potential of plywood varies widely depending on its handling and surface treatment. Typical reuse values for plywood, assuming reasonable care in handling, are given in Table 11.2.

Table 11.2 Reuse values for plywood formwork

Type	Reuses (up to)
Untreated	10
Treated with barrier paint or varnish	20
Resin film surface	50
GRP overlay	100

Profiled or textured surface plywoods are also obtainable. These impart a featured surface to the concrete. The reuse potential of these textured panels is limited as they become progressively more difficult to clean and maintain after each use.

As plywood is a laminated construction it has more layers parallel to the face grain than across it. This makes it stronger parallel to the face grain. Therefore for maximum strength the face grain should be placed at right angles to the panel supports. Where this is impracticable extra supports can be used.

Chipboard

Only exterior-grade chipboard or special-quality boards should be used as a structural sheathing material. Standard grades are unsuitable as they are not sufficiently resistant to moisture absorption; this leads to a large amount of moisture movement and a substantial loss in strength.

Special film-faced chipboards for formwork are also available. These have a fairly high reuse potential if properly treated and handled. In general, the number of reuses that can be expected from chipboard is less than that from the equivalent grade of plywood.

Fibreboard

This is used as a form lining material and is normally nailed or glued to a supporting backing of timber or plywood. The two main qualities used for formwork are standard hardboard and oil-tempered hardboard.

The oil-tempered boards have a smoother surface and are less absorbent. They produce a better concrete finish. The oil-tempered boards are also stronger and stiffer but tend to be brittle; therefore the standard boards are more suitable for curved formwork as they can be bent to a smaller radius.

Hardboard's reuse potential is fairly limited. Standard boards are considered suitable for one use only, whereas oil-tempered boards, if properly treated, may give up to 10 uses. However, the quality of the concrete finish will rapidly deteriorate with each reuse.

Other form lining materials

A variety of other materials is available for lining forms. These include building paper, corrugated cardboard, plastic laminate, moulded rubber, expanded polystyrene and so on. They are applied to the face of the form in order to obtain a specific concrete surface finish, for example, smooth, textured or patterned.

Medium-density and softboards. These are sometimes used as absorbent form liners, so reducing the risk of blow holes in the concrete finish and at the same time creating a fine surface texture. These boards are only suitable for one use.

Metal formwork

Steel and aluminium are normally used for formwork in circumstances where there is a high degree of repetitive work. The two main types of metal formwork available are:

- proprietary forms
- special purpose-made forms.

Proprietary forms

These consist of a wide range of equipment, including the following:

- Steel-framed panels with plywood or metal sheathing, which may be used for walls, columns and slabs.
- Aluminium beams for use as ledgers, joists and walings.
- Adjustable props, table forms and other support systems.
- Column cramps, beam cramps, wall ties and a variety of other ironmongery designed to secure formwork.

These proprietary systems are available for sale or hire from a number of specialist formwork suppliers.

The main advantages in using metal forms are their high strength, accuracy, smooth concrete finish, ease of erection and economy for repetitive use. Sometimes they have the disadvantages of being not readily adaptable, not readily acceptable of inserts, box-outs and fixing blocks, and generally heavier than other forms, so requiring crane handling.

Special purpose-made forms

These are normally manufactured for one specific job. They tend to be used more in civil engineering rather than construction work. Typical applications include tunnelling, retaining walls, sea walls, slip forming and so on.

In addition, the use of special purpose-made forms may be considered where a large number of identical components are required, for example, pre-cast moulds, column forms or complicated shapes such as mushroom-headed columns and so on.

The reuse potential of metal forms is very high and, provided they are handled with care and thoroughly cleaned, oiled and maintained after each use, between 100 and 600 reuses can be expected. Where the forms have become damaged, distorted or otherwise defective, for both accuracy and safety reasons they should not be used. As the repair of these forms is a highly specialised operation they should either be discarded or returned to the manufacturer for repair or replacement.

Plastic formwork

Sheet or foamed plastic can be an economical formwork material where repetitive use, high-quality finish and complicated shapes are required.

GRP is widely used for:

- waffle and trough moulds for floor construction
- profiled form liners, mainly for wall construction, to provide a patterned finish
- special purpose-made forms for complex shapes, for example, circular columns with mushroom heads and so on.

The reuse potential of plastic is, in common with other formwork materials, dependent on its treatment and handling. In general, one can expect up to 100 or more uses.

Plastics are particularly susceptible to impact damage and surface scratching. Therefore extra care must be taken to avoid hitting the surface. Some plastic forms are specially designed to be removed by compressed air. After striking, the form face should be immediately cleaned to remove all traces of cement dust, using a special cleaning agent, clean soapy water or an oil-soaked cloth. Any hardened cement paste can be removed with a wooden scraper. The use of metal scrapers is not recommended as they damage the surface.

In addition to their high reuse value and their ability to form complex shapes and produce a high-quality finish, they also have the advantage of being low weight.

Release agents

The correct use of release agents has a significant effect on the quality of the concrete finish and the reuse potential of the formwork. Their main functions are to:

- provide a quick release from the moulding
- prevent adhesion between the form face and concrete surface
- aid the production of a high-quality blemish-free concrete finish
- enable the maximum reuse of formwork.

There are various types of release agent available. The choice is dependent on the form face material and quality of finish required.

It is, therefore, essential that care is taken in choosing the correct release agent and that it is compatible with not only the form face but also any other formwork preparation such as barrier paints, fillers, surface retarders, waxes and so on.

The main categories of release agent and surface coatings in general use, together with their main characteristics and suitable form faces, are given in Table 11.3.

Table 11.3 Release agents

Type	Characteristic	Method of application			Suitable form face					
		Brush	Squeegee	Spray	Sawn	Planed	Plywood (Unsealed)	Plywood (Sealed)	Metal	Plastic
Neat oil	Encourages the formation of blow holes and so not recommended									
Neat oils with surfactant	Contains a surface wetting agent to minimise the formation of blow holes	*		*	*	*	*	*	*	*
Mould cream emulsion	General-purpose release agent particularly suitable for absorbent surfaces	*	*		*	*	*			
Water-soluble emulsion	Causes discoloration and retardation of surface cement and so is not recommended for visual concrete	*	*		*	*	*			
Chemical release agent	Causes a chemical reaction at the form face, so minutely retarding the concrete surface. Recommended for all high-quality work and particularly pretreated surfaces			*	*	*		*	*	*
Sealers and coatings, including barrier paints, waxes and other impermeable coatings	Mainly used as pretreatments. Not recommended for use without release agent. They prevent the absorption of release agents into the form face. Wax treatments are particularly useful for filling and sealing small imperfections in the form face	Often factory applied or as directed by the manufacturer			*	*	*	*	*	*
Other specialised release agents including wax emulsion	Formulated for specialised applications such as heated formwork systems and concrete forms	As directed by the manufacturer								
Surface retarders	Although these are not release agents they have the same effect as they retard the setting of the surface concrete. Surface retarders are in fact used to produce an exposed aggregate finish. The surface cement is brushed or washed away after striking the formwork.	*	*	*	*	*	*	*	*	

Precautions in use

Whatever the category of release agent being used, the best results are obtained with a very thin film. Excessive application causes a poor concrete finish, presents a greater risk of it getting on to the reinforcement and also creates a safety hazard because of the risk of people slipping on horizontal surfaces.

Release agents contain substances that can be harmful to your health. They must therefore always be handled and used with care.

- Always follow the manufacturer's instructions with regard to use and storage.
- Avoid contact with skin, eyes or clothing.
- Avoid breathing in the fumes, particularly when spraying.
- Keep away from foodstuffs to avoid contamination.
- Always use the correct personal protective equipment (PPE):
  - Barrier cream or disposable protective gloves.
  - Respirator when spraying.
- Do not smoke or use near a source of ignition.
- Ensure adequate ventilation when used on internal work.
- Thoroughly wash your hands before eating or smoking and after work with soap and water or an appropriate hand cleanser.
- In the case of accidental inhalation, swallowing or contact with the eyes, medical advice should be sought immediately.

Loads on formwork

The loads on formwork may be classified as follows:

- **Dead loads.** These consist of the self-weight of the formwork and the deadweight of the concrete and reinforcing steel.
- **Imposed loads.** These can be considered as temporary loads and include the impact load of concrete being placed and the live loads of people working and concreting equipment, in addition to the later storage of materials and forces from the permanent structure.
- **Environmental loading.** This includes wind loading and snow loading as well as accidental loading.
- **Hydrostatic pressure.** Caused by the fluid concrete acting against the sides of vertical or steeply sloping formwork. The amount of pressure is dependent on the height of the pour; as the concrete sets the pressure gradually disappears. The maximum hydrostatic pressure (P_{max}) at any point within a form is equal to the amount of concrete above it. Therefore, there will be minimum pressure at the top of the form, which will steadily increase to the maximum at the bottom. Factors that affect the development of pressure within formwork are given in Table 11.4.

For practical purposes these loads can be considered as wall-type loadings, deck-type loadings and hydrostatic pressure, see Figure 11.47.

Basic formwork construction

Formwork plays a very important part in the finished appearance of a concrete structure. Therefore workmanship of a very high standard is required to produce high-quality concrete.

When making formwork, its construction depends on the following basic requirements:

- **Containment.** It must be capable of shaping and supporting the wet concrete until it cures.
- **Strength.** It must be capable of safely withstanding the dead, imposed and environmental loadings without distortion or danger.

Table 11.4 Factors affecting the development of pressure

Factor	Effect on pressure
Density of concrete	Pressure increases in proportion to density
Workability	Pressure increases in proportion to increases in the 'slump' of the mix
Rate of placing	Slow rates of placing enable lower levels of concrete to commence setting before the pour is complete, so reducing pressure
Method of discharge	Discharge from a height causes a surge or impact loading
Concrete temperature	High temperatures quicken the setting and so reduce pressure, whereas low temperatures have the opposite effect
Vibration	This makes the concrete flow and increases the pressure up to P_{max}
Height of pour	In theory P_{max} increases with the height of the pour, although in practice during the time taken to fill the form the lower levels will have started setting, and so P_{max} over the full height is unlikely to be realised
Dimension of section	Pressure in walls and columns below 500 mm thick is reduced because of the arching or support effect of the aggregate between the formwork sides
Reinforcement detail	Heavy steel reinforcement tends to provide support for the concrete and so reduces pressure

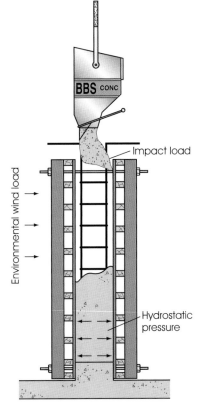

Wall and column loadings

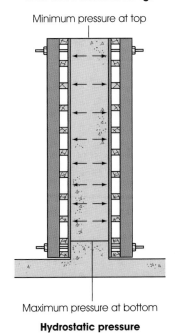

Hydrostatic pressure

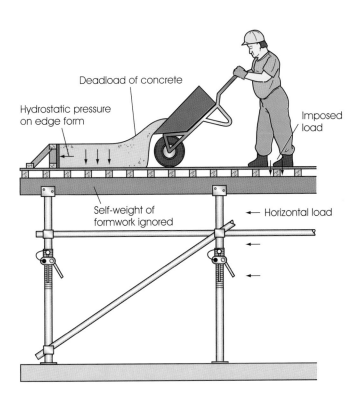

Deck and beam loadings

▲ **Figure 11.47** Loads on formwork

- **Resistance to leakage.** All joints in the form face must be close fitting or taped to prevent grout leakage.
- **Accuracy.** The accuracy of the formwork must be consistent with the item being cast. Formwork for foundations is not normally required to be as accurate as formwork for the superstructure.
- **Ease of handling.** All formwork should be designed and constructed to include facilities for adjustment, levelling, easing and striking without damage to the formwork or concrete.
- **Finish/reuse potential.** The form face must be capable of consistently imparting the desired concrete finish and at the same time achieving the required number of reuses.
- **Access for concrete.** The formwork arrangement must provide access for the placing of the concrete.

Formwork for foundations, kickers and columns

Foundations

In firm soil it is often possible to excavate foundations and cast the concrete against the excavated sides. In circumstances where this is not possible, formwork will be required.

Shallow foundations. Figure 11.48 shows typical details of a shallow foundation for a concrete strip, raft or ring beam. It consists of made-up form panels wedged, levelled and strutted in position. Strutting may be from stakes in the level ground or from a sole plate against the bank in deeper

▶ **Figure 11.48** Shallow foundation formwork

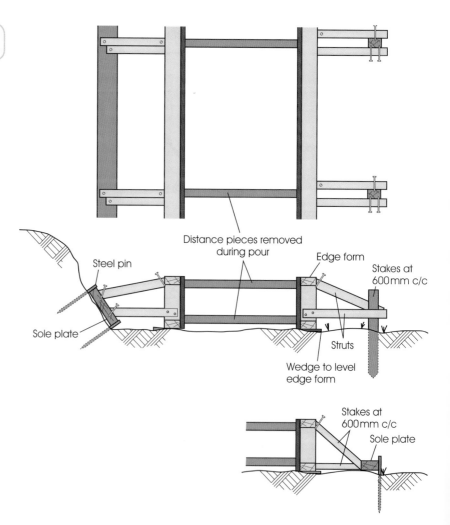

Distance pieces removed during pour

Steel pin

Edge form

Stakes at 600mm c/c

Sole plate

Struts

Wedge to level edge form

Stakes at 600mm c/c

Sole plate

excavations. The distance pieces should be removed during pouring as the concrete reaches them. The form may be lined in a polythene sheet to prevent grout leakage in porous sub-soil.

Pad bases and pile caps. Figure 11.49 shows typical foundation formwork suitable for deeper column pad bases or pile caps. Framed form panels are used for both the sides and the ends. They are wedged to level and then firmly strutted in position, again from either stakes, a sole plate or off the bank. Also shown is a method that is used to support and locate the column steel reinforcement starter bars. This consists of a simple framework suspended across the form sides and notched around the vertical reinforcement. If care is taken during the laying of the blinding concrete to form it to the required plan shape it can be used as a kicker, so positioning the form and preventing grout leakage. Alternatively, as before, the form sides and base may be lined with polythene sheeting.

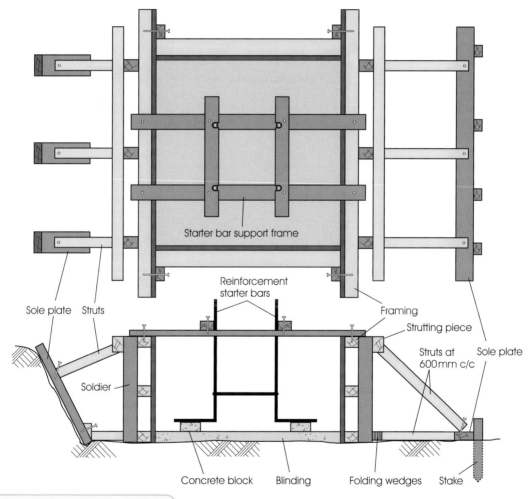

▶ **Figure 11.49** Column pad base formwork

The strutting shown in the previous details has been either cut square at both ends and tightened with folding wedges (do not forget to drive a nail through the wedges to prevent them slackening off), or birdsmouthed around a member at one end and cut square at the other. The form is moved into line or plumb by hammering the square end forward, causing the timbers to bite into each other and create a satisfactory bearing.

Bolt pockets. Often pad bases have to be provided with protruding hold-down bolts or bolt pockets for the later fixing of structural columns or portal frames. A simple framework similar to that used for locating starter bars can also be used for supporting and locating these hold-down bolts. The bolts are normally required to have a certain amount of tolerance movement in order to permit the later accurate positioning of the structural framework. This can be achieved, as shown in Figure 11.50, by the use of tapered polystyrene sleeves around the bolts that can be burnt or melted out after the concrete has been cast.

▶ **Figure 11.50** Locating hold-down bolts

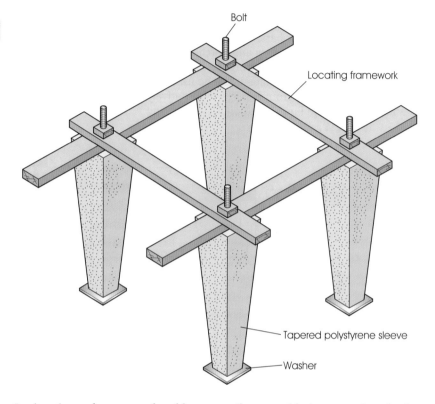

In situations where several pad bases or pile caps with the same plan size but of different depth are required, the formwork can be constructed to the greatest depth. When these are subsequently used for the shallower pours, battens fixed to the sides can be used to indicate the concrete casting level. A line of nails along the form sides is often used, but this is frowned on by the concreters who may damage their hands on them when trowelling off.

Kickers

A kicker is a small upstand of concrete the same plan size as the finished wall or column but only up to 150 mm in height. Its purpose is to locate the bottom of the form and prevent grout loss. It is an important component in concrete work as far as accuracy is concerned.

Kickers for walls and columns may be formed together with the base or slab as part of the same pour, or alternatively it may be formed later as a separate operation.

The ideal size for a kicker cast in a separate operation is about 150 mm in height; any more than this and accuracy is lost owing to the difficulty in ensuring that its sides are plumb. Much less than this and the concrete cannot be compacted properly, so later causing movement at the intersection.

Figure 11.51 shows typical kickers. These are normally held in position by tying them off the reinforcement starter bars with notched battens.

Kickers that are cast along with the slab are known as floating kickers or upstands. These are often more difficult to form than separate kickers. Depending on their height, this type of kicker either may be simply

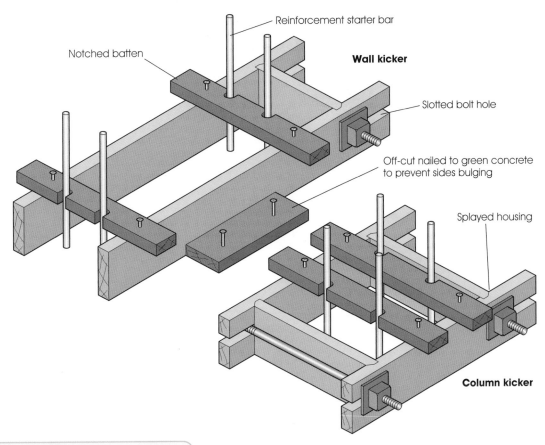

Reinforcement starter bar

Notched batten

Wall kicker

Slotted bolt hole

Off-cut nailed to green concrete
to prevent sides bulging

Splayed housing

Column kicker

▲ **Figure 11.51** Wall and column kicker boxes

supported by previously cast concrete blocks and also tied at intervals back
to the edge form with distance pieces, or may be suspended from the edge
form with the aid of plywood diaphragm brackets. Concrete blocks may still
be used underneath at intervals to reduce any tendency to sag. See Figure
11.52 for details of floating kickers.

In certain situations, provision must be made in the kicker design for the
later securing of formwork. This normally takes the form of a cast-in form
tie. Figure 11.53 shows a coil tie being cast in a wall kicker. As well as
providing a later securing point it ties the kicker sides together, keeping them
parallel. Also shown is a loop coil-tie cast in a kicker, where a later
anchorage point is only required on one side.

For reasons mainly of speed, kickers are sometimes not used, in which case
the form may be accurately positioned with the aid of plywood templates,
either nailed on to the green concrete, as shown in Figure 11.54, or fixed by
cartridge tool.

Columns

A plan view of a typical square column box is shown in Figure 11.55. This is
made up in four separate sections; two of these are the same width as the
column, whereas the other two are made twice the thickness of the ply over
size. Each side is made up so that the backing timbers overhang the edge of
the plywood by its thickness, so that the box will be self-aligning when it is
cramped up.

In order to prevent grout leakage at corners it is essential that the meeting
edges of the plywood are straight and square. As an additional precaution
Neoprene or foamed rubber self-adhesive strips can be applied between the
mating surfaces.

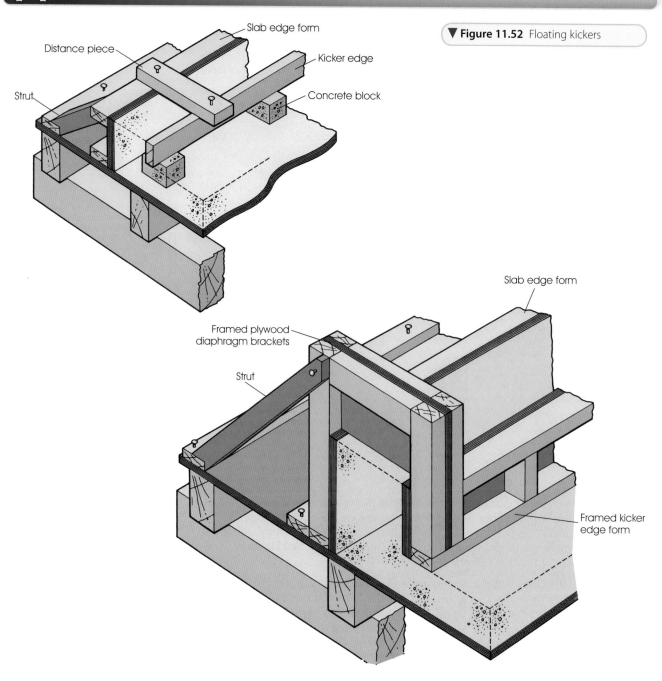

Slab edge form

Distance piece

Kicker edge

Strut

Concrete block

▼ **Figure 11.52** Floating kickers

Slab edge form

Framed plywood diaphragm brackets

Strut

Framed kicker edge form

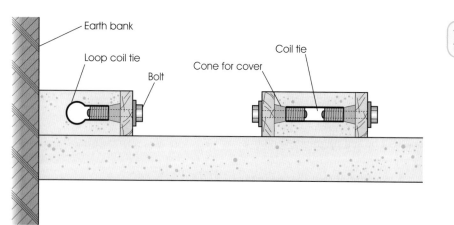

Earth bank

Loop coil tie

Bolt

Cone for cover

Coil tie

◀ **Figure 11.53** Kickers with provision for subsequent fixings

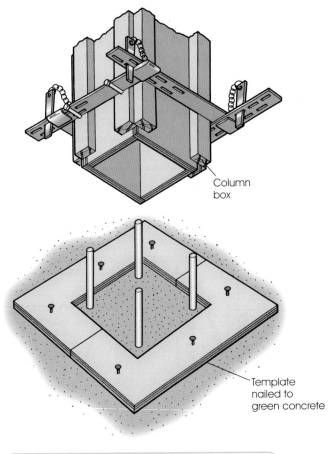

Column box

Template nailed to green concrete

▲ **Figure 11.54** Plywood template to position column

50 mm × 100 mm backing timber

18 mm plywood formface

Neoprene seal to prevent grout leakage

Screw hole filled for fairface work

Slight gap to ensure tight clamping

Ply fixed with either nails or screws

Additional backing timber for wide columns

▲ **Figure 11.55** Square column box plan

Large columns will require an additional backing timber placed centrally on each side, or noggins fixed between the backing timbers under each cramp. This is to prevent the ply bending.

An alternative to the use of proprietary column clamps, which is particularly useful for very large column boxes, is shown in Figure 11.56. This involves the use of yokes made from walings and tie bolts. The spacings of the yokes in common with proprietary clamps will be dependent on the hydrostatic pressure of the design.

Timber yokes and bolts are also particularly suited for use where L-shaped or rebated columns are being constructed, see Figure 11.57.

▼ **Figure 11.56** Use of yokes

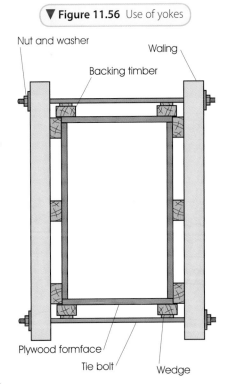

Nut and washer

Waling

Backing timber

Plywood formface

Tie bolt

Wedge

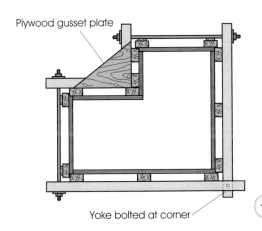

Plywood gusset plate

Yoke bolted at corner

◀ **Figure 11.57** L-shaped column

Plumbing columns. The four column sections are held together with cramps or yokes and the whole column box is plumbed up and held in position by an adjustable steel prop on each side. The column must be plumbed in both directions. This can be done with the aid of a plumb bob or straightedge and level.

Figure 11.58 shows both of these methods. The props must be adjusted until the column is plumb. This operation must be repeated on the adjacent edge so that the column is plumb in both directions. When tightening any prop its opposite one must be slackened otherwise the whole column box will be lifted from the kicker. Immediately after pouring, the column must be rechecked for plumb and carefully adjusted if required, as it may have been knocked during the pouring process.

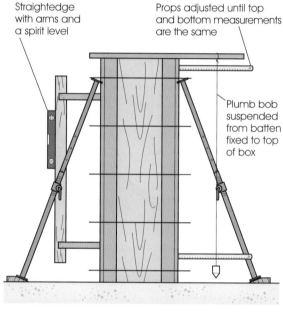

▶ **Figure 11.58** Plumbing a column box

Straightedge with arms and a spirit level

Props adjusted until top and bottom measurements are the same

Plumb bob suspended from batten fixed to top of box

Proprietary steel-framed panels. Either steel- or plywood-faced panels can be used to form square and rectangular columns. These are connected at their external corners with L-shaped angles that are secured with keys and wedges, see Figure 11.59.

Mushroom heads. Occasionally columns are topped with a splayed or mushroom-shaped head. Although they form a distinctive decorative feature, their main purpose is in fact to carry the load of large-span floor slabs and transfer these loads on to the columns. The more common method of supporting and transferring loads is to use a system of beams.

Where large numbers of mushroom heads are involved, the use of purpose-made steel or glass-reinforced plastic forms would be an economical consideration. A method for forming mushroom heads using timber and plywood is shown in Figure 11.60. The column is first cast up to the level of the splay. A timber yoke is bolted on to the column shaft at its top to form a platform onto which the head formwork can be erected. This consists of two long plywood sides fixed to shaped plywood formers and bolted together on top of the yoke platform. The other two sides are cut and fixed onto the first two.

Circular columns. These can be constructed using cardboard tube formers, proprietary glass-reinforced plastic formers, proprietary steel formers, or made-up timber and plywood forms.

Figure 11.61 shows a section through a cardboard tube former. These require backing timbers and either yokes or metal banding to prevent distortion from the circular during casting. The example shown uses metal banding of the

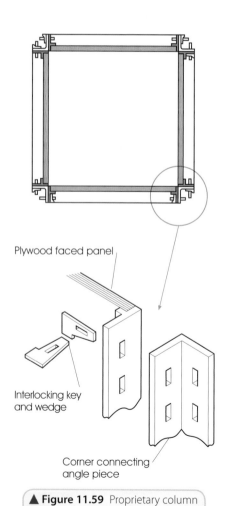

Plywood faced panel

Interlocking key and wedge

Corner connecting angle piece

▲ **Figure 11.59** Proprietary column form

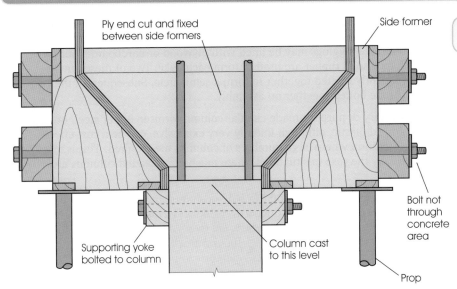

Ply end cut and fixed between side formers

Side former

Supporting yoke bolted to column

Column cast to this level

Bolt not through concrete area

Prop

▶ **Figure 11.60** Plywood mushroom-head form

type used in heavy packaging. The bands are tightened around the forms using a special tightening and crimping device. On striking, the bands are simply cut with shears to release the backing timbers. It is customary to leave the cardboard tube in position until occupation in order to provide physical protection for the concrete. Being spirally wound, it is then simply unwound from the column. This method has the advantage of being fairly economical, but the tubes are only suitable for one-off usage.

A method of forming a circular column in timber is shown in Figure 11.62. It is made up of plywood formers, timber lagging and vertical studs in two separate halves, which are later bolted together. Adjustable steel props are used for holding and plumbing in the same way as square columns.

▼ **Figure 11.61** Circular column using a cardboard tube

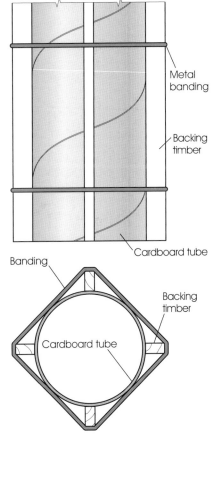

Metal banding

Backing timber

Cardboard tube

Banding

Backing timber

Cardboard tube

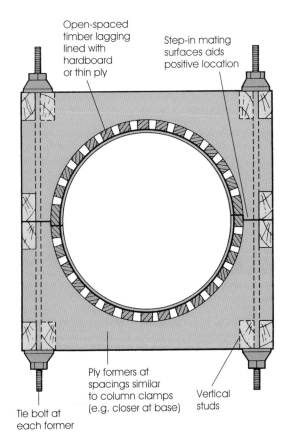

Open-spaced timber lagging lined with hardboard or thin ply

Step-in mating surfaces aids positive location

Tie bolt at each former

Ply formers at spacings similar to column clamps (e.g. closer at base)

Vertical studs

◀ **Figure 11.62** Circular column form

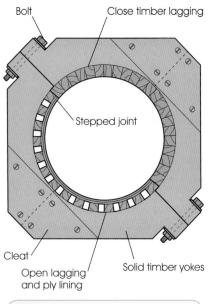

Bolt Close timber lagging

Stepped joint

Cleat

Open lagging and ply lining

Solid timber yokes

▲ **Figure 11.63** Circular column using timber yokes

Figure 11.63 shows an alternative method of constructing circular columns. This uses solid timber yokes and laggings. Where a fair-faced finish is required the laggings may be lined with plywood or hardboard. Four pieces of shaped solid timber are used to form each yoke. Two of these are permanently cleated together to form a semi-circle. Subsequently, the two halves are bolted together on assembly.

Proprietary or purpose-made circular column forms in either steel- or glass-reinforced plastic, although initially very expensive, can be a cost-effective consideration where large numbers of columns are required. This is more especially the case in situations where mushroom-headed columns are required.

Figure 11.64 shows a section through a column box with shaped ends and a box-out to create a service duct. This is in fact a combination of a square and circular column. It has been made up in two halves using plywood formers, vertical studs and timber laggings lined with either plywood or hardboard. The rectangular service duct has been formed using a box-out that incorporates a false draw. The plywood former and laggings method of column box construction is very versatile; it can be adapted to create any shape of column box.

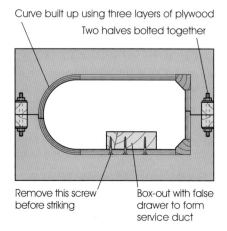

Curve built up using three layers of plywood

Two halves bolted together

Remove this screw before striking

Box-out with false drawer to form service duct

▶ **Figure 11.64** Column with shaped end and service duct

Formwork for walls

Wall formwork normally consists of a number of standard wall forms bolted together. These are then tied over their backs with either walings or soldiers. The choice of using the form panels vertically with horizontal walings, or of using horizontal panels and vertical soldiers, depends on the nature of the work and the designer. It is general practice for high lifts, where substantial pressures are expected, to use horizontal panels and vertical soldiers.

Figure 11.65 shows typical details of a wall form using standard panels and form ties. These are erected against the previously cast kicker. The tie bolt positions are dependent on the wall height, thickness, method and rate of pour, as these factors greatly affect the development of concrete pressure within the forms.

Adjustable steel props are used at about 1 m centres on both sides of the wall to plumb and hold the forms in position. A prop with an overly steep angle should be avoided, as this tends to lift the forms off the kicker. An allowance for shrinkage must be made on long walls. They are often, therefore, cast in a hit-and-miss pattern so that shrinkage movement can take place before alternate bays are cast. An alternative to adjustable steel props is push-pull props that bolt to the form and the substructure, enabling them to be used on one side of the form only.

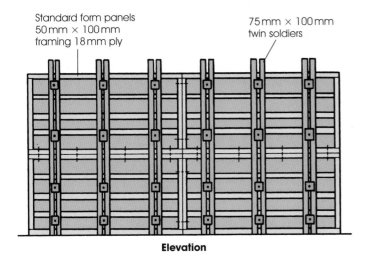

Standard form panels
50 mm × 100 mm
framing 18 mm ply

75 mm × 100 mm
twin soldiers

Elevation

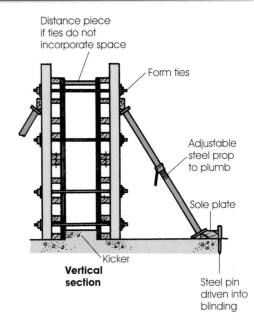

Distance piece
if ties do not
incorporate space

Form ties

Adjustable
steel prop
to plumb

Sole plate

Kicker

**Vertical
section**

Steel pin
driven into
blinding

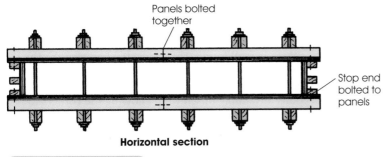

Panels bolted
together

Stop end
bolted to
panels

Horizontal section

▲ **Figure 11.65** Wall forms

Stop ends. Where provision is required in the stop end for the reinforcement starter bars to protrude into the next bay, the detail shown in Figure 11.66 can be used. This accommodates the starter bars and eases removal after striking the main panels as the two halves of the ply stop end simply slide out from either side.

Corners and piers. Careful formwork detailing of corners and attached piers is required in order to resist the increased pressures at these points. Typical details are shown in Figure 11.67.

▼ **Figure 11.66** Wall form stop end

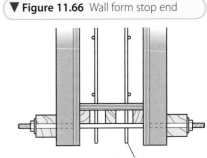

Starter bars projecting through stop end

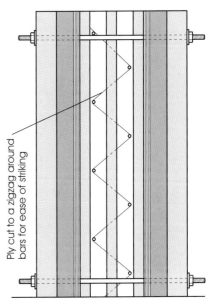

Ply cut to a zigzag around
bars for ease of striking

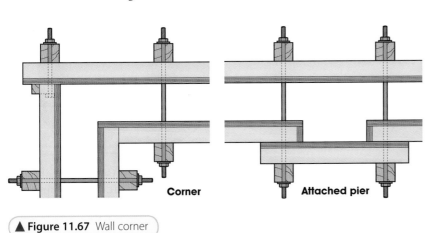

Corner

Attached pier

▲ **Figure 11.67** Wall corner

Proprietary systems

These are available for wall formwork. Most of these consist of steel-framed panels jointed together using keys and wedges. Figure 11.68 shows details of one such system. Scaffold tubes are used as walings to tie the panels and align the wall.

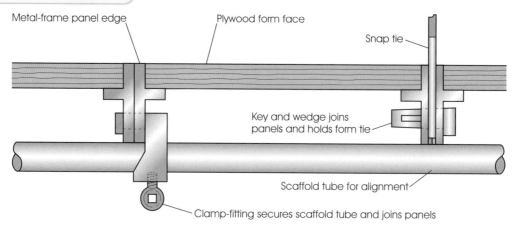

▼ **Figure 11.68** Proprietary wall form details

Metal-frame panel edge

Plywood form face

Snap tie

Key and wedge joins panels and holds form tie

Scaffold tube for alignment

Clamp-fitting secures scaffold tube and joins panels

▼ **Figure 11.69** Rectangular box-out

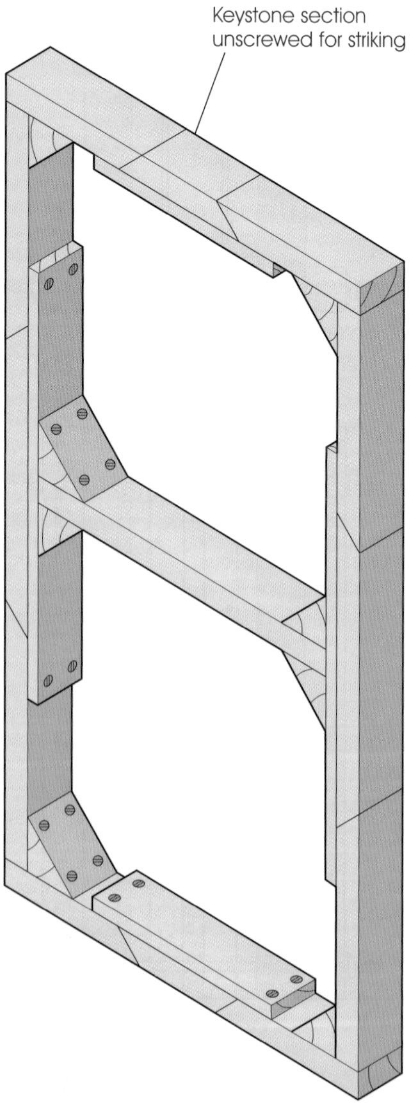

Keystone section unscrewed for striking

Rectangular box-outs. Figure 11.69 shows a box-out suitable for forming door and window openings in wall forms. This is located between the two form faces of the wall and allows the formwork on both sides to be continuous. Keystone-shaped sections are included in each edge of the box. On striking, these sections are simply unscrewed, allowing the corners to be removed separately.

Circular or curved box-outs can be formed using plywood fixed around shaped ribs, as shown in Figure 11.70. To assist removal the circular box-out has been made in three sections joined using dovetail keys. On striking, the keys are simply unscrewed to enable the sections to be withdrawn with ease into the hole formed.

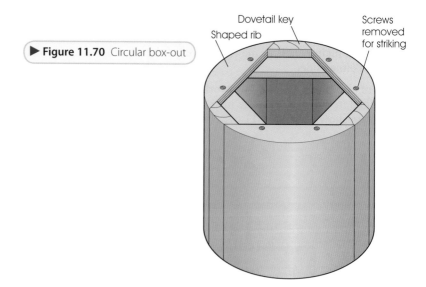

Dovetail key

Shaped rib

Screws removed for striking

▶ **Figure 11.70** Circular box-out

Climbing wall forms. Walls are sometimes cast in a number of vertical lifts. This is either to reduce the concrete pressures of a high lift, or to obtain the maximum reuse from a minimum number of form panels. The formwork for this method is known as climbing wall forms.

Figure 11.71 shows the basic arrangement for climbing forms. After casting and allowing the first lift to cure sufficiently, the form panels are removed and raised to form the next lift, which is then cast. This process can be repeated until the required wall height is reached.

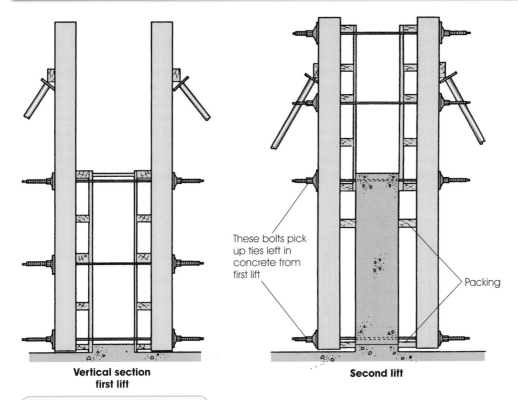

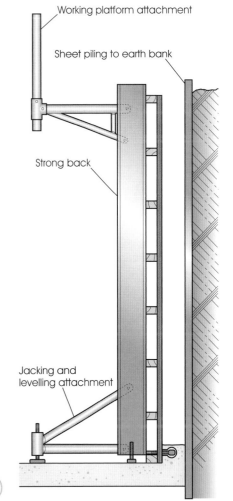

These bolts pick up ties left in concrete from first lift

Packing

Vertical section first lift

Second lift

▲ **Figure 11.71** Climbing wall forms

On fair-faced work the joint line created between lifts may be unsightly. This can be masked by adding a fillet to the panel, which forms a definite feature joint (see Figure 11.72). This detail is particularly useful on vertical joints in sawn board-marked feature forms.

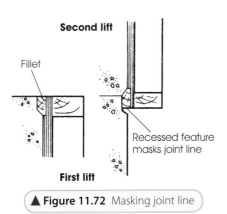

Second lift

Fillet

Recessed feature masks joint line

First lift

▲ **Figure 11.72** Masking joint line

Single-sided wall forms. In situations where a wall has to be cast against a bank or other structure, the single-sided wall form arrangement shown in Figure 11.73 is used. This has standard form panels and proprietary strong backs. Because of the lack of ties it is essential that each strong back is anchored firmly to the kicker. In the example, this has been achieved by bolting the strong backs to loop coil ties, which were previously cast in the kicker. Where an additional lift is required, loop coil ties could be cast at the top of the wall for subsequent anchorage.

Working platform attachment

Sheet piling to earth bank

Strong back

Jacking and levelling attachment

▶ **Figure 11.73** Single-sided wall form

Plumbing and lining wall forms

Walls are plumbed up and held in position by adjustable steel props on each side. The props at the two ends are adjusted until the wall is plumb using a similar method to that used to plumb columns. Remember, when tightening any prop its opposite one must be slackened, otherwise the whole wall form will be lifted from its kicker. The top of the wall can then be 'lined in' using the method shown in Figure 11.74. Drive nails near the top of the two ends and strain a line between them. Cut three identical pieces of packing, say 18 mm plywood; place one each end under the line. Use the third piece to check the distance between the form face and the line at each support position. Adjust the prop so that the packing piece just fits between the form face and the line.

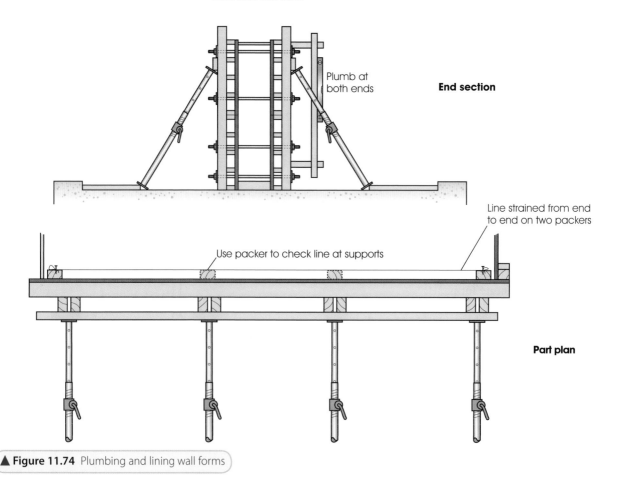

Figure 11.74 Plumbing and lining wall forms

Curved wall forms

The construction of curved concrete walls may be carried out using panels made up from shaped horizontal ribs, vertical laggings and plywood form face linings, as shown in Figure 11.75. Two different-shaped form panels are required – one set to the concrete face internal radius and the other to the external radius. Twin vertical soldiers and form ties are used to secure the panels where the ribs overlap.

Shown in Figure 11.76 is a proprietary system for forming curved wall shapes. This utilises strong backs attached to the plywood form face. The desired internal and external curvatures are formed by adjusting the turn buckles, which connect the edge angles that are attached to both ends of the plywood sheets. Adjacent forms may be joined by bolting through these edge angles.

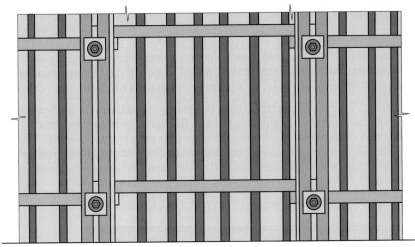

► **Figure 11.75** Constructing a curved concrete wall

Elevation

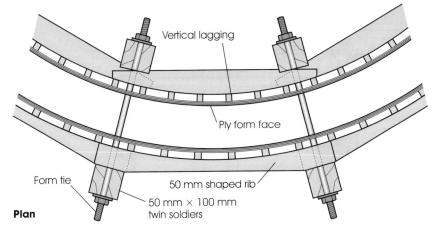

Vertical lagging

Ply form face

Form tie

50 mm shaped rib

Plan

50 mm × 100 mm
twin soldiers

▼ **Figure 11.76** Proprietary system for curved wall shapes

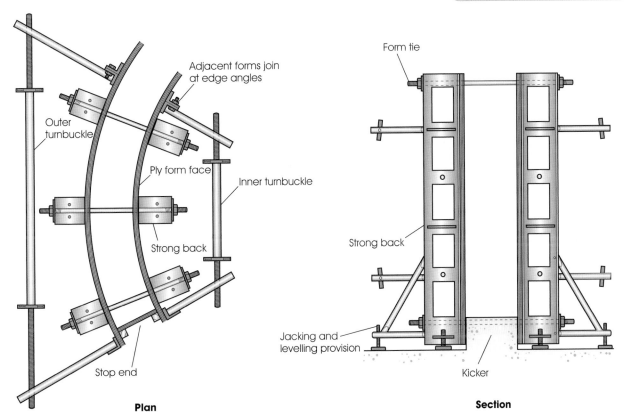

Adjacent forms join
at edge angles

Outer
turnbuckle

Ply form face

Inner turnbuckle

Strong back

Stop end

Plan

Form tie

Strong back

Jacking and
levelling provision

Kicker

Section

Slabs and beams

Floor and roof slabs

The decking for floor or roof slabs usually consists of plywood, joists and ledgers supported on adjustable steel props, or one of the various proprietary formwork systems available, or a partial combination of both.

Figure 11.77 shows the arrangement of timbers and adjustable steel props for a typical floor slab. Scaffold tubes and fittings have been used for the lacing and diagonal sway bracing. These permit greater prop loadings and prevent sideways movement of the formwork. The recommended positioning for the lacing members is up one-third of the extended inner tube height.

▼ **Figure 11.77** Floor slab decking

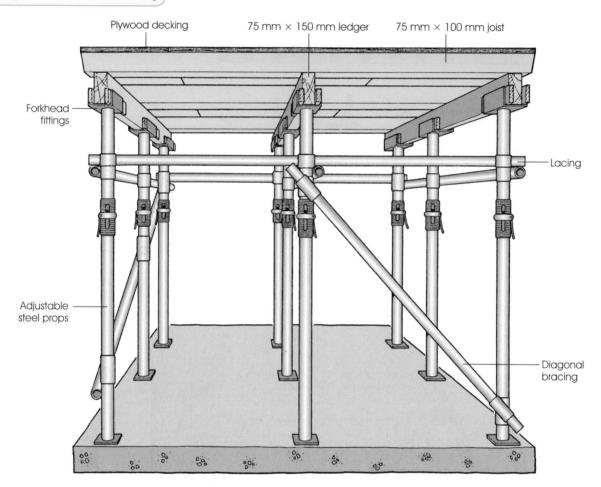

Plywood decking · 75 mm × 150 mm ledger · 75 mm × 100 mm joist · Forkhead fittings · Lacing · Adjustable steel props · Diagonal bracing

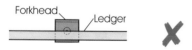

Eccentric loading

Forkhead / Ledger

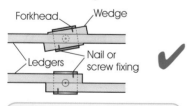

Concentric loading of forkheads

Forkhead / Wedge / Ledgers / Nail or screw fixing

▲ **Figure 11.78** Forkhead loading

Forkhead fittings are located in the top of the adjustable steel props to provide a secure bearing for the ledgers. Most forkheads are of a size to permit the side-by-side lapping of ledgers. Concentric (through the centre) loading of props is essential if they are to support their design loading. This can be achieved when fork heads are being used for single ledgers by aligning the square head diagonally across the timber and packing out with wedges (see Figure 11.78).

The spacing of formwork members will vary depending on the anticipated loads and pressures. As a general guide the following can be used:

- Props   1.200 m centres
- Ledgers   1.500 m centres
- Joists   400 mm centres

Figure 11.79 shows how a slab may be finished at an open edge. The edge, which sits on top of the decking, is framed up from 50 mm × 75 mm timbers

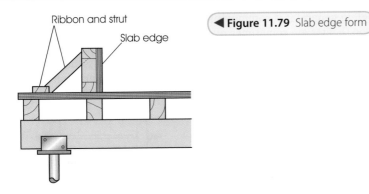

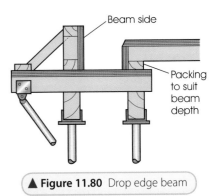

▲ **Figure 11.80** Drop edge beam

and strutted in position off the 50 mm × 75 mm ribbon with 50 mm × 50 mm struts. The ends of the struts are shown birdsmouthed over the edge framing and ribbon, but in practice these are often cut square.

Drop beams

These are sometimes incorporated at slab edges as shown in Figure 11.80. The beam soffit has been made up using the same method as for slab formwork. The ends of the main slab joists are supported by packing off the beam soffit. The joists supporting the beam soffit have been cantilevered out to enable strutting of the beam side. The cantilevered joists should in turn be strutted back to the adjustable steel props for support. This cantilevered arrangement is often necessary at upper floor levels where there is no direct bearing available for the adjustable steel props directly below the support position.

Figure 11.81 shows formwork details where a beam is incorporated in the middle of a floor slab. Where the beam is deep the beam sides may be held in position by the use of form ties spaced periodically along the beam and positioned towards their top edge.

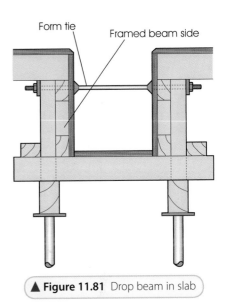

▲ **Figure 11.81** Drop beam in slab

Careful consideration must be given when determining the layout of plywood soffits where they abut previously cast work or existing structures. This is to avoid any trapping after casting. Figure 11.82 shows how the plywood should be cut around a previously cast column to enable striking. For grout-tight joints against the column, any gap can be filled with a compressible foam or rubber strip.

Figure 11.83 shows how narrow striking strips may be included along the edges of slabs where they abut beams or walls. On striking, the splayed edge main plywood sheet comes away first, leaving the narrow striking strip adhering to the soffit. This is then simply removed by the insertion of a wedge if required. The ends of the joists are also splayed to prevent binding when one end is lowered.

▼ **Figure 11.82** Cutting plywood around columns

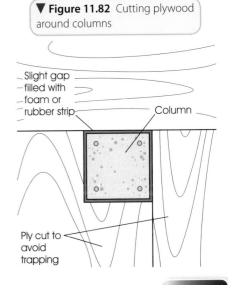

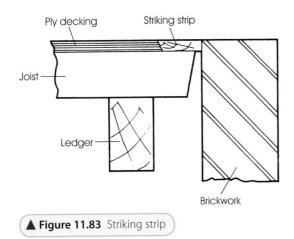

▲ **Figure 11.83** Striking strip

Although in general it is common practice to cast beams and slabs in one operation, beams are sometimes required to span between two supports on their own. Figure 11.84 shows the formwork for a beam spanning between two columns. The runners under the beam soffit are not essential, but their use enables wider spacing of the joists and permits easy variation of the beam depth.

► **Figure 11.84** Alternative design of beam boxes

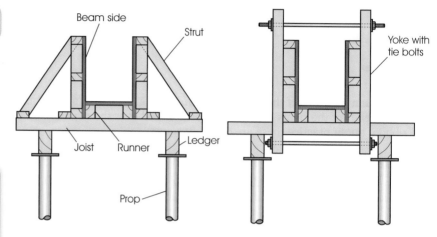

Related information

Also see Chapter 7 in Book 1, page 326, which deals with the setting up and use of quickset levels.

Levelling slab formwork and lining edge forms

Datum marks on adjacent walls or columns are used for the initial positioning of slab formwork. Having erected the slab formwork to the correct height, it should be checked for level and adjusted as required. This operation can be carried out with the aid of a quickset level using the following procedure (see Figure 11.85):

1. Set the instrument up and adjust ready for use, adjacent to the slab formwork and at the same level. This may be either on a previously cast section of the same slab or on an adjacent scaffold. Never set up the level on the actual formwork as the instrument will move along with the formwork as it is adjusted.

2. Two assistants are required, one on the decking holding the staff, the other below the decking to adjust the props. It is unlikely that the person below will hear verbal instructions, so prearrange a code system, for example one stamp with the foot on the decking to wind up the prop, two stamps to lower it, three stamps for okay.

3. The levels are checked along the lines of each ledger, since it is these that support the joists and decking.

4. Starting at the ledger adjacent to the columns or wall, adjust the props until the top of the decking coincides with the previously established mark. Take a staff reading at either end and midway along this ledger. Each reading should be the same. Record this, as it is the reading required to level the whole deck.

5. Position the staff at one end of the next ledger; adjust the prop until the same reading is achieved. Reposition the staff, first at the other end of the ledger, followed by midway along; adjust props as required to achieve the desired reading. Long ledgers or those formed in two lengths will require additional intermediate checks.

6. Repeat the previous stage for each remaining ledger.

7. After the levelling operation is complete and before loading the formwork, check to ensure that all the prop collars are adjusted tight up to the pin, as the process of adjusting various props can cause the inner tubes of intermediate props to lift up. If the collars were not so adjusted the formwork would sag and distort under load.

Slab edge forms will require lining. This is carried out using a similar method to that used for lining wall forms.

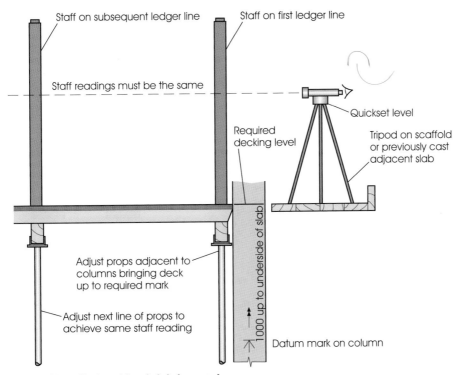

Staff on subsequent ledger line

Staff on first ledger line

Staff readings must be the same

Quickset level

Tripod on scaffold or previously cast adjacent slab

Required decking level

1000 up to underside of slab

Adjust props adjacent to columns bringing deck up to required mark

Adjust next line of props to achieve same staff reading

Datum mark on column

Using a quickset level to adjust and level slab formwork

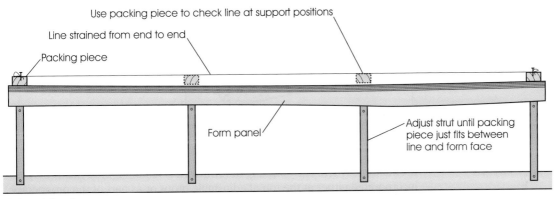

Use packing piece to check line at support positions

Line strained from end to end

Packing piece

Form panel

Adjust strut until packing piece just fits between line and form face

Lining slab edge

▲ **Figure 11.85** Levelling and lining

Steel-framed buildings have to be encased to protect against fire and corrosion. Where they are to be encased in concrete it is possible to suspend the formwork from the universal beams using hangers (see Figure 11.86).

Proprietary adjustable floor centres are useful for decking as they reduce not only the amount of timber but also the number of adjustable props required, so easing the levelling operations and creating a less 'cluttered' under-slab area. Figure 11.87 shows a typical example of the formwork for a floor slab using centres and standard form panels.

▶ **Figure 11.86** Suspended formwork in steel-framed buildings

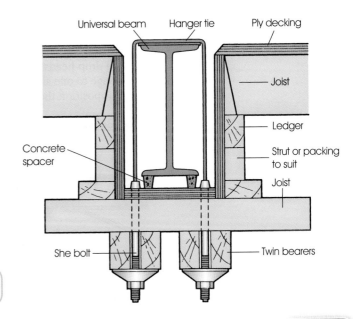

Universal beam

Hanger tie

Ply decking

Joist

Ledger

Strut or packing to suit

Joist

Concrete spacer

She bolt

Twin bearers

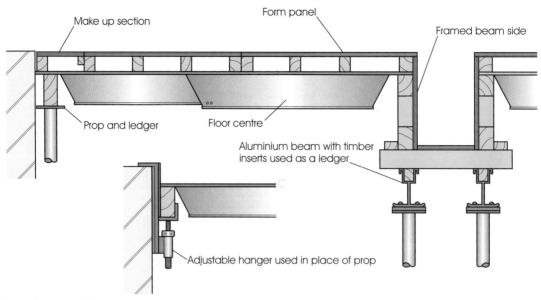

Use of floor centres

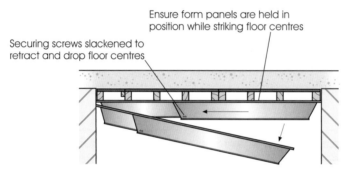

Striking floor centres

▲ **Figure 11.87** Proprietary floor centres

Drop-heads

In situations where early striking of formwork is required, drop-heads may be incorporated either at the top of adjustable props or as part of the proprietary support system. Their use enables the decking to be removed three to four days after curing the slab; the support structure remains undisturbed until the concrete has gained sufficient strength to be self-supporting over its full span. Figure 11.88 shows drop-heads fixed to the top of adjustable props. These are used to support proprietary steel beams and standard form panels. On striking, the drop-heads are released using a hammer blow to move the supporting flange to one side. The panels and beams will drop 100 mm or so, and from this position they can easily be handled and lowered.

▶ **Figure 11.88** Use of drop-heads

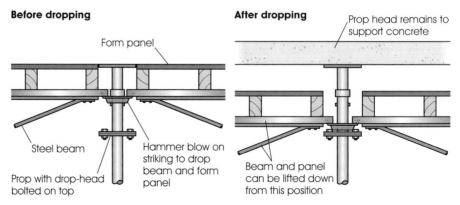

Table forms

These are popular for the repetitive casting of floor slabs in medium- to high-rise structures. This is because they greatly reduce the time taken to erect, strike and re-erect, due to the fact that these operations can be carried out without any dismantling.

Figure 11.89 shows a floor slab and edge beam detail using table forms as the support system. Where edge beams are concerned, one essential requirement is that the table must be capable of being lowered sufficiently to fit under the edge beam.

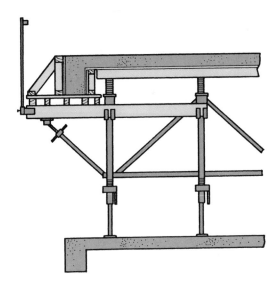

◀ **Figure 11.89** Table form edge beam detail

Slabs with sloping soffits

These are for use as projecting canopies and so on and may be formed using framed-up ribs acting as ledgers to support the joists and plywood decking, as shown in Figure 11.90. Cross-braces are fixed to the struts of adjacent ribs to prevent overturning.

Alternatively, sloping soffits may be formed, see Figure 11.91. Here the straight ledgers are raked using differing prop lengths. Wedges cut to the required slope are positioned in the forkhead to permit a flat bearing.

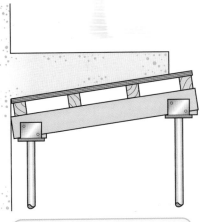

▲ **Figure 11.90** Sloping slab soffit

Stairs

When setting out formwork for stairs great care is necessary, as the finished result must obviously comply with all the requirements of the Building Regulations, see Chapter 5.

In situ formwork for stairs is possibly one of the most expensive operations; therefore, wherever possible, designers tend to use pre-cast stairs.

Stair soffits can be considered as a sloping slab; so the supporting formwork will be similar to that of floor slabs. The **step profile** is formed using cut strings and riser boards. Risers may be formed from timber, plywood or pressed steel, see Figure 11.92. Timber or plywood risers should have a splayed bottom edge to permit the entire tread surface to be trowelled off. The radius along the edges of the pressed steel risers acts as a stiffener and also forms a neat pencil round to the finished step profile.

Two alternative formwork arrangements are shown in Figure 11.93. These are both suitable for casting stairs abutting a wall on one side. For stairs that are built between two walls the strutted open edge form can be omitted and substituted by either a wall plate and hangers or a cut wall string. Likewise, free-standing stairs open on both sides can be formed using strutted edge forms on both sides.

▲ **Figure 11.91** Alternative sloping soffit detail

Related information

Also see Chapter 5, page 123, which covers the requirements of the Building Regulations related to stairs.

▶ **Figure 11.92** Stair risers

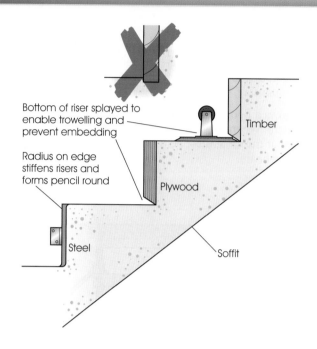

Bottom of riser splayed to enable trowelling and prevent embedding

Radius on edge stiffens risers and forms pencil round

Timber

Plywood

Steel

Soffit

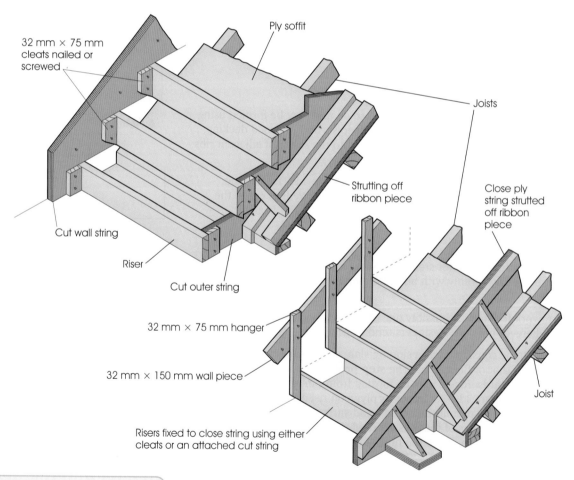

Ply soffit

32 mm × 75 mm cleats nailed or screwed

Joists

Cut wall string

Strutting off ribbon piece

Close ply string strutted off ribbon piece

Riser

Cut outer string

32 mm × 75 mm hanger

32 mm × 150 mm wall piece

Joist

Risers fixed to close string using either cleats or an attached cut string

▲ **Figure 11.93** Stair formwork details

Cut strings may be marked out using a tread/riser template that has been cut to the desired step profile, see Figure 11.94. This is stepped along the string edge the required number of times and marked around with a pencil each time. The outline of steps is best marked out on an abutting wall from the nosing/pitch line. This is an imaginary line that connects the edge of each step. A chalk line is used to mark this on the wall, and the pitch line length (taken from the template) for each step is marked along it. Using a spirit level the horizontal and vertical lines of the step profile can be marked out from these positions. The thickness of the slab concrete is termed the waste (see Figure 11.95).

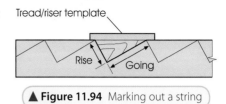

▲ **Figure 11.94** Marking out a string

◄ **Figure 11.95** Marking stair details on abutting wall

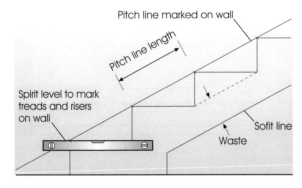

An inverted carriage piece strutted off the main structure should be fixed down the centre of wide flights to resist the tendency of long slender risers to bulge under concrete pressure (see Figure 11.96). Sectional details showing formwork arrangements for short flights either up to or from landings are shown in Figure 11.97.

◄ **Figure 11.96** An inverted carriage piece for wide stairs

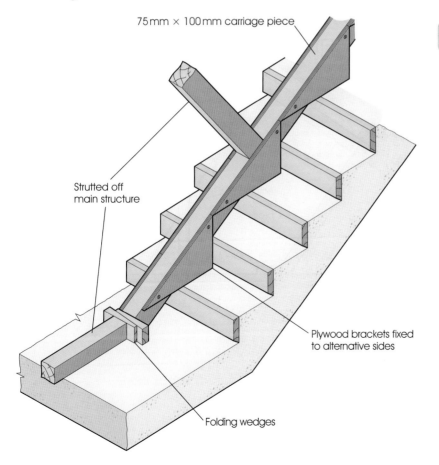

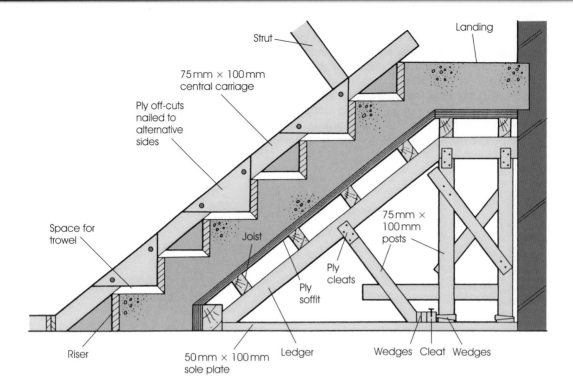

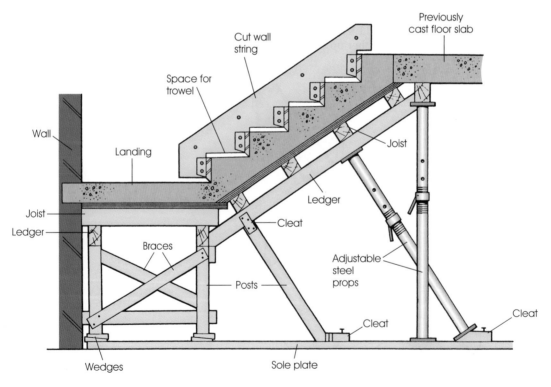

▲ **Figure 11.97** Formwork for short flights

Pre-cast concrete moulds

The moulds for pre-cast concrete products may be constructed from steel, glass-fibre-reinforced plastics, or timber and plywood. The first two are mainly the province of pre-casting factories, while timber and plywood are the common materials for use on site.

Construction

Single mould boxes

The basis of mould box construction is a level base or bed on which edge forms can be assembled and cast. The jointing of edge forms is best carried out using bevelled housings. These tighten on bolting, ensuring a grout-tight joint. In addition, this detail is less likely to result in damage at the corner of the concrete when striking.

Where not required for wedging purposes, bolt holes for tightening forms are best slotted, since half a turn or so of the nut is all that is required to permit bolt removal. Figure 11.98 shows a simple mould box suitable for casting blocks, slabs or wall panels.

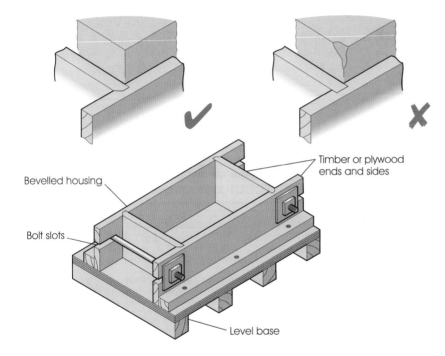

◀ **Figure 11.98** Basic mould box construction

Bevelled housing

Bolt slots

Timber or plywood ends and sides

Level base

Gang moulds

These are used to cast more than one concrete unit at the same time. Figure 11.99 shows a typical gang mould for the pre-casting of tapering fence posts. Placing well-greased steel rods through the distance pieces into the moulds at the required positions permits holes in the posts for bolts or straining wire. These should be removed shortly after the concrete has commenced stiffening.

Shaped pre-cast items can be formed by adaptation of the basic mould box design. Figure 11.100 shows how infill pieces are inserted to a rectangular mould box shape to form a weathered pre-cast sill. For speed of erection and striking, long folding wedges have been used on one side of the box instead of the continuous batten.

Figure 11.101 shows how a typical variety of concrete shapes can be formed using the infill technique.

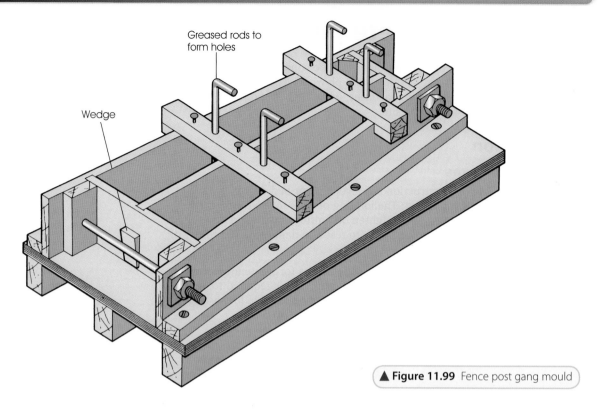

Greased rods to form holes

Wedge

▲ **Figure 11.99** Fence post gang mould

◀ **Figure 11.100** Mould for pre-cast sill

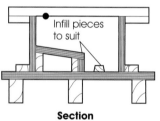

Infill pieces to suit

Section

Steel rod to form drip

Long wedges

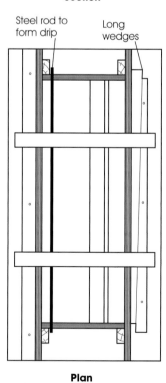

Plan

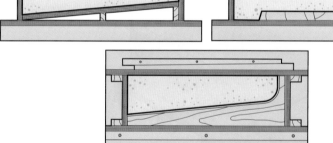

▲ **Figure 11.101** Infill technique for basic moulds

In order to facilitate striking, the bed for pre-cast moulds may be pivoted on one edge. After striking the mould box, the bed is tilted to a near vertical position. This causes the cast component to slip a few millimetres on to a restraint batten and, in doing so, to break the bond between the concrete and bed face, see Figure 11.102. The concrete component can them be simply lifted off and transported elsewhere for further curing.

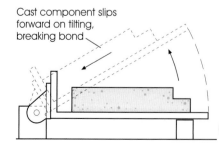

Cast component slips forward on tilting, breaking bond

◀ **Figure 11.102** Tilting pre-cast bed

Pre-cast stairs

Stairs are an expensive and time-consuming *in situ* casting operation and so are often specified to be pre-cast. A typical pre-cast stair detail is shown in Figure 11.103. This incorporates rebates at either end, which locate into similar rebates formed at slab landings. Reinforcement loops or cast-in sockets have been provided on the tread surface near the top and bottom of the flight to act as slinging points for crane handling.

Three main casting aspects are possible with stair moulds, see Figure 11.104.

- ◼ The inclined casting aspect is really an *in situ* casting technique applied to a pre-cast unit.
- ◼ Inverted flat casting allows the upper surfaces and edges of the flight to be of high-class mould quality, while the soffit can be trowelled to an acceptable finish. Inverted flat castings are particularly suitable where tiled nosings or non-slip tread surfaces are to be incorporated, as these may simply be placed in the mould prior to filling with concrete.
- ◼ Edge casting is useful for fair-faced flights, as all but one edge can have a high-quality moulded surface.

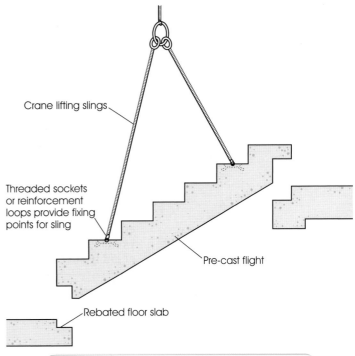

Crane lifting slings

Threaded sockets or reinforcement loops provide fixing points for sling

Pre-cast flight

Rebated floor slab

▲ **Figure 11.103** Dropping in a pre-cast flight of stairs

◀ **Figure 11.104** Casting aspects for pre-cast stairs

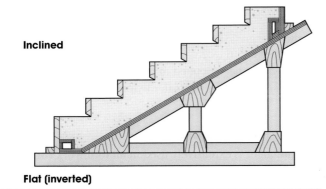

Inclined

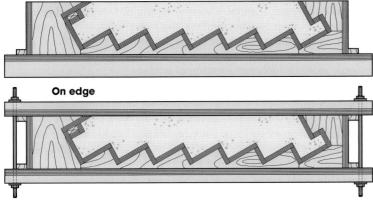

Flat (inverted)

On edge

Formwork inspection and safety

At all times during the erection of formwork and just prior to and during concreting, the formwork supervisor must keep a close eye on the work. This is to ensure that the formwork is to the engineer's specification, capable of producing the required quality (dimensions, tolerances and concrete finish) and, most important of all, safe. The following checklist is useful for this purpose. Owing to the wide range of formwork constructions, the list includes items that may not be relevant to a particular job; therefore, it must be amended to suit the work in hand. In addition to the formwork supervisor, similar checks should also be carried out by the general foreman, site engineer, clerk of the works and, where appointed, the false (temporary) work coordinator, one of whom will have the duty of giving permission to load or fill the formwork. During the actual pouring of the concrete, it is essential that a fully experienced form worker is on stand-by adjacent to the work to deal with minor adjustments during casting, re-plumbing, levelling or lining after casting and any other contingencies.

Before concreting

- Are all materials suitable and free from distortions, defects, rust, rot and so on?
- Is the formwork in accordance with the design; in the right position; of the correct dimensions; plumb, level and in line?
- Are all ties, supports, struts, cramps and braces, correctly positioned, tight and suitably tied or fixed to prevent later movement?
- Has provision been made for ease of striking without causing damage to the concrete?
- Have all the stop ends, features, fillets, inserts, box-outs and so on been correctly located and fixed?
- Has a release agent been applied to the form face?
- Have all holes, fixings and joints in the form face been suitably filled or taped?

Checking proprietary equipment

Proprietary equipment is extensively used in all formwork construction. A special check to look out for the following common errors should be made during all site inspections:

- **Column and beam cramps.** Over-tightening of wedges and bolts, causing crushing of backing timbers and subsequent loss of strength; no support under cramps (nails or blocks) to prevent them from slipping; projecting end of cramps not painted white or bright to make them obvious; loose or incorrect location of wedges.
- **Floor centres.** Incorrect location of, or failure to tighten, locking nuts or wedges; bearing plates inadequately seated; centres used at their minimum span (no provision to retract for striking).
- **Form ties.** Too small or incorrectly positioned washer plates; over-tightening of bolts; unequal size of soldiers or walings; damaged or substandard threads (see Figure 11.105).
- **Adjustable steel props.** Use of visibly bent, damaged or corroded props; use of nails, bolts or steel reinforcement in place of the high-tensile steel pins supplied by the manufacturers; missing or inadequate bracing and lacing; props out of plumb (check with a spirit level); props eccentrically loaded; inadequate bearing at either top or bottom (see Figure 11.106).

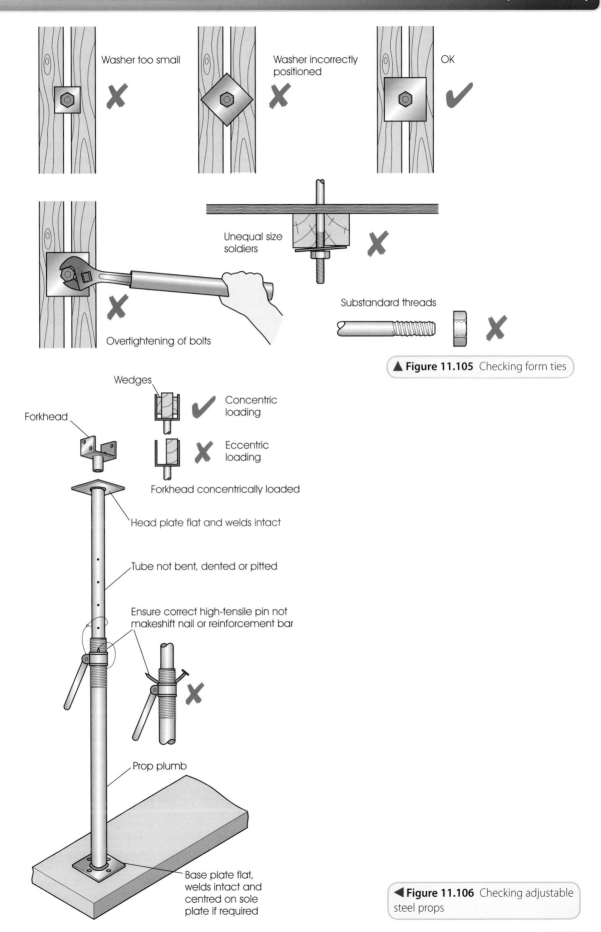

Washer too small ✗

Washer incorrectly positioned ✗

OK ✔

Unequal size soldiers ✗

Overtightening of bolts ✗

Substandard threads ✗

▲ **Figure 11.105** Checking form ties

Wedges

Concentric loading ✔

Eccentric loading ✗

Forkhead

Forkhead concentrically loaded

Head plate flat and welds intact

Tube not bent, dented or pitted

Ensure correct high-tensile pin not makeshift nail or reinforcement bar ✗

Prop plumb

Base plate flat, welds intact and centred on sole plate if required

◀ **Figure 11.106** Checking adjustable steel props

During concreting

Look out, listen for and bring to the attention of the engineer any of the following:

- Formwork straining noises (ties snapping, timber creaking and so on).
- Any movement or distortion in the formwork.
- Any joints opening or grout leakage.

Each of these factors may be an indication of a possible formwork blow-out or collapse. Common failures commence with grout leakage followed by distortion and bulging in the formwork. In cases where a form tie fails, its neighbouring ties become overloaded and rapid progressive failure can result. The immediate action is to stop the pouring and vibration immediately, evacuate personnel from potentially dangerous situations, and then take remedial action to relieve and control the excessive pressures.

Where appropriate, check that distance pieces have been removed and fixing blocks and so on inserted.

Note: There are other items, which must be checked but are not usually the responsibility of the formwork carpenter such as: safe access; steel reinforcement; key to previously poured surface; availability of concrete and equipment; method and rate of fill; curing and so on.

Striking

The time between the pouring of the concrete and the striking of the formwork varies widely. This depends on the following four factors:

- The type of formwork, for example, wall, column, slab or beam.
- The concrete used and the type of surface finish required.
- The weather conditions and exposure of the site.
- The requirements of the specification.

Minimum striking times are shown in Table 11.5 for ordinary Portland cement (OPC).

Table 11.5 Minimum striking times

Location of formwork	Temperature (t)		
	7 °C	16 °C and over	Formula for t between 0 °C and 25 °C
Vertical surfaces			
Walls, columns and beam sides	18 h	12 h	$300 \div (t + 10)$ h
Horizontal surfaces			
Slab soffits	6 days	4 days	$100 \div (t + 10)$ days
Slab support structure	15 days	10 days	$250 \div (t + 10)$ days
Beam soffits	15 days	10 days	$250 \div (t + 10)$ days
Beam support structure	21 days	14 days	$360 \div (t + 10)$ days

Example

To determine the minimum striking time for slab formwork, assuming a temperature of 10 °C during curing.

Slab soffit

$$\text{Minimum time} = \frac{100}{t + 10}$$

$$= \frac{100}{10 + 10}$$

$$= 5 \text{ days}$$

Support structure

$$\text{Minimum time} = \frac{250}{t + 10}$$

$$= \frac{250}{10 + 10}$$

$$= 12.5 \text{ days}$$

- No formwork must be removed until the concrete has cured sufficiently for it to be self-supporting and capable of carrying any loads imposed upon it.
- Vertical surfaces such as walls, columns, beam sides and so on can, in very good weather, be struck as little as nine hours after pouring.
- Horizontal surfaces normally require a much longer time and this can be up to 28 days in poor winter weather. It is usual after striking horizontal formwork to re-prop the concrete in strategic positions for up to a further 14 days.
- The final decision to strike must always rest with the site engineer, who has knowledge and experience of what is involved.
- The safe time to strike the formwork can be determined by tests on cubes made from the same mix as the concrete and stored under similar conditions.

Safety

Striking is an important operation. The procedure for striking must be carefully considered to avoid any safety risk to personnel, stress or damage to the completed building and damage to the formwork itself.

In general, the procedure to adopt is the reverse of that used for erection. Observe the following precautions:

- Always wear appropriate PPE (personal protective equipment) – safety boots, helmet, gloves and so on.
- Never stand or allow anyone else to stand close to the striking area.
- Use hardwood wedges to ease the forms away from the concrete surface (the use of nail bars results in both form and concrete damage).
- Always release (ease) loaded member gradually. For example, props for floor slabs must be released in small stages (half a turn of the handle) starting from the centre and working out, then repeating the operation.
- Always control the striking by removing a section at a time.
- Never allow the formwork to fall (crash striking).
- Always clean, de-nail, repair and store under cover all formwork material, ready for reuse. In addition, all props, cramps, bolts and so on should be cleaned, lubricated and coated with a rust inhibitor.

> **Related information**
>
> Also see Chapter 10 in Book 1, which deals with health and safety procedures in depth.

Timbering to excavations

Timbering is the temporary support given to the sides of excavations, in order to prevent soil collapse while construction work is taking place. The term 'timbering' is normally applied irrespective of the materials used.

The sides of excavations are prone to collapse in on themselves unless they are either battered back or supported with timbering. The angle of battering is dependent on the type of subsoil, the presence of ground water, the time of year and the weather conditions. An angle of up to 45° is suitable in most circumstances, see Figure 11.107.

▼ **Figure 11.107** Battering or timbering to foundation trenches

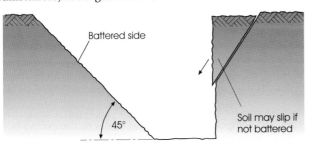

Open timbering

Open timbering is where the sides of the excavation are supported at intervals and are not completely covered. This method is normally used for excavations in firm and moderate ground.

Figure 11.108 shows the stages undertaken when erecting open timbering to the sides of a foundation trench. As the trench is mechanically excavated, pairs of vertical poling boards are positioned in the trench and temporarily held tight to the sides with struts.

▶ **Figure 11.108** Open timbering

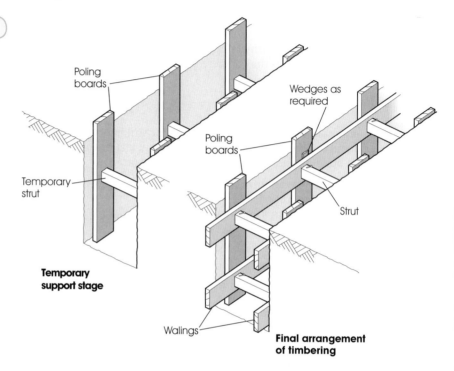

Horizontal pairs of walings are introduced once 2 m to 3 m has been excavated. These are held in position against the poling boards by struts at between 1.2 m and 1.8 m centres. Wedges may be required between the poling boards and walings to ensure each board is held tight up to the earth.

Finally, on completion, the temporary struts to the individual poling boards can be removed to give a better working space.

Closed timbering

Closed timbering used in loose, soft or waterlogged ground is where the sides of the excavation are completely covered.

Figure 11.109 shows the stages undertaken when erecting close timbering to the sides of a foundation trench. In this case, poling boards (either timber or metal) are driven into the ground before any excavation is undertaken.

As the ground is excavated between the poling boards, temporary short lengths of walings and struts are inserted for support. As before, when a sufficient length has been excavated, continuous pairs of walings and struts can be inserted.

Vertical puncheons are used at intervals between the walings to stop them slipping. The struts may be cleated to the walings for additional security using short lengths of batten, called strutlips. Again, wedges may be required between the poling boards and walings to ensure each board is held tight up to the earth.

Finally, on completion, the temporary walings and struts to the poling boards can be removed to give a better working space.

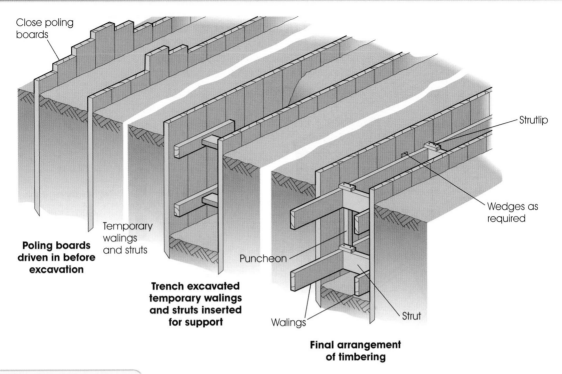

Close poling boards

Strutlip

Poling boards driven in before excavation

Temporary walings and struts

Wedges as required

Trench excavated temporary walings and struts inserted for support

Puncheon

Walings

Strut

Final arrangement of timbering

▲ **Figure 11.109** Closed timbering

Site hoardings and footpaths

Hoardings

Hoardings are, in effect, screens that shield areas such as wastelands, allotments, demolition sites and building sites from public view for various reasons, for example, protection, safety, security and aesthetics.

Figure 11.110 shows typical hoarding details. They are often constructed using panels that are framed up in sections from 50 mm × 100 mm timbers, and sheeted with 18 mm WBP (weather- and boil-proof) plywood. These panels are secured at their base by 50 mm × 50 mm stakes, which are driven approximately 600 mm into the ground. Braces are fixed to the tops of the panels to hold them plumb. These braces are also fixed at their bottom end to 50 mm × 50 mm stakes. Where the hoarding is adjacent to a scaffold, the stakes and braces can be omitted and the panels fixed back to the scaffold using hooks and bolts (providing the additional wind loading has been allowed for in the scaffold design).

Viewing screens are sometimes incorporated in the hoarding to give the public a view of the building works that are going on behind it. Where there is the possibility of a large amount of public interest in the building, which could result in congestion on the pavement, a short set-back in the hoarding could be formed to create a viewing area.

If personnel access is required through the hoarding, a wicket gate in the panel as shown can provide this. For vehicular access, several panels could be hinged to form a large pair of gates. Temporary walkways for public protection may be required where a hoarding partially blocks a pavement (see Figure 11.110).

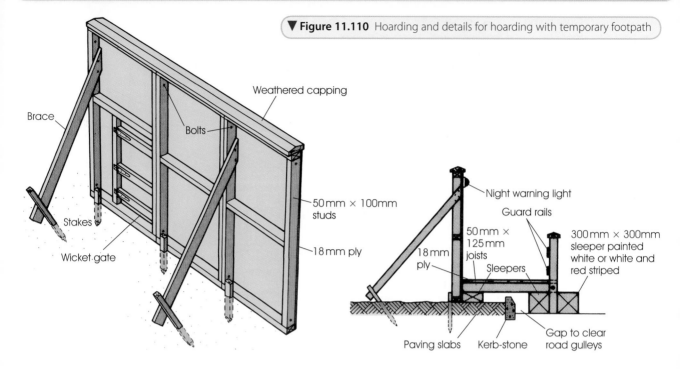

▼ **Figure 11.110** Hoarding and details for hoarding with temporary footpath

Brace

Weathered capping

Bolts

Stakes

Wicket gate

50 mm × 100 mm studs

18 mm ply

Night warning light

Guard rails

50 mm × 125 mm joists

18 mm ply

Sleepers

300 mm × 300 mm sleeper painted white or white and red striped

Paving slabs

Kerb-stone

Gap to clear road gulleys

Boundary fences and gates

Timber fences can be classified into two main types.

- **Open-boarded fences.** These define the boundary by providing a visual barrier.
- **Close-boarded fences.** These, as well as defining the boundary, also provide a certain amount of privacy and security.

Open-boarded fences

These are usually constructed from loose fencing materials in a variety of designs.

Ranch-style fence, see Figure 11.111, consists of 25 mm × 150 mm boards, screwed to the face of the 75 mm × 75 mm posts. For improved durability the boards can be partly recessed into the posts. When it is necessary to lengthen the boards they can be butted midway over a post. These joints should be staggered so that adjacent boards are not butted on the same post. The matching gate is made in a similar way, using two stiles and a brace.

▶ **Figure 11.111** Ranch-style fence

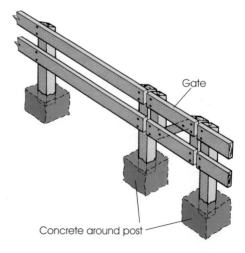

Gate

Concrete around post

Picket or palisade fence, Figure 11.112, shows a simple method that consists of 75 mm × 75 mm posts, 50 mm × 75 mm weathered edge rails and 25 mm × 75 mm shaped top pickets or paling. The rails can either be mortised and pinned into the posts or fixed using a metal bracket. Again, a matching gate can be made in a similar way, using two rails and a brace, with the pickets nailed on the face.

Stockade or interlap fence, see Figure 11.113. The stockade is a taller version of the picket fence. For increased privacy, boards can be fixed alternately to both sides of the rails to form an interlap fence.

The gap between the boards is a matter of personnel preference; narrow gaps are used for privacy and wider ones simply to define the boundary with a decorative visual barrier. This type of fence is ideal for use in exposed situations as the gaps between the boards allow strong winds to pass through, reducing the amount of wind pressure exerted on the fence.

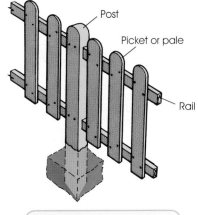

▲ **Figure 11.112** Picket fence

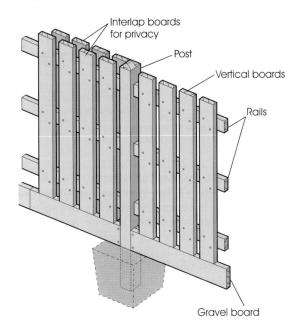

◄ **Figure 11.113** Stockade or interlap fence

Close-boarded fences

These may be constructed from either loose fencing materials, such as featheredge boarding, or using prefabricated fencing panels.

Featheredge fence. Figure 11.114 shows a close-boarded featheredge fence. This is made up from 100 mm × 100 mm or 100 mm × 125 mm posts, 75 mm triangular arris rails and 19 mm × 100 mm featheredge boarding. Fences up to 1.2 m in height require two arris rails, each positioned about 150 mm above and below the intended ends of the featheredge boards. For taller fences, up to 1.8 m in height, an additional centre rail should be fitted. The ends of the arris rails are best tenoned into the posts and secured in position using dowels. As an alternative they can be fitted using metal brackets. The capping and gravel board are often omitted on cheaper work, but this is a false economy as their purpose is to prevent moisture absorption into the end grain of the featheredge boarding, which if allowed to take place would quickly result in decay. Each piece of featheredge should overlap by about 20 mm and be fixed by one nail through the thick edge at each rail. Care must be taken to ensure the nails do not go through the thin edge of the board below where they overlap, as it will split when subjected to moisture movement. The entire run of the posts and rails should be erected and secured before boarding. Gates for close-boarded fences are usually made up in the same way as ledged and braced doors.

> ### Related information
>
> Also see Chapter 6, Door frames and linings, page 147, which deals with the construction of ledged and braced doors.

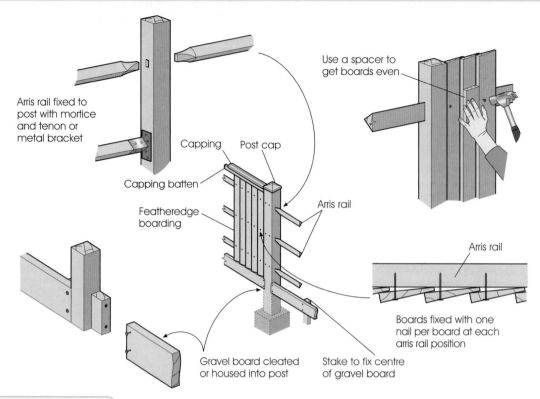

Arris rail fixed to post with mortice and tenon or metal bracket

Use a spacer to get boards even

Capping
Post cap

Capping batten

Featheredge boarding

Arris rail

Arris rail

Boards fixed with one nail per board at each arris rail position

Gravel board cleated or housed into post

Stake to fix centre of gravel board

▲ **Figure 11.114** Featheredge close-boarded fence

Prefabricated fence panels. Standard fence panels are 1.8 m wide and range in height from 600 mm to 1.8 m, increasing in 300 mm intervals. Most fence panels are made from horizontal overlapping thin strips of timber, often waney edged, which are sandwiched between a sawn timber frame. These panels are sold as lap or waney lap panels, see Figure 11.115. Again, the use of capping and gravel boards is recommended for extended fence life. After positioning and securing the end post, the first panel is fixed to it using nails through the edge frame or metal clips. The entire fence is constructed by erecting posts and panels alternately. Matching gates are available from many suppliers.

▶ **Figure 11.115** Prefabricated fence panel

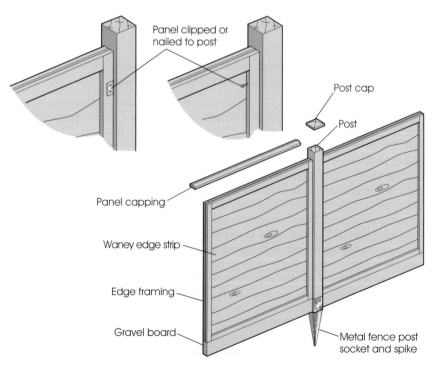

Panel clipped or nailed to post

Post cap

Post

Panel capping

Waney edge strip

Edge framing

Gravel board

Metal fence post socket and spike

Fence posts

Timber posts are traditionally used for most types of fence either 75 mm or 100 mm square. Hardwood (oak) posts are more durable, but preservative-treated softwood posts are cheaper. All timber posts should be either weathered at the top or capped, to shed rainwater so prolonging their life, see Figure 11.116.

Concrete posts are made in a variety of styles from pre-cast concrete. Mortised posts are used with arris rails and grooved posts for panel fences, see Figure 11.117. They are considered to be stronger than timber posts and are also rot free. Concrete gravel boards are available for use with grooved posts.

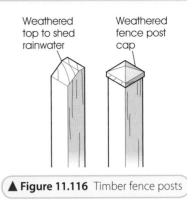

▲ **Figure 11.116** Timber fence posts

Erection of fence posts

The type of fence will determine whether all of the posts are erected and secured first, or the posts are put up one at a time.

The first task is to establish the line of the fence. Drive pegs into the ground at either end along the boundary and stretch a builder's line between them to align the row of posts. These should be spaced at between 1.8 m and 3 m centres, depending on the size and type of fence. Where the fence runs up to a property, the first post can be fixed to the wall using expanding masonry bolts.

Fence posts may be secured in the ground using either hardcore and concrete or proprietary metal spiked sockets.

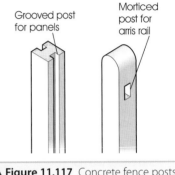

▲ **Figure 11.117** Concrete fence posts

Hardcore and concrete footings. Use the following sequence of operations, see Figure 11.118:

1. Prepare hole for the post. The depth of the post into the ground should be 600 mm for fences up to 1.2 m in height and 750 mm for fences up to 1.8 m.
2. Ram some hardcore (broken brick or stones) in the bottom of each hole, to give a firm base for the post ends and provide drainage.
3. Plumb the post with a sprit level and temporarily brace to hold in position.
4. Ram more hardcore around the post up to a depth of 300 mm.
5. Fill around the post with concrete, tamping with a batten to ensure compaction.
6. Finish the top of the concrete just above ground level and smooth it off so that it slopes away from the post, to shed rainwater and reduce the risk of rotting in timber posts.
7. Allow the concrete to set before removing the temporary braces.

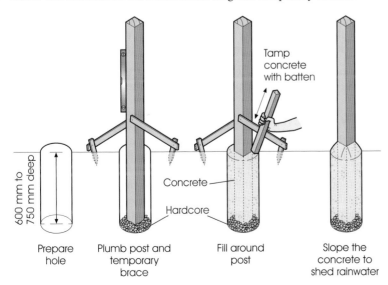

◀ **Figure 11.118** Concreting in a fence post

Proprietary metal spiked sockets. These do not require any holes digging; they are simply driven into firm ground. Use 600 mm spikes for fences up to 1.2 m high and 750 mm ones for those up to 1.8 m. Bolt-down sockets are also available for use where posts are to be erected on to previously concreted surfaces, see Figure 11.119.

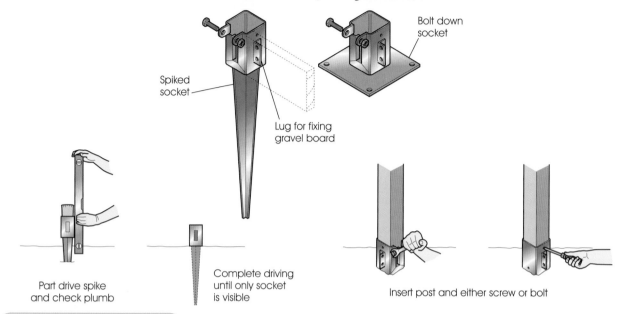

Spiked socket

Bolt down socket

Lug for fixing gravel board

Part drive spike and check plumb

Complete driving until only socket is visible

Insert post and either screw or bolt

▲ **Figure 11.119** Metal fence post socket

The following sequence of operations is used to insert metal fence post spikes:

1. Insert an off-cut of post or a proprietary driving block into the spike socket.
2. Drive the spike partway into the ground using a sledgehammer.
3. Check the socket for plumb using a spirit level.
4. Continue driving the spike into the ground until only the socket is visible. Re-check for plumb at intervals to ensure you are still driving the spike in vertically.
5. Insert post into socket and secure it by either screwing or tightening the bolt, depending on the make and type of spike.

Hanging gates

Most gates are simply hung between two posts, using strap hinges, see Figure 11.120. Low gates are held closed by an automatic gate latch; taller gates are fitted with a thumb or ring latch. A coil spring may be added to provide a self-closing action. Where security is required, tower bolts or a padlock, hasp and stable may be fitted to the inside of the gate. Alternatively, a rim or mortise lock may be used, depending on the gate construction.

Door frames are sometimes used for side-entry gates, these should be fitted with a weathered capping for protection, see Figure 11.121.

Preserving fences

All timber that is used for fencing must be preservative treated. This can either be pressure treated or done by steeping the fencing material in preservative. Where the fence is to have a painted finish, for example, small picket or ranch-type fences, the timber should be treated with a preservative that is capable of receiving a paint finish. The preservative treatment of the posts is particularly important, as untreated or ineffectively treated timber that is in contact with the ground will rot in a very short space of time. It was often for this reason that traditionally durable oak posts were used. Coating the bottom part of the post with bitumen, thereby forming a weatherproof skin, can prolong their lifespan.

> ### Related information
>
> Also see Chapter 1 in Book 1, page 40, which covers the methods used to preservative treat timber.

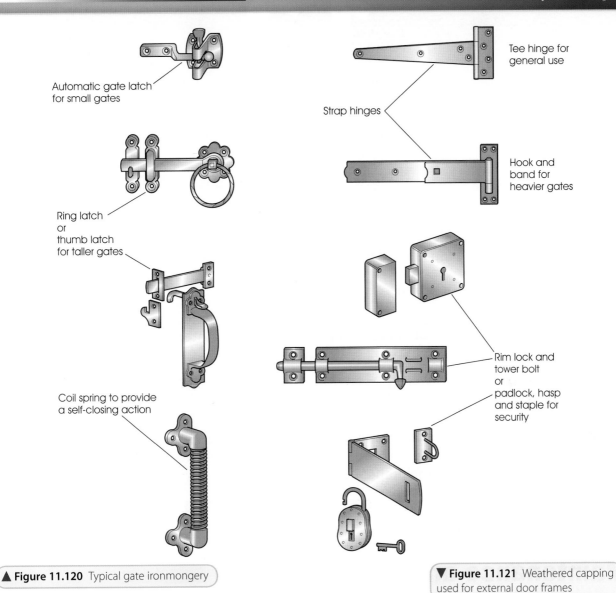

▲ Figure 11.120 Typical gate ironmongery

- Automatic gate latch for small gates
- Ring latch or thumb latch for taller gates
- Coil spring to provide a self-closing action
- Tee hinge for general use
- Strap hinges
- Hook and band for heavier gates
- Rim lock and tower bolt or padlock, hasp and staple for security

In order to keep the timber posts clear of the ground, or where the posts of existing fences have rotted through at ground level, they may be repaired by bolting to a short concrete spur post that has been concreted behind the timber post (see Figure 11.122).

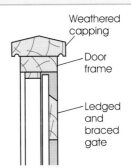

▼ Figure 11.121 Weathered capping used for external door frames

- Weathered capping
- Door frame
- Ledged and braced gate

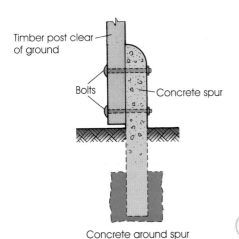

- Timber post clear of ground
- Bolts
- Concrete spur
- Concrete around spur

◀ Figure 11.122 Concrete spur

Repairs and maintenance

This chapter covers the work of both the site carpenter and bench joiner. It is concerned with principles of undertaking repairs and maintenance to buildings. It includes the following:

- Maintenance and repair routines.
- Causes of failure.
- Repairs and maintenance of timber components and associated fittings.
- Re-glazing and painting as part of a maintenance programme.
- Multi-skill repairs to brickwork, plasterwork, tiling, guttering and rainwater pipes as part of a maintenance programme.

Maintenance is taken to mean the keeping, holding, sustaining, or preserving of a building and its services to an acceptable standard. This may take one of two forms: planned maintenance and unplanned maintenance or repairs.

Maintenance and repair routines

There are a variety of reasons why products and processes fail. The impact of many of these failures can be reduced or contained by maintenance routines.

Planned or routine maintenance

This is a definite programme of work aimed at reducing to a minimum the need for often costly unplanned repair work. It includes the following:

- The annual inspection and servicing of general plumbing, heating equipment, electrical and other services.
- The periodic inspection and cleaning out of gutters, gullies, rainwater pipes, airbricks and so on.
- Periodic redecoration, both internally and externally.
- The routine general inspection/observation of the building fabric and moving parts.

Also included under this heading is what is known as preventative maintenance. Basically this is any work carried out as a result of any of the previous inspections in anticipation of a failure, for example, the early replacement of an item, on the assumption that minor faults almost certainly lead onto bigger and more costly faults unless preventative work is carried out.

Unplanned emergency or corrective maintenance

This is work that is left until the efficiency of the element or service falls well below the acceptable level or even fails altogether. This is the most expensive form of maintenance, making inefficient use of both labour and materials and often creating serious health/safety risks. Yet it is the type most often carried out. This is because the allocation of money to enable maintenance work to be planned is often given low priority.

Causes of failure in products, processes and materials

Apart from the natural ageing process of all buildings during their anticipated life (however well maintained), early deterioration of buildings can be attributed directly to one or often a combination of the following agents:

- Design and construction factors.
- Chemical attack: effect on surface finish of chemical pollution in the atmosphere.
- Weathering and damage: the effect of frost, rain, snow, sun and wind.
- Dampness: the effect of water when allowed to penetrate into the building.
- Biological attack: fungal decay and insect attack.
- Fire: the effect of extremely high temperatures on building structures and different materials.

Design and construction factors

Faulty design and construction methods can lead to rapid deterioration of a building. In fact, approximately one third of all maintenance/repair work could be avoided if sufficient care was taken at the design and construction stages.

Faulty design

This results from inadequate knowledge or attention to detail on the part of the architect or designer leading to, for example, poor specification of materials/components, structural movement, moisture penetration, biological attack and the inefficient operation of the building services.

Faulty construction

Inadequate supervision during the construction process can result in poor workmanship, the use of inferior materials and the lack of attention to details/specifications. These can all lead to similar problems as those stated for faulty design, resulting in subsequent corrective measures and expense for the building owner.

Dampness

In buildings this is the biggest single source of trouble. It causes the rapid deterioration of most building materials, can assist chemical attack and creates conditions that are favourable for biological attack. Dampness can arise from three main sources: rain penetration, rising damp and condensation. In addition, leaking plumbing and heating systems and spillage of water in use are also significant causes of dampness.

Rain penetration

This is rain penetrating the external structure, either through the walls, windows or the roof, and appearing on the inside of the building as damp patches. After periods of heavy rain these patches will tend to spread and

▲ Figure 12.1 Efflorescence stains on brickwork

then dry out during prolonged periods of dry weather. They will, however, never completely disappear, as a moisture stain and in some cases even efflorescence (crystallised mineral salts) will be left on the surface, see Figure 12.1.

Mould growth (fungi resulting in dark-green or black patchy spots) may occur in damp areas particularly behind furniture, in corners and other poorly ventilated locations. The main causes of rain penetration are shown in Figure 12.2. It can be seen that penetration takes place through gaps, cracks, holes and joints in, around or between components and elements.

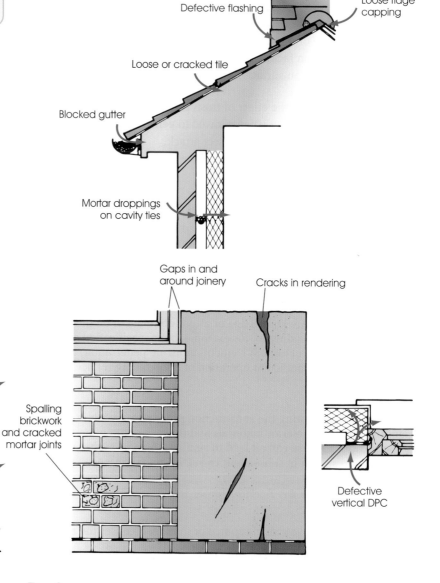

▲ Figure 12.2 Rain penetration

Roofs

Loose or missing tiles or slates, including the hip and ridge capping tiles, will allow rainwater to run down rafters, causing damp patches on the ceilings and tops of walls. These patches may appear some distance away from the defective area as the water spreads along timbers, across the ceiling, and so on. This dampness will also saturate any thermal insulation material making it ineffective. If left unrepaired, saturation of the roof timbers will occur leading to fungal decay in due course. Another major area of penetration is around the chimneystack and other roof-to-wall junctions; this may be due to cracked chimney pots, cracked or deteriorating flaunchings (the sloping mortar into which the pots are set), or corroded or pitted metal flashings (these cover the joint between the stack or wall and the roof), which may be cracked or deteriorated. Poor pointing to the stack can also be a cause of penetration. Any of these defects can cause large patches of damp on an internal wall.

Walls

Clearly rainwater travels downwards and when assisted by high winds it will travel sideways through gaps. However, depending on the nature of the material, it can often move unassisted both sideways or upwards because of capillary attraction. **Capillarity** is the phenomenon whereby water can travel against the force of gravity in fine spaces or between two surfaces that are close together; the smaller the space the greater the attraction.

A simple experiment that demonstrates capillary action is shown in Figure 12.3. It shows that the water rises highest where the two glass plates touch and reduces as the gap widens. Measures taken to prevent or reduce the effect of capillary action are called 'anti-capillary' measures. The grooves that run around the edges of window frames to widen the gap are an example of an anti-capillary measure.

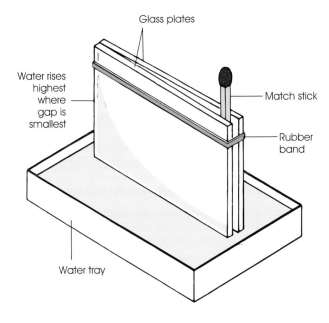

Glass plates

Water rises highest where gap is smallest

Match stick

Rubber band

Water tray

◀ Figure 12.3 Capillary action experiment

There are two main conditions that promote capillarity in the external envelope. The fine cellular structure of some materials provides the interconnecting pores through which water can travel. Also the fine joints between components, for example, wall to door or window frame, mortar joints between brickwork, close joints between overlapping components. The risk of capillarity is reduced or avoided by either:

- physically separating the inside and outside surface by introducing a gap (for example, cavity wall construction)
- introducing an impervious (waterproof) barrier between components (for example, mastic pointing, DPCs, DPMs, moisture barriers, flashing and so on).

Over time, the water resistance of brick and stonework and their mortar joints will deteriorate. This deterioration can be accelerated by the action of frost. Rainwater may accumulate below the surface and freeze. Ice expands causing the brickwork and stonework and their mortar joints to spall

▲ Figure 12.4 Spalling brickwork

(crumble away), see Figure 12.4. The wall then offers little resistance to the weather and should be replaced. This entails either:

- cutting the surface of the spalled components back and replacing with matching thin components (half bricks) and finally re-pointing the whole wall
- 'hacking off' the entire wall (cutting back to remove spalling) and covering with one of the standard wall finishes, cement rendering, rough cast, pebble dash, Tyrolean or silicone-nylon fibre.

Cracks in cement rendering and other wall finishes can be caused by shrinkage on drying, building movement or chemical attack. Once opened up, deterioration is accelerated by frost action. Small cracks may be enlarged and filled with cement slurry. Large areas that may have 'blown' (come away) from the surface will require hacking back to a sound (or firmly adhering) work surface and replaced. With cavity wall construction, rainwater that does penetrate the outer leaf should simply run down inside the cavity and not reach the internal leaf. The vertical mortar joints of the outer leaf are sometimes raked out at intervals along the bottom of the cavity to provide weep holes through which the water can escape.

However, when the cavity is bridged by a porous material (for example, the collection of mortar droppings on the wall ties during construction), the water will reach the inner leaf causing small isolated damp patches on the internal wall surface. The remedy for this fault is to remove one or two bricks of the outer leaf near the suspected bridge and either clean out or replace the tie as necessary.

Dampness around door and window frames is likely to be caused by wind-assisted rain entering the joint between the wall and frame by the action of capillarity or by a defective vertical DPC (damp-proof course) used around openings in cavity walls, where the inner and outer leaf join. Exterior mastic can be used to seal the joints but where DPCs are defective they will require cutting out and replacing. A check should be made at the sill level of frames. Cracked sills allow water to penetrate and, therefore, should be filled. The drip groove on the underside of the sill should be cleaned out as it often collects dirt/dust and is filled by repeated painting. The purpose of the drip groove is to break the under surface of the sill making the water drip off at this point and not run back underneath into the building.

Blocked or cracked gutters and down-pipes, dripping outside taps and constantly running overflows can cause an excessive concentration of water in one place that will be almost permanently damp. This will result in an accelerated deterioration of the wall and subsequent internal damp patches. Repairing or replacing the defective component can easily rectify the immediate fault. However, if left unattended the resulting damage to the building structure has most serious and costly implications.

Rising damp

This is normally moisture from below ground level rising and spreading up walls and through floors by capillarity. This most often occurs in older buildings. Many of these were built without DPCs and DPMs (damp-proof membranes), or, where they were incorporated they have broken down, possibly with age (for example, slate, once a popular DPC material, is not flexible and will crack with building movement, so allowing capillarity). The visual result on the walls is a band of dampness and staining spreading up from the skirting level, wallpaper peeling from the surface and signs of efflorescence. The skirting, joists and floorboards adjacent to the missing or failed DPC are almost certain to be subject to fungal attack. Solid floors may be almost permanently damp causing considerable damage to floor coverings and adjacent timber, furniture and so on.

Rising damp can still occur in buildings that have been equipped with DPCs and DPMs. One of the main reasons for this is the bridging of DPCs; in the case of cavity walls, builders' mortar droppings or rubble may have collected at the bottom of the cavity, allowing moisture to rise above the DPC level; or earth in a flowerbed being too high above the DPC. DPCs are normally located at least two courses of brickwork (150 mm) above the adjacent ground level. This is because even very heavy rain is unlikely to bounce up and splash the walls much more than 100 mm from the surrounding surface. So the splashed rainwater is still prevented from rising above the DPC. Where the surrounding surface is later raised, these splashes might bypass the DPC and result in rising damp. Weak porous rendering that has been continued over the DPC is another means by which the DPC may be bypassed (see Figure 12.5).

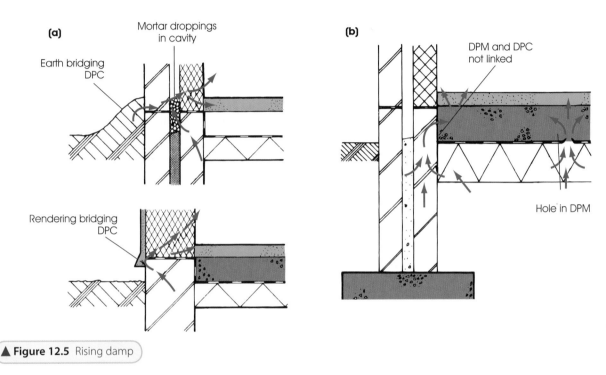

▲ **Figure 12.5** Rising damp

In solid floors with a DPM, rising damp can only occur if this is defective. For example, it may have been penetrated by jagged hardcore during the pouring of the concrete, it may have been inadequately lapped (permitting capillarity between the lapped joint), or it may not have been linked in with the DPC in the surrounding walls (allowing moisture to bypass at this point).

The remedy to rising damp faults will of course vary. Bridged or bypassed DPCs can be rectified by simply removing the cause, for example, lowering the ground level or removing mortar and rubble from the cavity and so on. Where the DPC itself is faulty or missing altogether, one can be inserted by either cutting out a few bricks at a time to allow the positioning of a new DPC, or sawing away the mortar joint a section at a time and inserting a new one.

Alternatively, liquid silicone may be injected near the bottom of the wall. This soaks into the lower courses, which then acts as a moisture barrier preventing capillarity.

Localised faults in DPMs can be remedied by cutting out a section of the floor larger than the damp patch, down to the DPM, taking care not to cut through it. This should reveal the holed or badly lapped portion, which can be repaired with a self-adhesive DPM. An alternative method, which can also be used in floors without any DPM, is to cover the existing concrete floor with a liquid bituminous membrane or a sheet of heavy-duty polythene sheeting before laying a new floor finish, although, to be effective it should be joined into the DPC.

Condensation

The results of this form of dampness are often mistakenly attributed to rain penetration or rising damp, as they can all cause damp patches, staining, mould growth, peeling wallpaper, efflorescence, the fungal attack of timber and generally damp, unhealthy living conditions. The water or moisture for most condensation actually comes from within the building. People breathing, kettles boiling, food cooking, clothes washing and drying, bath water running and so on, are all processes that add more moisture to the air in the form of vapour, which can lead to **surface condensation**.

Air is always capable of holding a certain amount of water vapour. The warmer the air, the more vapour it can hold but when air cools, the excess vapour will revert to water. This process is known as condensation. So whenever warm moist air meets a cool surface condensation will occur (see Figure 12.6). This can only be controlled effectively by achieving a proper balance between heating, ventilation and insulation. The building should be kept well heated but windows should be opened or mechanical ventilators used, especially in kitchens and bathrooms, to allow the vapour-laden air to escape outside and not spread through the building. External walls need thermally insulating to remove their cold surfaces. Both cavity-wall insulation and lining the walls with a thin polystyrene veneer help a great deal. Double-glazed windows also help reduce condensation by preventing the warm moist air coming into direct contact with the cold outside pane of glass.

▶ **Figure 12.6** Surface condensation

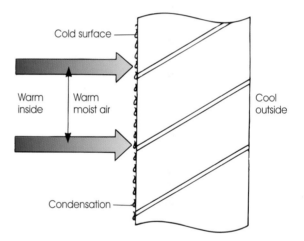

In addition to this surface condensation, there is another condensation problem that occurs when wall surfaces are warm. This is known as **interstitial** or **internal condensation**. This is illustrated in Figure 12.7. The warm moist air passes into the permeable structure until it cools, at which point it condenses leading to the same problems associated with penetrated and rising dampness.

▶ **Figure 12.7** Interstitial condensation

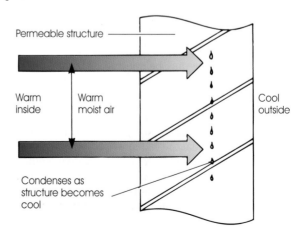

Interstitial condensation can be dealt with by:

- using a **vapour barrier** (this prevents the passage of water vapour) on the warm inside of the wall, for example, a polythene sheet or foil-backed plasterboard
- allowing this water vapour to pass through the structure into a cavity where it can be dispersed by ventilation.

Movement

The visual effects of movement in buildings may apparently be of a minor nature (e.g. windows and doors that jamb or bind in their frames; fine cracks externally along mortar joints and rendering; fine cracks internally in plastered walls, along the ceiling line and so on). They can, however, be the first signs of serious structural weakness. Movement in buildings takes two main forms: ground movement and movement of materials.

Ground movement

Any movement in the ground will cause settlement in the building. When it is slight and spread evenly over the building it may be acceptable, although when more than slight or differential (more in one area than another), it can have serious consequences for the building's foundations and load-bearing members, requiring expensive temporary support (shoring) and subsequently, permanent underpinning (new foundations constructed under existing ones).

Ground movement is caused mainly by its expansion and shrinkage near the surface, owing to wet and dry conditions. Compact granular ground suffers little movement, whereas clay ground is at high risk. Tree roots cause ground shrinkage owing to the considerable amounts of water they extract from it (see Figure 12.8). Tree roots can extend out in all directions from the tree base, greater than its height. In addition, overloading of the structure beyond its original design load can also result in ground movement.

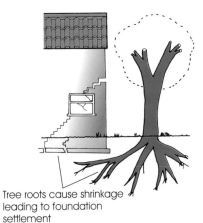

Tree roots cause shrinkage leading to foundation settlement

▲ **Figure 12.8** Ground movement

Frost also causes ground movement. Water in the ground expands on freezing. Where this is allowed to expand on the undersides of foundations it has a tendency to lift the building (known as frost heave) and drop it again on thawing. This repeated action often results in serious cracking. Freezing of ground water is limited in the UK to about the top 600 mm in soil.

Movement in materials

All building materials will move to some extent owing to one or more of the following reasons: temperature changes, moisture-content changes and chemical changes. Provided the building is designed and constructed to accommodate these movements, or steps are taken to prevent them, they should not lead to serious defects.

- **Temperature changes** cause expansion on heating and shrinkage on cooling; particularly affected are metals and plastics, although concrete, stonework, brickwork and timber can be affected also.

Large areas of brickwork, stonework and concrete may start to show cracks due to thermal expansion and contraction, see Figure 12.9. Movement joints should be included during the construction or can be retro fitted. These take the form of a straight joint that is filled with a compressible board, such as bitumen-impregnated fibreboard, with the outer surface sealed with polysulphide to prevent water penetration, see Figure 12.10.

- **Moisture changes** cause expansion when wetted and shrinkage on drying. This is known as moisture movement. The greatest amount of moisture movement takes place in timber, which should be painted or treated to seal its surface. Moisture movement can also affect brickwork, cement rendering and concrete. Rapid drying of wetted brickwork in the hot sun can result in cracks, particularly around window and door openings.

▲ **Figure 12.9** Cracking due to thermal movement

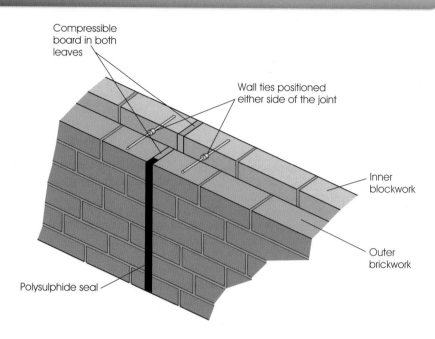

▶ **Figure 12.10** Brickwork movement joint

Chemical and biological attack

Chemical

This consists of the corrosion of metals and the sulphate attack of cement. Corrosion causes metals to expand and lose strength. Corrosion of steel beams can lift brickwork causing cracks in the mortar joint. Corroded wall ties may cause bulges in cavity brickwork. The sulphate attack of cement occurs either in the ground or through products of combustion in chimneys. The sulphate mixes with water and causes cement to expand. Sulphate-resisting Portland cement (SRPC) should be used in conditions where high levels of sulphate are expected.

Smoke containing chemicals is given off into the atmosphere as a result of many manufacturing processes. This mixes with water vapour and rainwater to form dilute or weak acid solutions known as acid rain. These solutions corrode iron and steel, break down paint films and erode the surfaces of brickwork, stonework and tiles. Acid rain also attacks decorative metal and plastic finishes. Regular cleaning to remove the contamination can prolong the useful life of materials in these environments.

Ageing. Exposure to sunlight can cause bleaching, colour fading of materials and even decomposition owing to solar radiation, especially ultra-violet light. Particularly affected are bituminous products, plastics and painted surfaces.

Biological

Timber, including structural, non-structural and timber-based manufactured items, are the targets for biological attack. The agents of this are fungi and infestation of wood-boring insects. Given the right conditions an attack by one or both agents is almost inevitable. There are two main types of fungi that cause decay in building timbers, these being dry rot and wet rot.

Dry rot is the more serious and is more difficult to eradicate than wet rot. It is caused by a fungus that feeds on the cellulose found mainly in sapwood (the outer layers of a growing tree). This causes timber to lose strength and weight, develop cracks in brick-shape patterns and, finally, to become so dry and powdery that it can easily be crumbled in the hand. Two initial factors for an attack are damp timber in excess of about 20 per cent moisture content (MC) and bad or non-existent ventilation.

Wet rot does not spread to the same extent as dry rot. It feeds on wet timber (30 per cent to 50 per cent MC) and is most often found in cellars, neglected external joinery, ends of rafters, under leaking sinks or baths and under impervious (waterproof) floor coverings. During an attack, the timber becomes soft, darkens to a blackish colour and develops cracks along the grain. Very often timber decays internally with a fairly thin skin of apparently sound timber remaining on the surface.

Infestation of wood-boring insects commonly called woodworm, after the larvae that are able to feed on and digest the substance of wood. The female adult beetle lays eggs during the summer months, usually in the end grain, open joints, or cracks in the timber. This affords the eggs a certain amount of protection until the larvae hatch. The larvae then start their damaging journey by boring into the timber, consuming it and then excreting it as fine dust. The duration of this stage varies between six months and 10 years, depending on the species. During the early spring, at the close of this stage, the larvae bore out a pupal chamber near the timber surface, where they undergo the transformation into adult beetles. This takes a short period after which the beetles bite a way out of the timber leaving characteristic flight holes. The presence of flight holes is often the first external sign of an attack. After emerging from the timber the beetle's instinct is to mate, lay eggs and then die, so completing one life cycle and starting another.

Fire

All materials can be classified as either of the following:

- Capable of burning solids such as timber (termed combustible) and capable of burning liquids such as petrol (termed flammable).
- Incapable of burning solids such as brickwork, concrete and steel (termed non-combustible) and capable of burning liquids such as water-based paints (termed non-flammable).

It is not the actual material that burns when combustible or flammable materials are subjected to a fire, it is the vapour they give off when heated. The lowest temperature at which a material gives off a flammable vapour is known as its ignition temperature or flashpoint. These will vary widely – timber, for example, has an ignition temperature of around 300 °C, whereas many flammable liquids have a flashpoint well below normal internal temperatures.

Most building fires are started from a relatively small source of heat, for example, a lighted match, a cigarette end or an overloaded electrical circuit. Any of these can supply the initial heat required to ignite a small fuel source, such as an armchair or waste paper in a rubbish bin. These heat sources in turn preheat other materials. This cycle of events will continue to escalate, with more flammable vapours being given off as each different material in a room is heated above its ignition temperature. It is at this point that a 'flashover' will occur. The term 'flashover' is used to describe the simultaneous ignition of the flammable vapours, causing an intense fire involving all the contents of a room or building.

The rate of burning depends upon the supply of oxygen. It is possible for a fire in a room or small building to burn itself out owing to the depletion of the oxygen supply. However, if the hot gases produced as a result of the fire are allowed to escape, then a fresh oxygen supply will be drawn in causing a rapid fire growth.

If a fire is put out at an early stage, the only damage may be that caused by smoke. This will blacken ceiling, floor and wall surfaces. It can to some extent be washed off but may require pressure washing externally and repainting internally. Timber discoloured by smoke will also require washing down and re-decoration.

> ### Related information
> Also see Chapter 1 in Book 1, pages 32–9, which covers the recognition, eradication and treatment of timber affected by both fungi and wood-boring insects.

Extended exposure to fire can cause damage to the buildings structure. Just because a material is non-combustible it does not mean that it will not be affected:

- Bricks, blocks and stone will expand in extremely high temperatures, causing cracking, spalling and buckling,

- Concrete can crack, spall and even explode when subjected to prolonged high temperatures in a building fire. This is due mainly to the heating up of the steel reinforcement that expands and forces off the surface concrete.

- Metal rapidly conducts heat, buckles, expands and will lose its structural strength at fairly low temperatures. This may cause damage to other structural elements such as walls where a steel beam or lintel has been built-in, leading to structural collapse in the worst cases. Metals can be protected to some extent from fire by coating with an intumescing paint or by encasing with a non-combustible material such as plasterboard or brickwork.

- Timber is combustible, but large structural sections have a slow rate of burning and the charring caused helps to insulate it from the heat. Eventually, after extended exposure, it will lose its structural integrity, leading to collapse. Increased fire protection can be given by coating with an intumescing paint; pressure treating with a chemical that does not give off a flammable vapour when heated; or by encasing with a non-combustible material such as plasterboard.

- Plastics will melt when exposed to fire and may give off toxic fumes.

> **Related information**
>
> Also see Chapter 11 in Book 1, which covers material properties and the causes of failure.

Inspections and repairs

Whenever a building firm undertakes major repairs or maintenance it is desirable to undertake a survey of the building. The extent of the measurements, sketches and details taken will depend on the nature and extent of the work. Clearly a survey prior to replacing windows in a house will be very different to that of one involving structural movement.

Asbestos is particularly harmful to your health, especially if its dust or fibres are inhaled. You may come across asbestos when working on maintenance jobs. Contractors should have an asbestos survey undertaken by a specialist surveyor before undertaking major refurbishment or maintenance jobs. The removal of asbestos or working in an area of asbestos must only be carried out by specialist companies.

In many cases there will information already existing that can assist in a general survey for maintenance purposes. The building's occupants may explain the apparent major problems to you or you may be guided by another survey report, such as a valuation survey or schedule of remedial work, see Figure 12.11. These are used as the starting point for notes and sketches.

Internal survey

Notes are best complied in the form of a table listing the defects in each area, see Table 12.1, which identifies the defects found when undertaking an internal survey of an early 1900s-built semi-detached house. These should be backed up by internal sketches, see Figure 12.12.

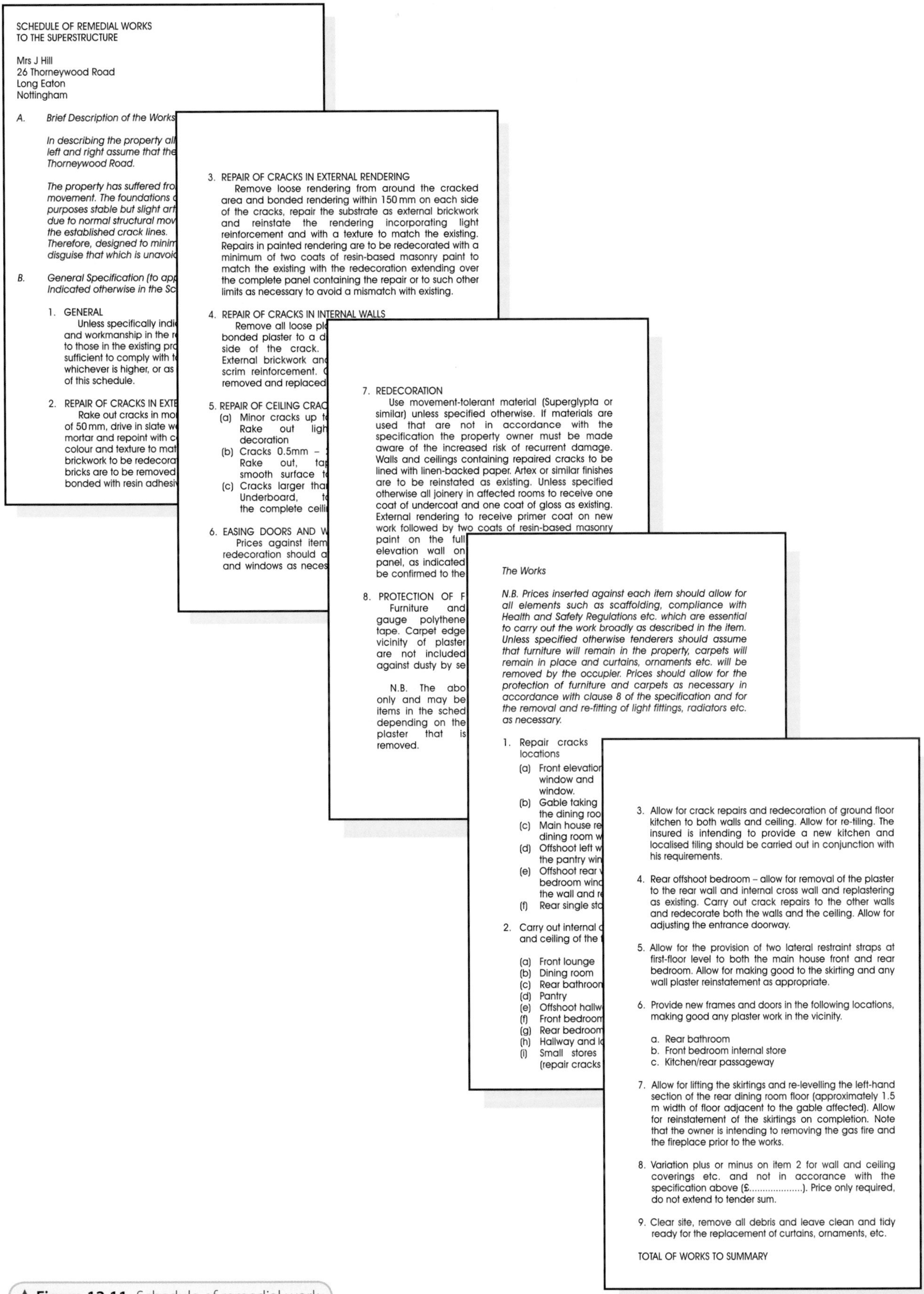

SCHEDULE OF REMEDIAL WORKS
TO THE SUPERSTRUCTURE

Mrs J Hill
26 Thorneywood Road
Long Eaton
Nottingham

A.   *Brief Description of the Works*

*In describing the property all
left and right assume that the
Thorneywood Road.*

*The property has suffered fro
movement. The foundations
purposes stable but slight art
due to normal structural mov
the established crack lines.
Therefore, designed to minim
disguise that which is unavoi*

B.   *General Specification (to app
Indicated otherwise in the Sc*

1.   GENERAL
        Unless specifically indic
and workmanship in the re
to those in the existing pro
sufficient to comply with t
whichever is higher, or as
of this schedule.

2.   REPAIR OF CRACKS IN EXTE
        Rake out cracks in mo
of 50 mm, drive in slate w
mortar and repoint with c
colour and texture to mat
brickwork to be redecora
bricks are to be removed
bonded with resin adhesiv

3.   REPAIR OF CRACKS IN EXTERNAL RENDERING
        Remove loose rendering from around the cracked
area and bonded rendering within 150 mm on each side
of the cracks, repair the substrate as external brickwork
and reinstate the rendering incorporating light
reinforcement and with a texture to match the existing.
Repairs in painted rendering are to be redecorated with a
minimum of two coats of resin-based masonry paint to
match the existing with the redecoration extending over
the complete panel containing the repair or to such other
limits as necessary to avoid a mismatch with existing.

4.   REPAIR OF CRACKS IN INTERNAL WALLS
        Remove all loose pl
bonded plaster to a d
side of the crack.
External brickwork an
scrim reinforcement. C
removed and replaced

5.   REPAIR OF CEILING CRAC
     (a)   Minor cracks up t
              Rake   out   ligh
              decoration
     (b)   Cracks 0.5mm –
              Rake   out,   ta
              smooth surface t
     (c)   Cracks larger than
              Underboard,   te
              the complete ceili

6.   EASING DOORS AND W
        Prices against item
redecoration should a
and windows as neces

7.   REDECORATION
        Use movement-tolerant material (Superglypta or
similar) unless specified otherwise. If materials are
used that are not in accordance with the
specification the property owner must be made
aware of the increased risk of recurrent damage.
Walls and ceilings containing repaired cracks to be
lined with linen-backed paper. Artex or similar finishes
are to be reinstated as existing. Unless specified
otherwise all joinery in affected rooms to receive one
coat of undercoat and one coat of gloss as existing.
External rendering to receive primer coat on new
work followed by two coats of resin-based masonry
paint on the full
elevation wall on
panel, as indicated
be confirmed to the

8.   PROTECTION OF F
        Furniture   and
gauge polythene
tape. Carpet edge
vicinity of plaster
are not included
against dusty by se

     N.B.   The abo
only and may be
items in the sched
depending on the
plaster   that   is
removed.

The Works

*N.B. Prices inserted against each item should allow for
all elements such as scaffolding, compliance with
Health and Safety Regulations etc. which are essential
to carry out the work broadly as described in the item.
Unless specified otherwise tenderers should assume
that furniture will remain in the property, carpets will
remain in place and curtains, ornaments etc. will be
removed by the occupier. Prices should allow for the
protection of furniture and carpets as necessary in
accordance with clause 8 of the specification and for
the removal and re-fitting of light fittings, radiators etc.
as necessary.*

1.   Repair cracks
     locations
     (a)   Front elevation
              window and
              window.
     (b)   Gable taking
              the dining roo
     (c)   Main house re
              dining room w
     (d)   Offshoot left w
              the pantry win
     (e)   Offshoot rear
              bedroom wind
              the wall and r
     (f)   Rear single st

2.   Carry out internal
     and ceiling of the

     (a)   Front lounge
     (b)   Dining room
     (c)   Rear bathroom
     (d)   Pantry
     (e)   Offshoot hallw
     (f)   Front bedroom
     (g)   Rear bedroom
     (h)   Hallway and l
     (i)   Small stores
              (repair cracks

3.   Allow for crack repairs and redecoration of ground floor
     kitchen to both walls and ceiling. Allow for re-tiling. The
     insured is intending to provide a new kitchen and
     localised tiling should be carried out in conjunction with
     his requirements.

4.   Rear offshoot bedroom – allow for removal of the plaster
     to the rear wall and internal cross wall and replastering
     as existing. Carry out crack repairs to the other walls
     and redecorate both the walls and the ceiling. Allow for
     adjusting the entrance doorway.

5.   Allow for the provision of two lateral restraint straps at
     first-floor level to both the main house front and rear
     bedroom. Allow for making good to the skirting and any
     wall plaster reinstatement as appropriate.

6.   Provide new frames and doors in the following locations,
     making good any plaster work in the vicinity.

     a.   Rear bathroom
     b.   Front bedroom internal store
     c.   Kitchen/rear passageway

7.   Allow for lifting the skirtings and re-levelling the left-hand
     section of the rear dining room floor (approximately 1.5
     m width of floor adjacent to the gable affected). Allow
     for reinstatement of the skirtings on completion. Note
     that the owner is intending to removing the gas fire and
     the fireplace prior to the works.

8.   Variation plus or minus on item 2 for wall and ceiling
     coverings etc. and not in accordance with the
     specification above (£...................). Price only required,
     do not extend to tender sum.

9.   Clear site, remove all debris and leave clean and tidy
     ready for the replacement of curtains, ornaments, etc.

TOTAL OF WORKS TO SUMMARY

▲ **Figure 12.11** Schedule of remedial work

Table 12.1 Internal defects schedule

Location	Defect	Possible cause	Remedial action
Lounge	• Flight holes in floorboards	• Woodworm	• Expose under-floor space and check remainder of house to determine the extent of attack, then rectify
Dining room	• Uneven floor	• Possible structural movement (subsidence) or fungal attack	• Consult structural engineer • Expose under-floor space to determine extent, then rectify
Lower hall	• Entrance door sagging	• Joints failed	• Dismantle and re-assemble door using external WBP (weather- and boil-proof) adhesive
Stairs	• Creaking treads • Gap between wall and string	• Shrinkage between tread and riser, glue blocks loose • Moisture movement and/or failure of fixing	• Screw treads to risers. Re-fix glue blocks • Re-fix string, cover gap with decorative moulding
Kitchen	• External door rotten • Cracked tiling	• Wet rot • Movement or accident	• Replace with new door • Replace tiles
Bathroom/larder	• Doors will not close	• Door twisted • Hinge bound • Defective ironmongery	• Ease rebates or replace door • Scrape off paint, pack out hinge • Adjust or replace tiles
Bedroom 1	• Cracks in plaster • Door will not close	• Structural movement/shrinkage • As above	• Consult structural engineer • Cut out and repair • As above
Bedroom 2	• Damp patches to ceiling and upper wall	• Condensation • Defective slates • Deflective flashing	• Provide roof space ventilation • Replace • Replace
Bedroom 3	• Decayed window	• Wet rot due to lack of repainting	• Replace window

External survey

Sketch outline elevations of the building and add any defects found. Photographs of the elevation may be taken as a back up to your sketches, especially where intricate details have to be replaced. Often defects found on the internal survey are a result of external defects. These should be your starting point. As an example, a damp mouldy patch in the corner of a kitchen might be the result of surface condensation due to poor ventilation. Alternatively, it may be due to a leaking rainwater gutter or downpipe. Binoculars are useful for viewing higher levels of a building; closer observation might involve the use of a ladder or the erection of a scaffold. Figure 12.13 shows a typical set of external survey sketches produced during the inspection of the early 1900s semi-detached house.

The focus of attention when carrying out the external survey should include the following points:

Walls
■ Signs of structural movement: cracks in brickwork joints, rendering and missing pointing.
■ Staining: particularly just below the roof eaves, above the DPC and behind rainwater downpipes.
■ Height of DPC above ground level: should be a minimum of 150 mm.
■ Air bricks: ensure they are clear and not blocked by overgrown vegetation.

Windows and doors
■ Condition of woodwork: look out for poorly fitting doors and casements.
■ Condition of paintwork: cracked paintwork at joints will allow water penetration and may lead to wet rot. The easy insertion of a penknife or bradawl may confirm your suspicions.
■ Condition and operation of ironmongery.
■ Condition of glass, putty and glazing beads.

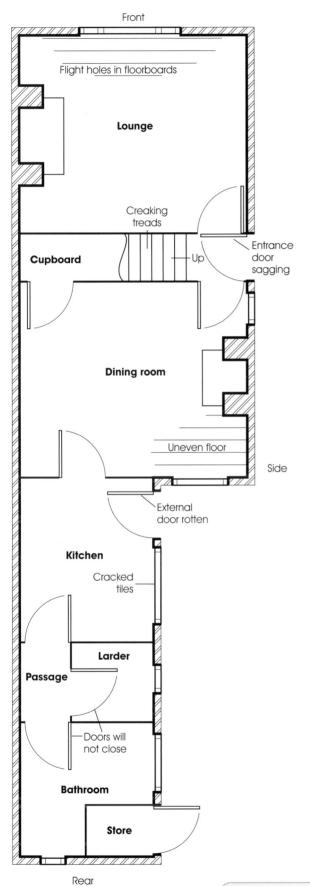

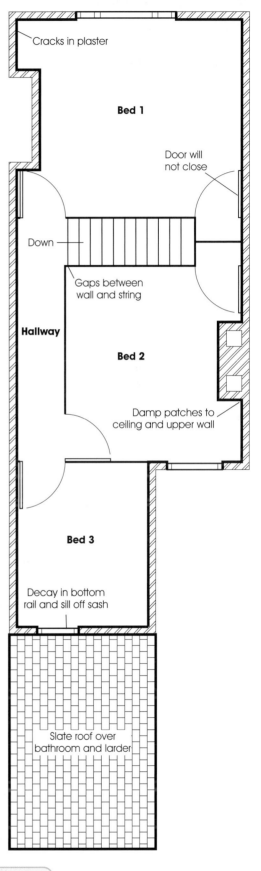

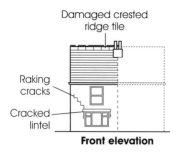

Damaged crested ridge tile

Raking cracks

Cracked lintel

Front elevation

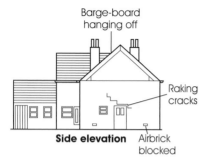

Barge-board hanging off

Raking cracks

Side elevation

Airbrick blocked

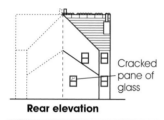

Cracked pane of glass

Rear elevation

▲ **Figure 12.13** External survey sketches showing defects

Roofs

■ Missing, displaced or damaged tiles, slates and flat roof coverings.

■ Missing, displaced or damaged flaunching, flashings, valley gutters and verges.

■ Bargeboards, fascia and soffits – condition of paintwork: look out for signs of decay and distortion. Also check soffits for the presence of any ventilation gaps or grills, these may require cleaning as collected debris or repeated painting may have blocked them.

Guttering, downpipes and drains

■ Check joints are sealed: look out for signs of water staining and moss growth.

■ Feel behind cast iron downpipes for corrosion damage due to lack of paint protection.

■ Check all brackets and clips are secure.

■ Check drain gullies are clear: look out for signs of them discharging water over their edges onto a path or house wall.

■ Check drain gullies are retaining their water seal. If no water is seen in the 'U' bend, it may be cracked or broken and discharging water to undermine the foundations and also causing dampness.

Outside areas

■ Check condition of garden walls, paths and driveways: look out for signs of structural movement.

■ Wooden fences, post and gates: look out for signs of decay. They are particularly susceptible at ground level and joints, where water can be retained.

■ Note position of trees and other large plants.

Table 12.2 shows the results of the external survey in schedule form.

Repairs to timber

When considering timber repairs, a great deal of judgement and negotiation with the client is often required. Can it be repaired cost-effectively or is it cheaper in the long run to replace it? Each job is considered on its own merits, with cost against future service life being the main consideration.

Always wear a dust mask and eye protection when working on, rubbing down or scraping off old paintwork. Surfaces painted prior to the 1960s may contain harmful lead within the paint. In these circumstances, rub down using a wet process (with wet and dry paper) to minimise the potential risks.

Table 12.2 External defects schedule

Location	Defect	Possible cause	Remedial action
Front elevation	• Ridge tile damaged • Cracking to ground floor lintel and raking cracks above	• Wind damage/uncertain • Structural movement	• Replace and make good • Consult structural engineer
Side elevation	• Air bricks blocked • Raking cracks above entrance door • Bargeboard hanging off	• Build up of dirt, soil and vegetation • Structural movement • Fixings failed	• Clean out, reduce ground level to at least 150 mm below DPC • Consult structural engineer • Re-fix and make good slates if required
Rear elevation	• Cracked pane of glass to bathroom	• Accidental/unknown	• Re-glaze window
Outbuildings and structures	(Not viewed)		

Doors

There are many defects associated with doors. Remedial action will depend on the type and location and may range from a simple adjustment through to complete replacement. Table 12.3 and Figure 12.14 cover the most common door defects, the causes and the recommended remedial action. In all but minor cases, consider/discuss with clients the possibility of renewing the door.

Table 12.3 Door defects

Defect	Sketch	Causes	Remedial action
Will not close, door sticking at head, sill or stile	Stile hits against jamb	• Build-up of paint over time or swelling due to intake of moisture	• Ease (plane) edge or top of door, remove extra material to allow for clearance; repaint/seal exposed edges on completion
Large gaps between door and frame	Large gap	• Shrinkage due to reduction in moisture content	• Add lipping to one or preferably both edges of the door as these should be at least 10 mm thick. It may be necessary to further reduce the door width before fitting them; repaint/seal exposed edges on completion
Sagging, tapered gap at top of door; may be touching floor	Large gap at top / Touching floor	• Uneven settlement (downwards movement) of hanging stile, wall and door frame • Loose joints in framed doors	• Remove top bevel, lift up door to fit frame; add new piece to the bottom of the door • Dismantle door, re-glue joints and reassemble square
Hinge bound or binding door springs or resists as it is being opened and closed	Paint buildup / Protuding screwheads / Gap / Distorted stile / Hinge recess too deep	• Hinge recess cut too deep • Build-up of paint over hinges • Protruding screw heads • Distorted hanging stile	• Pack out hinge recess with a piece of thin card (can be cut from screw box) • Scrape off paint build-up • Remove screws; plug holes and replace with smaller-headed screws • Small distortions can be corrected by adding an extra hinge in the centre, otherwise renew door
Twisted door, does not close evenly to stop	Not closing at bottom	• Door distorted often due to irregular grained timber and variable moisture conditions particularly if not completely painted/sealed • Door frame fixed out of line	• For small gaps, stops can be adjusted or rebates eased to mask situation, otherwise renew door • As for small gaps or re-fix frame in-line

continued

Table 12.3 continued

Defect	Sketch	Causes	Remedial action
Decayed or damaged	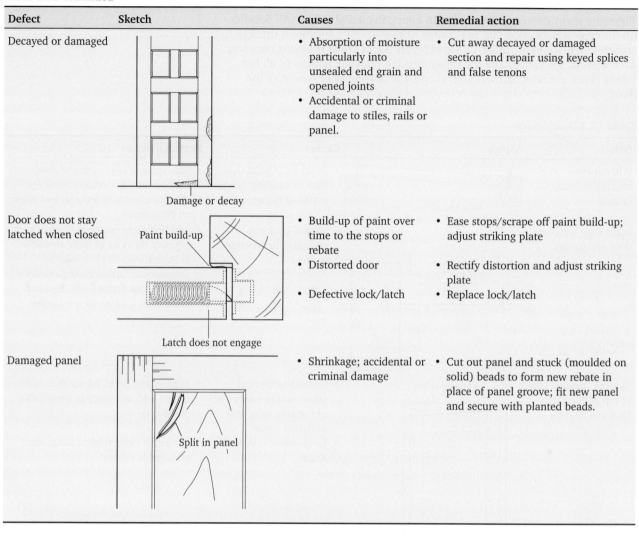	• Absorption of moisture particularly into unsealed end grain and opened joints • Accidental or criminal damage to stiles, rails or panel.	• Cut away decayed or damaged section and repair using keyed splices and false tenons
Door does not stay latched when closed		• Build-up of paint over time to the stops or rebate • Distorted door • Defective lock/latch	• Ease stops/scrape off paint build-up; adjust striking plate • Rectify distortion and adjust striking plate • Replace lock/latch
Damaged panel		• Shrinkage; accidental or criminal damage	• Cut out panel and stuck (moulded on solid) beads to form new rebate in place of panel groove; fit new panel and secure with planted beads.

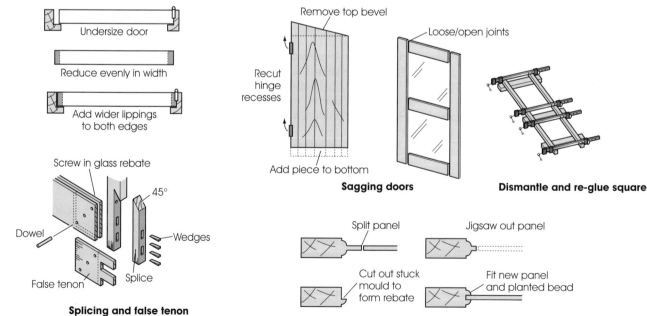

▲ **Figure 12.14** Remedial treatment to doors

Door frames

Defects to external door frames can normally be attributed to either wet rot to the lower end of the jambs or breakout damage in the lock striking plate area as a result of an attempt to force the door. Both of these can normally be resolved by splicing in new timber to the area of damage, see Figure 12.15. In the worst cases a new frame may be required. However, this option will cause the most disturbances to the internal plasterwork and decoration.

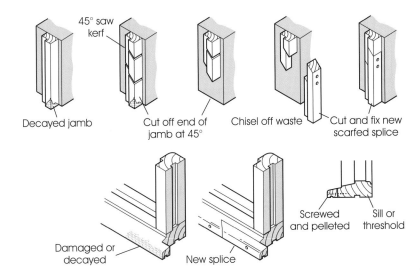

◀ **Figure 12.15** Repairs to door frames and window frames

Decayed jamb — 45° saw kerf — Cut off end of jamb at 45° — Chisel off waste — Cut and fix new scarfed splice

Damaged or decayed — New splice — Screwed and pelleted — Sill or threshold

Windows

Defects to wooden window frames are similar to those of doors and door frames. They will either be associated with poorly fitting opening parts (sashes and casements) or decayed/damaged frames.

These can be rectified using the methods previously outlined for doors; scraping off the paint build-up, easing leading edges, dismantling and re-assembly of sagging casements and splicing of new timber to decayed or damaged areas. In the worst cases, the installation of a new window should be considered.

Boxed-frame sash windows

This type of window is the traditional pattern of sliding sashes and for many years has been superseded by casements and solid-frame sash windows. This was mainly due to the high manufacturing and assembly costs of the large number of component parts. An understanding of their construction and operation is essential, as they will be met with frequently in renovation and maintenance work.

The double-hung boxed window consists of two sliding sashes suspended on cords that run over pulleys and are attached to counter-balanced weights inside the boxed frame.

Figure 12.16 consists of an elevation, horizontal and vertical section of a boxed-frame sliding sash window. It shows the make-up of this type of window and names the component parts.

Re-cording sashes

The maintenance carpenter is often called upon to renew a broken sash cord (see Figure 12.17). It is good practice to renew all four cords at the same time since the remaining old cords will be liable to break in the near future.

> **Related information**
>
> Also see Chapters 5, 6 and 7, which cover stairs, windows and doors in depth.

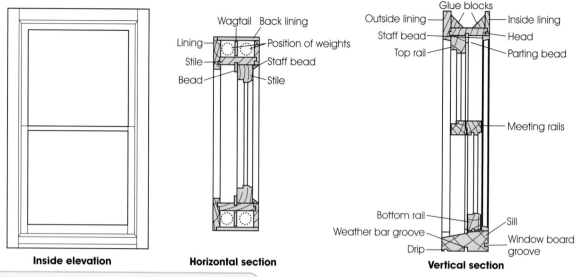

Inside elevation   **Horizontal section**   **Vertical section**

▲ **Figure 12.16** Boxed-frame sliding sash window details

The sequence of operations for renewing sash cords is as follows:

1. Carefully remove staff beads.
2. Carefully remove pockets (access pieces cut towards the bottom of pulley stiles).
3. Take out bottom sash. The sash cords should be wedged at the pulley and removed from the groove in the sash.
4. Carefully remove parting beads. Break paint joints first and carefully prise out with a chisel.
5. Take out top sash in a similar manner to the bottom sash.
6. Remove the weights and cords through the pockets. The wagtail will move to one side to give access to the outside weights. Note: the weights may not all be the same, so ensure they are returned to their original positions.
7. Thread new cords over pulleys and down to the pockets. A 'mouse' can be used to thread the cords easily. A 'mouse' is a small lead weight that is attached to a 2 m length of string that in turn is tied to the cord. The mouse is inserted over the pulley and drops to the bottom of the frame. The sash cord can now be pulled through. Many carpenters use a length of small chain instead of a mouse.
8. Fasten sash cords to weights. To obtain the length of cord for the top sash, rest the sash on the sill and mark on the pulley stile the end of the sash cord groove. Pull the weight up to almost the top and cut the cord to the position marked on the pulley stile. Wedge the cord in the pulley to prevent the weight from dropping. To obtain the length of cord for the bottom sash, place the sash up against the head of the frame and mark on the pulley stile the end of the sash cord groove. With the weight just clearing the bottom of the frame cut the cord to the position marked on the pulley stile. Wedge the cord in the pulley.
9. Fix sash cords to the top sash and insert the sash into the frame. The cords are normally attached to the sashes by nailing them into the cord grooves. Alternatively, the cord can pass through a closed groove and end in a knot.
10. Replace the parting beads. Where these have been damaged new ones should be used.
11. Fix the sash cords to the bottom sash and insert the sash into the frame.
12. Replace the staff beads and check the window for ease of operation. Candle wax can be applied to the pulley stiles and beads to assist smooth operation.

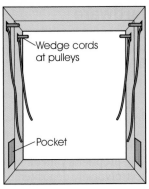

Working from inside, wedge cords and remove sashes

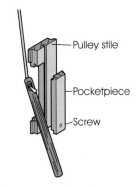

Remove cords and weights through pockets

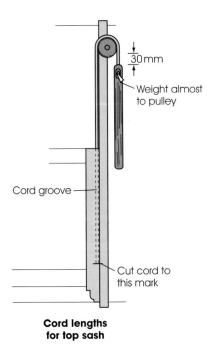

Cord lengths for top sash

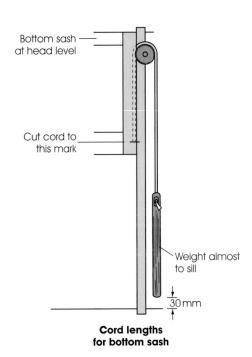

Cord lengths for bottom sash

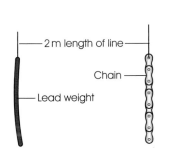

Types of mouse

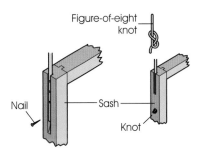

Cord nailed or knotted

▲ **Figure 12.17** Re-cording sashes

Door and window hardware

The moving parts of locks, latches, bolts and hinges require regular lubrication and cleaning and also need to be kept free from paint build-up in order to ensure their trouble-free operation.

Door furniture

Door furniture can be ruined by incorrect cleaning. It is important to ensure proper care is taken to keep them clean. Dust, contamination and moisture are the main hazards that can affect door furniture. Irreparable damage can be caused to the surface by using proprietary metal polishes, harsh abrasive cleaners or emery cloths on electrolytically deposited or special surface finishes. The correct cleaning procedures for the most common finishes are:

- **Anodised aluminium finishes** should be dusted regularly, washed periodically with weak detergent solutions and occasionally wiped with wax polish.
- **Stainless steel finishes** should be dusted regularly and occasionally washed with soap and water.
- **Plastic products** require only wiping with a damp cloth. Strong sunlight can cause surface deterioration that cannot be easily rectified.
- **Powder-coated finishes** should be cleaned with a soft cloth and any household furniture polish. Industrial solvents must not be used as these can destroy the finish.
- **Lacquered finishes** (polished brass and so on) should be cleaned occasionally with good-quality wax-free polish. Abrasives and metal polishes should not be used as these will remove the lacquer.

Hinges

A door that is not properly hinged can result in sagging, which can also have an effect on the lock side of the door and frame. An adequate number of hinges are, therefore, always recommended. External doors, heavy internal doors and doors in areas of high humidity should have three hinges. Heavy good-quality hinges should always be used on doors where high-frequency service is expected.

Tips on the fixing and maintenance of hinges:

- The leaf with the largest number of knuckles should be fixed to the door frame.
- The hinge pins must be in true alignment.
- The hinge recesses must be equal in depth, square and plumb.
- Check that all screws have been fitted, ensure they are tight and of the correct size.
- Ensure that metal hinges are lubricated with oil three times a year.

Locks and latches

The correct installation of locks and latches is paramount to their smooth operation. Alignment of the strike plate is essential to ensure correct location of both the latch bolt and deadbolt. Lock and latch cases that are mortised into the door too tightly will result in jamming of the key and bolt.

Tips on the fixing and maintenance of locks and latches:

- Check that all screws have been fitted and ensure they are tight.
- Check the alignment of lever furniture and keyhole plates.
- Check the springing on lever furniture and doorknobs.
- Apply graphite to the cylinder if the keys have a tendency to stick in the lock.
- Check the mortising of the lock case and the alignment of the striking plate, where the latch or deadbolt either sticks or will not engage.

Other items, such as door closers, floor springs, panic bars and so on, should be checked and maintained in accordance with the specific manufacturer's instructions. Periodically check:

- that all screw fixings are sufficiently tight
- the operation of the item is correct and smooth
- adjust and lubricate in accordance with the instructions.

Replacement of hardware

Inevitably, items will eventually begin to wear and require replacement. In general these should be replaced with like for like or the nearest alternative. However, proposed replacement does provide the opportunity to upgrade hardware, particularly locks for increased security.

When replacing or upgrading hardware it is often necessary to make good holes and recesses. Figure 12.18 shows a typical situation. In general all follow the same procedure:

1. Cut oversize filling pieces of a similar material. These are often bevelled to provide a key.
2. Mark and cut out enlarged hole and recesses to receive filling pieces.
3. Glue and pin filling pieces in place.
4. Punch in pins, plane off flush, fill pinholes with wood filler, sand off smooth and repaint or seal the area.

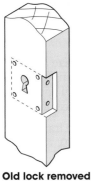

Old lock removed

May be bevelled to provide key

Filling pieces cut, holes and recesses enlarged

◄ **Figure 12.18** Making good holes and recesses

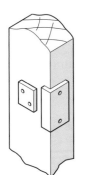

Oversize filling pieces glued and pinned in place

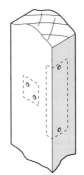

Filling pieces planed off flush, pin holes filled and sanded off flush

Replacement of door and window frames

When extensive repairs are required it is often more cost-effective in the long run to consider a complete replacement. This should be discussed with the building owner. It should be emphasised that although the replacement may be more expensive than the repair initially, the replacement would last a lot longer and also provide the opportunity of upgrading fittings and modification to suit new requirements.

Before removing frames, always check the wall above for signs of support, see Figure 12.19. Newer buildings may have a concrete or steel lintel, older properties a stone lintel, brick arch, or soldier course or sometimes none at all apart from the frame. Also look out for cracks in the brickwork mortar joints above the opening, which could be indications of structural movement, possibly leading to collapse on removing the frame.

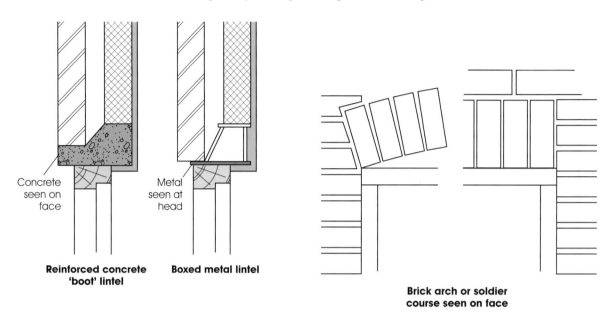

Concrete seen on face

Metal seen at head

Reinforced concrete 'boot' lintel

Boxed metal lintel

Brick arch or soldier course seen on face

▲ **Figure 12.19** Means of support over openings

Do not proceed if a means of support is not evident or there are signs of movement. Seek the advice of a structural expert, as arrangements may have to be made for temporary support, the insertion of a lintel and structural repairs to be carried out by others before the frame replacement is made.

Once you are sure the opening is correctly supported, the frame can be cut out and removed in sections in sequence, see Figure 12.20.

When the frame has been removed, clean off any projecting mortar and previous fixings. Make good any damaged brickwork and holes previously occupied by 'built-in' horn fixings. Finally, fix in position the new frame.

▶ **Figure 12.20** Removing a defective frame

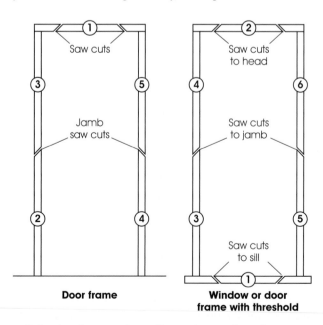

Saw cuts

① ②

Saw cuts to head

③ ⑤ ④ ⑥

Jamb saw cuts

Saw cuts to jamb

② ④ ③ ⑤

Saw cuts to sill

①

Door frame

Window or door frame with threshold

Cut out and remove in sections using numbered sequence

Repairing floors and roofs

Repairs to floor joists and rafters will require the localised or sometimes the total stripping out of floor boarding, ceilings and the tiles and felt, in order to gain access. A proper working platform is required when undertaking work on a roof; working off a ladder is not considered appropriate and may contravene the Working at Height Regulations. Employers should undertake a risk assessment before anyone starts to work at height.

If the defect is the result of fungal decay or wood boring insect damage, the procedures covered in Book 1 should be adopted. In these circumstances the details shown in the previous chapters can be used to replace the components on a like-for-like basis.

Defective joists and rafters in ground, upper floors and roofs can be repaired on a localised basis by cutting away the affected timber and bolting a new piece of timber alongside. Any new member inserted in this way should normally extend to another load-bearing member, such as a wall plate, binder, purlin and so on, as shown in Figure 12.21.

Related information

Also see Chapters 1 and 2, which cover floor and roof construction in depth.

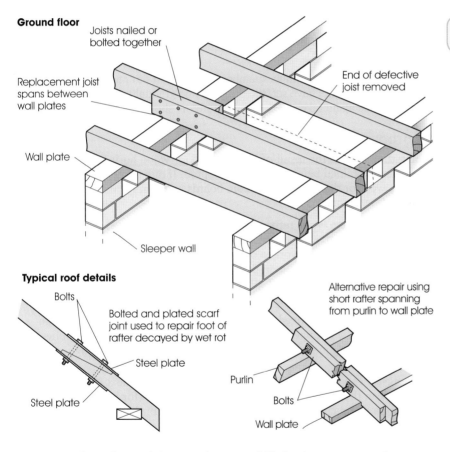

Figure 12.21 Replacing defective structural members

Ground floor
- Joists nailed or bolted together
- Replacement joist spans between wall plates
- End of defective joist removed
- Wall plate
- Sleeper wall

Typical roof details
- Bolts
- Bolted and plated scarf joint used to repair foot of rafter decayed by wet rot
- Steel plate
- Steel plate
- Alternative repair using short rafter spanning from purlin to wall plate
- Purlin
- Bolts
- Wall plate

However, where the work is extensive, especially in circumstances where structural timber is affected, such as upper floor joists, rafters, purlins and so on, it is often wiser and more cost-effective to have this work referred to a specialist contractor who will be fully equipped and experienced to undertake it.

When replacing entire upper floor joists and rafters, shoring may be required to provide temporary support. Use adjustable steel props to support floor and ceiling joists at either end; these should be used in conjunction with head and sole plates to spread the load. On hollow ground floors some floorboards will need to be removed so that the props can extend down to the oversite concrete see Figure 12.22. Props should not be over tightened as this may lift the structure causing further damage and cracking to floors, ceilings, roofs and

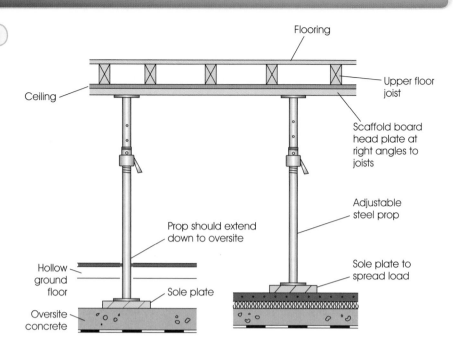

► **Figure 12.22** Shoring to upper floor

walls. Once the affected timber member has been cut out and replaced and any brickwork repairs have set, the props can be gradually eased so that the structure once again takes up the load before the shoring is removed.

Floorboards

Where the defect is the result of fungal attack or wood boring insects, the procedure explained before should be adopted. If due to movement (loose fixings), wear or damage, one or more boards can simply be refixed or replaced.

Care must be taken when refixing or replacing floorboards as there is always the possibility of services (water, gas and electricity) running below. The surface of the area to be worked on can be scanned before starting work, with a metal/live electric circuit detector, see Figure 12.23. As an added precaution it is wise to turn off all service supplies at the meter/stop valve before any refixing or cutting out operations commence.

Loose boards are best refixed by screwing down into the joists rather than nailing, especially at upper floor levels in older properties where significant amounts of nailing may damage the lath and plaster ceiling through vibration.

1. A jigsaw can be used to cut the ends of a floorboard next to a joist. The ends can then be chopped back to form a bearing (for heading joints) using a wood chisel, or fix a batten to the edge of the joist.
2. Alternatively, a small circular saw may be used to cut the heading joints over the joists. The blade should be set to the floorboard thickness.
3. A sharp knife, padsaw or jigsaw can be used along the length of the board to separate the tongue.
4. Punch the fixing nails through the board. Insert two wide-blade flooring chisels in one of the edge joints and lever up the board.
5. Where more than one adjacent board is to be replaced the heading joints should be staggered over different joists.
6. Finally, cut the new boards to length and refix in place using a folding technique.

Stairs

The most common defects encountered are shown in Figure 12.24.

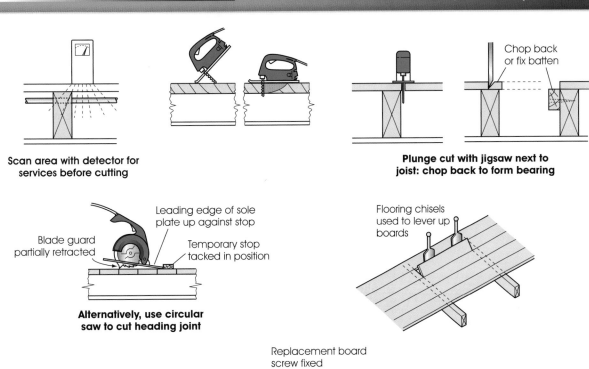

Scan area with detector for services before cutting

Plunge cut with jigsaw next to joist: chop back to form bearing

Chop back or fix batten

Leading edge of sole plate up against stop

Blade guard partially retracted

Temporary stop tacked in position

Alternatively, use circular saw to cut heading joint

Flooring chisels used to lever up boards

Replacement board screw fixed

Tongue removed

▲ **Figure 12.23** Removing and replacing floorboards

▼ **Figure 12.24** Common stair defects

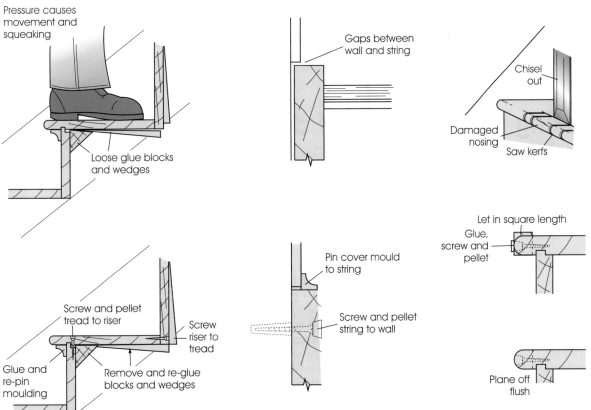

Pressure causes movement and squeaking

Loose glue blocks and wedges

Gaps between wall and string

Chisel out

Damaged nosing

Saw kerfs

Screw and pellet tread to riser

Screw riser to tread

Glue and re-pin moulding

Remove and re-glue blocks and wedges

Pin cover mould to string

Screw and pellet string to wall

Let in square length

Glue, screw and pellet

Plane off flush

413

Creaking treads

This results from movement between the tread and riser joint when walked on.

■ Where access to the underside of the flight is possible (in cases where it has not been plastered over) the creaking can be remedied by renewing the glue blocks at the tread-to-riser junction and re-gluing/re-wedging the treads and risers in their string housings.

■ Where access to the underside is not possible, gluing and screwing the tread down into the top of the riser can remedy the problem. Small gaps, normally the result of shrinkage, may be apparent at the tread-to-string housing. These should be filled by gluing in thin strips of timber veneer.

Gaps between the wall and string

This may be the result of either shrinkage or movement, or a combination of both.

■ Shrinkage gaps can be masked by the application of a cover mould.

■ Movement gaps are normally the result of the fixings between the wall and string becoming loose or failing, causing the string to move away from the wall when using the stairs. Refix the string back to the wall by plugging and screwing. A cover mould can again mask any gap or damage between the string and plasterwork.

Damaged nosings

Cutting out the damaged section and splicing in a new piece can repair these using the following procedure:

1. A square length should be let in. Cut at 45° at either end for additional support and a longer glue line.
2. Fix the let-in piece using glue and screws.
3. Finally, plane and rub down the let-in piece to match the nosing profile.
4. Where a Scotia mould is used at the underside of the tread-to-riser junction, it is best to renew the whole length, gluing and pinning the new one in place.

Glazing and painting of doors and windows

Maintenance of glazing work involving sealed double-glazing units and large window panes is best undertaken by specialist glaziers, who will be kitted up to undertake the work efficiently and safely.

Reglazing

The carpenter and joiner can undertake reglazing of small single-glazed panes. Wherever possible, glass should be pre-cut to size by the glass supplier. Measure the timber rebate sizes and order glass 3 mm undersize in both directions. For example, a piece of glass for a rebate opening size of 150 mm × 250 mm should be ordered as a cut size of 147 mm × 247 mm. If necessary, pieces of glass can be cut to size by scoring along the required line using a glass-cutting wheel. Lay the glass on a flat surface with matchsticks placed under the scored line at either end. Apply pressure on both sides to snap in two along the line. Never attempt to cut narrow strips and always wear eye protection and gauntlets.

The first stage is to remove the broken pane. Where practical, sash or casements should be removed from their frame so that broken glass can be removed and replaced with a new piece with relative safety at ground level.

When replacing broken glass at high level ensure that the area below is cordoned off so that no one can enter. Before starting to hack out the broken pane and hardened putty, ensure you are wearing eye-protection goggles and gauntlets to protect hands, wrists and lower arms. These should be worn during the whole process as shards of glass and fine splinters will inevitably be created as the pane is removed and the rebates cleaned up.

Start at the top of the pane by removing the old putty or glazing beads with a wood chisel or hacking knife. Remove glazing sprigs (flat or square nails) with pliers and then lever out remaining glass from behind again using a wood chisel. Finally, continue hacking out remaining back putty to rebates.

The procedure for reglazing is shown in Figure 12.25.

1. Check glass is correct size.
2. Ensure rebates are primed with paint.
3. Work a bead of putty around the back of the rebate.
4. Position plastic seating blocks in the bottom of the rebate to support the glass.
5. Position the bottom edge of the glass on seating blocks. Gently push the glass into the rebate, applying pressure evenly around the edge until a back putty thickness of 1 mm to 2 mm is achieved.
6. Use glazing sprigs to secure the glass in place. These may be driven in using a pin hammer or the edge of a firmer chisel. (Panel pins should not be used to secure the glass in place, as their round point of contact results in pressure points, which can lead to the formation of cracks.)
7. Work a bead of putty all around the rebate in front of the glass. Only a small bead is required when using glazing beads to secure the glass.

▼ **Figure 12.25** Reglazing procedure

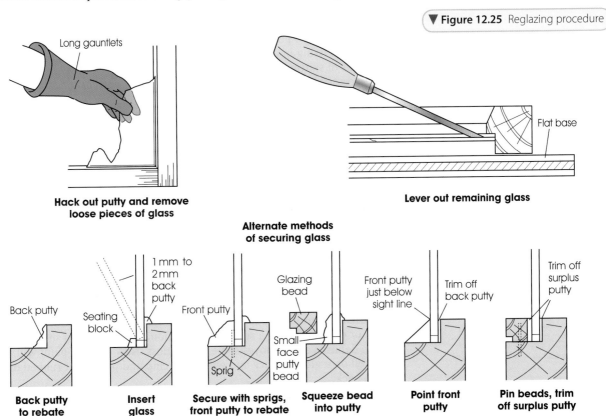

Long gauntlets

Hack out putty and remove loose pieces of glass

Flat base

Lever out remaining glass

Alternate methods of securing glass

Back putty

Seating block

1 mm to 2 mm back putty

Front putty

Sprig

Glazing bead

Small face putty bead

Front putty just below sight line

Trim off back putty

Trim off surplus putty

Back putty to rebate

Insert glass

Secure with sprigs, front putty to rebate

Squeeze bead into putty

Point front putty

Pin beads, trim off surplus putty

8. Replace glazing beads squeezing putty into glass, or point up the front putty bead using a putty knife or chisel to form a bevelled fillet. Its upper edge in contact with the glass should be just below the rebate sight line. Slight imperfections in the putty fillet can be improved by running over with a wet paintbrush.

9. Trim off the surplus putty to beads and then trim off the back bedding putty on the inside of the glass.

10. Clean the glass to remove oil and putty marks before it dries.

Painting woodwork

Woodwork is painted to provide a decorative finish. However, more importantly, it serves also to protect it from the elements. As a carpenter and joiner carrying out maintenance work, you may be required to paint newly replaced items as well as repaint existing items that have been eased or repaired. The paint system for wood normally consists of a primer to seal the surface and provide a bond for later coats, an undercoat that provides a smooth opaque covering coat and finally a decorative gloss or satin topcoat.

Preparation

The key to a successful paint system is careful preparation. Paint will not last long on a defective surface.

Bare wood. Normally only needs an initial rub down with glass paper to remove any roughness and sharp arrises. Knots in bare softwood can be full of resin and may later 'bleed' through to the finished paint surface if not sealed. First wipe over the surface with a cloth soaked in white spirit to remove any stickiness and excess resin. Then coat all knots with a knotting solution. On resinous hardwoods that are to be painted, wipe off excess resin using white spirit and seal the entire surface with an aluminium-based wood primer.

1. Apply wood primer to all surfaces and edges, taking particular care to achieve full penetration of any end grain. This is best undertaken prior to fixing in order to ensure full protection. For example, the backs of skirtings, architraves, doorframes, linings and so on are inaccessible when fixed. This is particularly important for external timber.

2. Fill any defects and open-end grain with a wood filler. Always select a waterproof type for external use. Rub down flush with the surface using glass paper.

3. As priming tends to raise the grain of woodwork resulting in a felt-like hairy surface, the whole job will require rubbing down (de-nibbing) prior to over-painting.

4. Wipe off surface with a 'low tack' cloth to remove any surface dust; apply the undercoat.

5. When the undercoat is dry, apply the topcoat. Refer to the paint manufacturer's information with regard to minimum and maximum over-coating times. If the topcoat is applied too soon, the undercoat will tend to bleed into the topcoat causing defects; too long and it may not bond successfully to the undercoat, resulting in early breakdown and peeling off.

Previously painted wood. If in good condition, lightly rub down with glass paper or clean off using a sugar soap solution. This cleans the surface dirt or grease deposits and removes some of the gloss. New coats of paint will not key well on a gloss surface and will easily chip and peel if not rubbed down or cut back.

1. Knot and prime any eased edges or repairs. When dry, rub down to blend in primer to existing paint surface.

2. Fill and rub down any minor defects and imperfections.

3. Remove surface dust; apply the undercoat followed by the topcoat within the recommended over-coating time.

When the old system has broken down, it is best to completely strip off the old paint, make good and start again from scratch using the same procedure as for bare wood.

Small areas showing signs of deterioration may be repaired without fully stripping the area.

1. Treat any minor areas of soft timber caused by wet rot with a wood hardener.

2. Rub down the surface to remove all loose defective paint. Note: Always wear a dust mask when rubbing down paintwork. Surfaces painted before the 1960s may contain harmful lead within the paint. In these circumstances it is best to rub down using a wet process (wet and dry paper) to minimise the potential risk.

3. Apply the paint system as before.

Painting procedure

Internal painting. Protect carpets and furniture with dustsheets. Doors to other rooms can be sealed with masking tape prior to any rubbing down. Open the window to ensure adequate ventilation. This is both for your own health and to help the paint to dry. Always wear a dust mask when rubbing down and eye-protection goggles when scraping off.

■ Primers and undercoats are applied by brushing out along the grain.

■ Topcoats are initially applied along the grain, brushed out across the grain and finally 'laid off' by finishing with gentle brush strokes along the grain. The aim is to produce a thin, even paint film, which does not 'sag' on vertical surfaces or 'pond' on horizontal ones.

External painting. In general the same procedure as for internal painting can be adopted, avoiding adverse weather conditions and taking extra care to ensure full paint coverage to avoid the possibility of moisture penetration.

■ Do not work in strong sunlight, as this prevents paints drying in the correct way and is likely to cause it to 'blister'. Wait until the area is in shade before painting.

■ Do not paint if rain is expected.

■ Do not paint first thing in the morning or last thing in the evening, when there might be dew. The resulting moisture will spoil the paint finish.

■ Do not paint when there is a risk of frost.

■ Do not use paint intended for interior use only.

It is best to aim to paint from mid morning to just after lunch. This will allow the air to dry before starting and the paint film to dry before the early evening dampness starts to form.

Sequence of operations. A logical approach is required when painting framed joinery. The aim is to keep a 'wet edge' blending in adjacent areas of paint, so that joints are not seen when the paint dries.

Figure 12.26 shows typical numbered sequences for painting doors and windows.

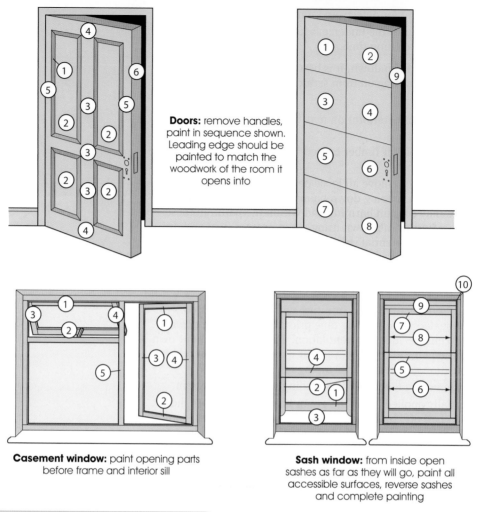

Doors: remove handles, paint in sequence shown. Leading edge should be painted to match the woodwork of the room it opens into

Casement window: paint opening parts before frame and interior sill

Sash window: from inside open sashes as far as they will go, paint all accessible surfaces, reverse sashes and complete painting

▲ **Figure 12.26** Sequence for painting doors and windows

Multi-skill maintenance

Building firms who specialise in maintenance work will often employ or give preference to operatives who possess multiple skills. They will be expected to carry out, in conjunction with their main craft skill, a range of basic skills in other crafts. In addition to painting and glazing of replaced doors and windows covered previously, they may also be expected to undertake minor brickwork, plaster repairs, tiling, other painting tasks and the replacement of gutters and rainwater pipes.

Replacing bricks

The first task is to attempt to match the pattern, size and make of the original. Measure the brick size and take a small piece to the brick supplier for them to identify. Brick sizes vary slightly due to the way they are made. New metric size bricks are a little smaller than the old imperial ones. When working on older property matching bricks may be difficult to obtain. Try suppliers who specialise in reclaimed building materials. If imperial bricks cannot be obtained, matching metric ones can be bonded into the existing work by slightly increasing the mortar joints.

Figure 12.27 shows the procedure to follow when replacing a single brick. In this case the brick previously cut around the built-in horn of a sill or threshold.

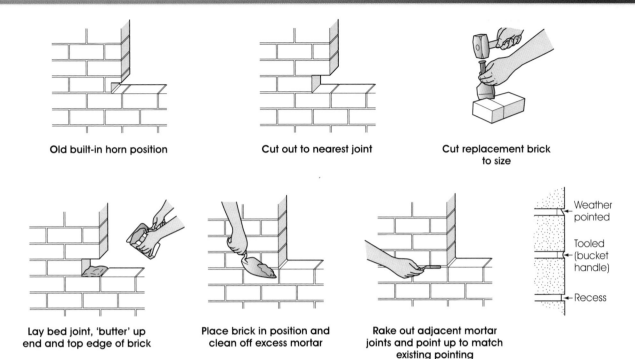

Old built-in horn position   Cut out to nearest joint   Cut replacement brick to size

Lay bed joint, 'butter' up end and top edge of brick   Place brick in position and clean off excess mortar   Rake out adjacent mortar joints and point up to match existing pointing

Weather pointed / Tooled (bucket handle) / Recess

▲ **Figure 12.27** Replacing a brick

1. Using a bolster and club hammer (do not forget to wear eye-protection goggles) cut out the old brick and clean away the old mortar joints. Brush out all dust particles, taking care to ensure nothing is allowed to enter the cavity.
2. Cut the brick to size if required. First gently score a line around the brick using a bolster and club hammer and finally use a heavier blow on the lines to sever it.
3. Prepare a mortar mix, typically 1:6 (one part Portland cement to six parts bricklaying sand). Sufficient water is added so that the mix has the consistency of soft butter (firm enough not to collapse when heaped, but easily compressed with a shovel). One part lime or a mortar plasticiser may be included in the mix for improved workability. Alternatively, pre-packed bricklaying mortar mixes are available. Note: The colour of the sand used should match the existing mortar joints. Red and yellow sands are commonly available.
4. Lay the bed of mortar into the prepared hole.
5. Butter up (apply mortar) to the end and the top edge of the brick. Place it in position and then remove the surplus mortar.
6. Fill any gaps in the mortar joint. When it has started to 'go off', rake the joints out below the brick surface.
7. After about 24 hours rake out the adjacent mortar joints.
8. Repoint the joints with a mortar mix that matches the original in colour and finish.

Repointing

The most common methods:

■ **Weather pointed**, which is done with a pointing trowel.

■ **Tooled**, a concave 'bucket handle' finish created by working along the drying mortar with a special jointing tool or, alternatively, a metal bucket handle (hence the name) or a piece of 15 mm copper pipe may be used.

■ **Recessed**, created by brushing out the drying mortar with a stiff bristle hand brush. Alternatively, a more consistent depth can be achieved by working the joint with a piece of timber having a protruding countersunk screw.

Relaying brick courses

Figure 12.28 shows the procedure to follow when relaying or building whole courses of bricks; in this case to under the sill of a reduced-height replacement window.

1. Cut out bricks at either end to form the bond between the adjacent vertical joints.

▶ Figure 12.28 Relaying brick courses

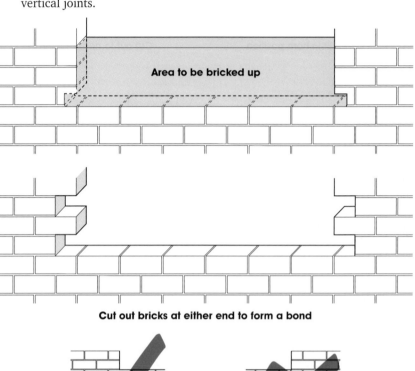

Area to be bricked up

Cut out bricks at either end to form a bond

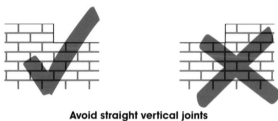

Avoid straight vertical joints

Lay bed mortar, butter up and lay bricks. Remove surplus mortar

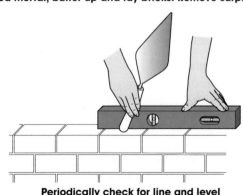

Periodically check for line and level

2. Clean away any old mortar and dust particles.

3. Dry lay the bricks to determine the pattern. Cut the bricks to size if required.

4. Prepare a mortar mix as before, using a 10 mm joint. Approximately 1 kg of mix is required for each brick.

5. Apply a bed of mortar to the existing brick course, approximately 10 mm thick.

6. Furrow the surface to a V-shaped groove with the point of a trowel. Butter up and lay bricks, removing surplus mortar as you go.

7. Repeat the process to lay the subsequent brick courses. Periodically check the bricks are being laid horizontal and in line with a spirit level. Use the end of the trowel to tap the bricks into place if required.

8. Complete the job by raking out the mortar joints and finally pointing them as before.

Repairs to plasterwork

In modern buildings the internal brick and blockwork walls will often have a hard plastered finish. This is normally applied in two layers, a 9 mm to 12 mm backing coat and a 2 mm to 3 mm finishing coat. Ceilings and stud partition walls are surfaced with sheets of plasterboard. These may be finished by a 2 mm to 3 mm coat of board finishing plaster applied onto the plasterboard that acts as the backing. Alternatively, the plasterboard joints may be taped up and filled to provide a dry-lined finish, ready for decoration. Dry-lined ceilings were often decorated using a textured coating and worked to create a repeating pattern or stipple finish.

In older buildings, the wall plaster may be much softer, but still normally two coats: a thick lime-based backing coat followed by a thin finishing coat. Ceilings and stud partition walls were then finished using **lath and plaster**. This is a system using thin timber laths nailed to the undersides of joists and faces of studs. Wet plaster was pressed up against them and allowed to squeeze between the gaps in the laths forming a key to hold this backing plaster in place. This was finished using a thin coat of finishing plaster.

The external walls of both modern and older buildings may be covered in **rendering**. This is a surface coat of sand and cement mortar applied to a wall for decorative and/or waterproofing purposes.

All of these finishes may crack due to structural movement, damage by accidental impact or disturbance during renovation work. They may therefore require repairs that the maintenance carpenter and joiner may be asked to undertake.

Patching plasterwork

The first thing to do when patching plasterwork on any background is to protect the floor by covering with a dustsheet. Figure 12.29 shows the procedure to follow when patching a 'blown' or damaged area of plasterwork to a brick or blockwork wall.

1. Tap plaster around the damaged area to 'see' (hear) if any part sounds hollow or loose.

2. Use a club hammer and bolster to hack off all existing loose plaster until the surface is sound.

3. Brush down the surface to remove all loose particles and dust, using a stiff bristle or wire brush.

4. Damp down the wall surface by brushing or spraying with water. This prevents the wall 'suction' from drying out the plaster too quickly, which could result in cracking on drying. Some surfaces such as concrete, shiny or glazed bricks and impervious engineering bricks do not help the plaster to stick. In these cases brush on a PVC bonding agent before plastering to ensure good adhesion.

Hack off existing
loose plaster

Brush off to remove
dust and loose particles

Damp down
with water

Mix plaster
in a bucket

Scoop up plaster from
hawk and apply to wall
in an upward sweep

Reinforce large
areas with repair
mesh or scrim

Comb or scratch area
to provide a key

Brush off and apply
finishing plaster

Rule off using a sideways
sawing action,
working upwards

Trowel up to a
smooth finish

Repeat trowelling up while
splashing with water

▲ **Figure 12.29** Patching plasterwork

5. Add a small amount of backing plaster into clean cold water in a bucket. Stir with a timber stick until a thick creamy consistency is achieved.

6. Transfer some of the plaster to the hawk. With the hawk tilted away from the wall scoop up a small amount of plaster on the edge of the steel trowel and press the plaster against the wall using an upward sweep of the trowel. The trowel should be used at an angle to the wall, with the angle reduced as you sweep it up the wall.

7. Take care not to allow the trowel to lay flat against the wall, as the suction will pull the fresh plaster off the wall.

8. Continue adding plaster to the wall until the whole area to be repaired is covered to within 2 mm to 3 mm below the surrounding wall finish. Pushing a repair mesh or scrim into the backing coat will reinforce larger areas.

9. Before the backing plaster is completely set, scratch the surface with a comb. This provides a key to help with the adhesion of the finishing coat.

10. After about 3 to 4 hours the backing coat surface will be hard, but not dry. It is then ready for finishing. If allowed to dry further, it will require damping down again with water before finishing.

11. Brush down the surface to remove any loose particles. Mix up a small batch of finishing plaster by adding to a little water as before, except this time it should be a runnier consistency. Trowel on the plaster, aiming to leave it slightly proud of the surrounding area.

12. Rule flat the surface using a timber or metal straightedge. Start at the bottom of the patch, move it up the wall with a side-to-side sawing action keeping it tight against the existing sound plaster as a guide. Trowel on more plaster to fill any hollows before ruling off again.

13. The plaster will start to set within 30 to 45 minutes. At this stage smooth the surface with a plastering trowel, again using upward sweeps with the trowel held at an angle. After 15 or so minutes lightly splash the surface

with clean cold water, while trowelling up and over the surface, to provide a smooth hard finish. Ensure the trowel is kept clean and damp during this process to prevent damaging the newly plastered surface. Regular brushing off in a bucket of cold water is ideal.

14. Small patches in rendering can be repaired in one or two coats using the above procedure, except that a sand and cement mix is used in place of plaster.

Patching damage to lath and plaster surfaces

Damage to lath and plaster partitions (see Figure 12.30) can be repaired using the following procedure:

1. Tap the surface to determine the extent of any loose plaster.
2. Score a deep knife line around the perimeter of the loose or blown area, remove the loose plaster and wire brush both the laths and the studs.
3. Damp down the laths and edges of the surrounding plaster and apply a coat of backing plaster to within 2 mm to 3 mm of the surrounding plaster. Sufficient pressure must be applied to force the plaster between the laths while still leaving a layer of plaster over them.
4. When the backing layer is partially set, scratch the surface to provide a key and leave to set further without fully drying out.
5. Finish the repair using steps 10 to 13 of the previous procedure.

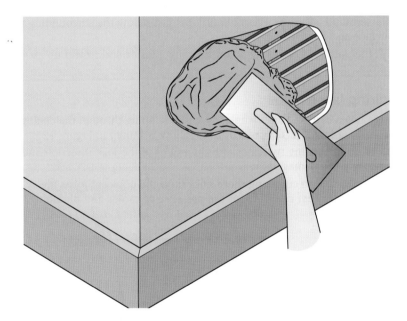

Figure 12.30 Hole in a lath and plaster partition

Patching damage to plasterboard

The procedure to repair small holes caused, for example, by striking the surface with the corner of a piece of furniture is shown in Figure 12.31.

1. Neaten up the jagged edges using a sharp knife or a pad saw.
2. Cut a strip of plasterboard about one and a half times the length of the neatened hole and just narrower in width. Make a hole in its centre, pass through a piece of string and knot it behind on a nail. (Remember, plasterboard is simply cut by scoring the face with a sharp knife and breaking along the line by applying pressure along from the scored side. Run the knife along the paper on the other side to separate.)
3. Mix up some plasterboard adhesive and apply to both ends of the strip. Feed the strip into the hole, using the string to pull it tight against the inner face of the board. Tie off the string to a scrap of timber positioned over the face of the hole.

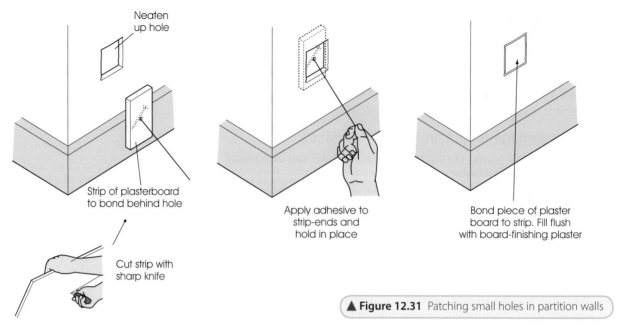

Neaten up hole

Strip of plasterboard to bond behind hole

Cut strip with sharp knife

Apply adhesive to strip-ends and hold in place

Bond piece of plaster board to strip. Fill flush with board-finishing plaster

▲ **Figure 12.31** Patching small holes in partition walls

4. When the adhesive has set, cut off the string. Cut another piece of plasterboard, this time to fit the hole and again bond in place using plasterboard adhesive. Press in until it is just below the surrounding wall surface and leave to set.

5. Fill the patched hole using board-finishing plaster and trowel up as before.

Repairing large holes

For large holes caused, for example, by a foot slipping through the ceiling when working in a loft, the procedure to follow is shown in Figure 12.32.

1. Mark on the ceiling two lines at right angles to the joist direction and enclosing the damaged area.

2. Use a pad saw to cut along these lines until the adjacent joists are reached after first checking for the presence of cables and plumbing.

▼ **Figure 12.32** Repairing large holes in plasterboard

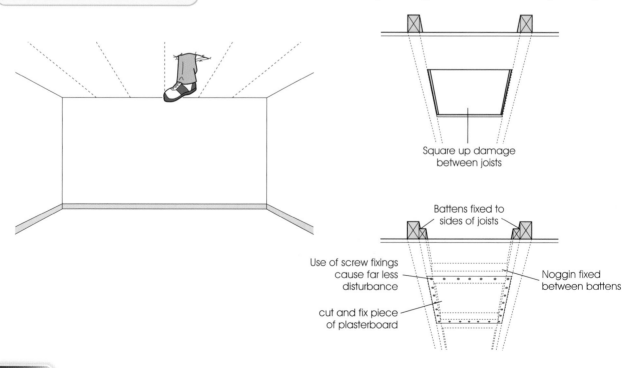

Square up damage between joists

Battens fixed to sides of joists

Use of screw fixings cause far less disturbance

Noggin fixed between battens

cut and fix piece of plasterboard

3. Again, using the pad saw, cut along the joist edges. Remove the damaged area, leaving a neat rectangular hole.

4. Cut battens; fix to the sides of the joists.

5. Cut and fix noggins between the battens at either end of the hole, ensuring the noggins' centre lines straddle the cut line to provide a bearing for both the existing sound ceiling and new plasterboard.

6. Cut a piece of plasterboard 2 mm to 3 mm smaller than the hole in both directions. Fix in place into the noggins and battens using plasterboard nails or plasterboard screws.

7. Use a sharp knife to cut away the finishing plaster about 25 mm all round the hole. Bed lengths of plasterer's scrim over the joints between the patch and existing sound ceiling, using a thin (runny) mix of board-finishing plaster as an adhesive. This is to reinforce the joint and reduce the risk of subsequent cracking.

8. Fill the patched area using board-finishing plaster and trowel up as before.

Damaged reveals

To make good the plaster work around the reveals of a replaced door or window: in most circumstances these can be patched using the two-coat backing and finishing plaster method as before.

Where there is extensive damage, or the plaster is loose, the entire reveal should be hacked off and replaced. Figure 12.33 shows the procedure to follow using plasterboard and board-finishing plaster:

1. Hack off plaster reveal back to the brick or blockwork surface and extending around the corner by about 75 mm to 100 mm.

2. Cut two strips of plasterboard, one for the reveal and the other for the return.

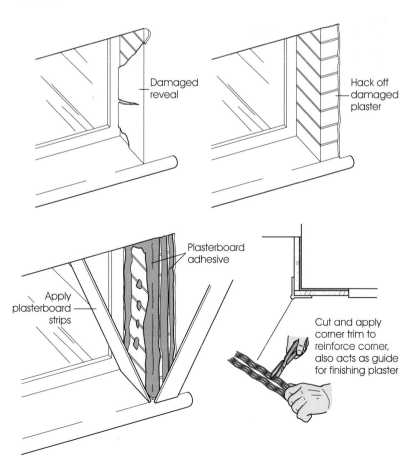

◄ **Figure 12.33** Making good plasterwork to damaged reveals

Damaged reveal

Hack off damaged plaster

Plasterboard adhesive

Apply plasterboard strips

Cut and apply corner trim to reinforce corner, also acts as guide for finishing plaster

3. Brush down the wall surfaces. Mix up some plasterboard adhesive. Using a trowel or special caulker, apply dabs of adhesive up the centre of the reveal and continuously around the perimeter.

4. Press plasterboard strips in place and check for plumb with a spirit level. The return strip should finish 2 mm to 3 mm below the adjacent plaster to allow for a coat of board finish.

5. Reinforce the corner with a length of metal plasterboard bead, which also acts as a guide for the subsequent finishing plaster coat. Apply a bed of plasterboard adhesive or board finish to the corner. Press the bead in place, check with spirit level for plumb. Also ensure it is in line with existing wall surface. Remove excess adhesive and allow too dry for 2 to 3 hours.

6. Reinforce the joint between return plasterboard strip and existing wall plaster using a length of scrim as before.

7. Complete repair by applying a coat of board-finishing plaster and trowel up in the normal way.

Replacing damaged tiles

Probably the hardest task is to find a good match for replacement. The building owner may have spares left over from the original work. Alternatively, take a damaged piece to a tile supplier for them to find a match. Figure 12.34 shows the procedure to follow:

1. Prepare the surrounding area and yourself. Cover the floor and surrounding units or bath and sanitary ware with dustsheets to protect from dust and possible scratching by the small sharp particles of broken tiles. Protect yourself by wearing eye-protection goggles, gloves and a dust mask.

2. Rake out grout joint around the damaged tile to relieve the perimeter stresses.

3. Drill a series of holes around the centre of the tile and break out using an old chisel, working progressively towards the edges. Do not try to break out the tile by trying to prise it off from the edge joints, as almost inevitably you will damage the adjacent tiles. Masking tape can be applied when drilling out the centre to prevent the masonry drill skidding across the ceramic surface.

▼ **Figure 12.34** Replacing a damaged tile

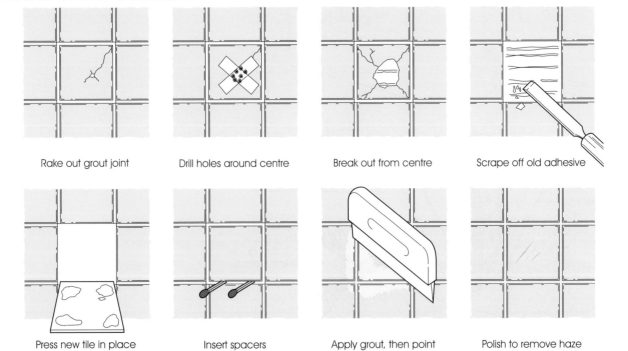

Rake out grout joint   Drill holes around centre   Break out from centre   Scrape off old adhesive

Press new tile in place   Insert spacers   Apply grout, then point   Polish to remove haze

4. Scrape or chip off the old tile adhesive back to the surface, taking care not to damage the plaster base.

5. Apply four dabs of tile adhesive to the back of the tile, or use a notched comb to provide a uniform ribbed layer.

6. Press the tile in place so that it lies flush with the surrounding ones. Insert tile spacers or matchsticks in the joints to position or support the tile. Adjust the tile as required to ensure a uniform gap all around.

7. Allow the adhesive to set for about 24 hours and then remove the spacers.

8. Fill the gap around the tile with grout, working in with a rubber squeegee. Point the joint with a fingertip or piece of wood dowel with a rounded end point.

9. Finally, when dry, polish up the surface to remove the grout haze using a clean dry cloth.

The long-term success of tiling depends on the adhesive and grout used. Most are available either in ready-mixed or powder form for mixing with water to a creamy paste. A standard type mix is only suitable for dry areas. It may also tolerate a little condensation or occasional splashing with water. In areas subject to more prolonged condensation or extensive wetting such as a shower area, always use waterproof products.

Cutting tiles

Tiles are best cut using a proprietary tile cutter, or diamond-tipped wet saw. However, small amounts can be cut by scoring the surface with a carbon-tipped tile-cutting point and snapping the tile along the scored line. Place two matchsticks under the scored line. Press down firmly on either side to snap in two, see Figure 12.35. Corners and curves can be cut out of tiles to fit around projections using a tile saw blade in a coping saw frame. Alternatively, the lines of the area to be removed may be scored and the waste nibbled away with a pair of pincers.

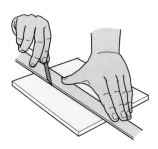

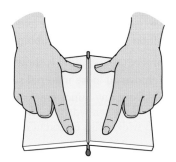

◀ **Figure 12.35** Cutting ceramic tiles

Score along line with tile cutting point

Place matchsticks under score, press down to snap tile in two

Finishing off

The joint between tiles and horizontal surfaces, such as kitchen worktops, baths and sanitary ware, will require sealing with a silicone sealant to prevent moisture penetration. Figure 12.36 shows the procedure to follow to ensure a neat bead of sealant in these locations.

1. Apply masking tape to both the vertical and horizontal surfaces, 2 mm to 3 mm away from the internal angle. Trim the cartridge nozzle off at an angle of 45° to give a bead just wide enough to fill the gap between the two taped edges.

2. Gently squeeze the cartridge gun trigger until the sealant is just seen at the tip.

3. Place the nozzle at one end of the angle to be filled, with the gun held at 45° to the wall.

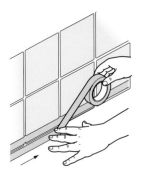
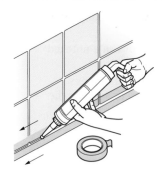
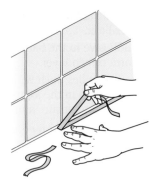

Apply masking tape
along angle to be sealed

Push the cartridge nozzle
along the angle

Allow sealant to skin
over, peel off tape

Complete
set bead

▲ **Figure 12.36** Sealing tiles to a worktop

4. Apply steady, even pressure to the trigger while pushing the gun along the angle. The nozzle will form the sealant into a neat concave curve. To stop the flow at the corners and on completion, release the metal tag adjacent to the trigger.
5. Any unevenness in the bead can be smoothed using a small paintbrush dipped in water. Leave sealant for a short while to skin over, and then peel off the tape to leave a well-formed neat bead.

Painting plasterwork

Walls and ceilings are normally painted using emulsion paint. Newly plastered and repaired surfaces should be left for 7 to 10 days to dry and then treated with a coat of plaster sealer before decoration. This prevents the new plaster showing through the paint finish in the form of 'patchiness'. Alternatively, a thinned emulsion can be applied as a primer before at least two full-strength coats are applied. The priming coat should be about one part water to about three parts emulsion paint.

Before starting work, arrange for any furniture in the room to be removed. Protect carpets and fixtures with dustsheets. Wear a dust mask and eye protection goggles when rubbing down and scraping off.

The procedure to follow is shown in Figure 12.37:

1. Remove all loose material such as dirt, dust and flaking paint.
2. Rake out any minor cracks in the plaster surface, using the end of an old slot-blade screwdriver. The raking out is to make the crack a little deeper and wider with undercut edges.

▼ **Figure 12.37** Painting plasterwork

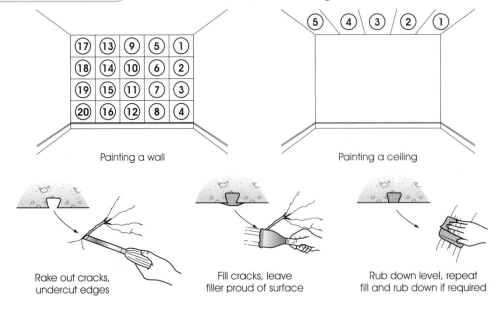

Painting a wall

Painting a ceiling

Rake out cracks,
undercut edges

Fill cracks, leave
filler proud of surface

Rub down level, repeat
fill and rub down if required

3. Fill cracks with plaster filler. Ensure filler is pressed well into the cracks in order for it to key on the undercut edges.

4. Leave filler slightly proud of surrounding surfaces. Rub down level when dry. Fill the area again and rub down if required.

5. Rub down the entire wall or ceiling surface.

6. Wash down the surface and allow it to dry.

7. Ceilings should be painted before walls. A small brush is used to cut into the corners and up to the frames, skirting and so on.

8. Use a roller or large brush to cover an area of about 1 m^2 at a time. Apply paint in one direction; spread it out by brushing or rollering diagonally. Finally, finish off using light pressure only, in the same direction as you started.

9. Using the numbered sequence, continue painting the subsequent squares or strips, blending in the paint application of one with another while the paint is still wet; otherwise if wet paint is applied over a drying one, pronounced lines will be apparent in the finished work.

10. Clean all brushes/rollers and equipment in water on completion.

Guttering and downpipes

Carpenters and joiners may be required to clean out fallen leaves and twigs from gutters or replace damaged or worn out guttering and down pipes while undertaking other maintenance activities.

Gutters are used to collect rainwater that falls on the roofs of buildings. It flows from the gutter into a down or rainwater pipe (RWP), see Figure 12.38, which is connected either directly to a back inlet gulley, or over an open-grated gulley. Hopper heads and branches may be used to allow several down pipes to flow into a single pipe part the way down. In situations where the rainwater pipe discharges into an open-grated gulley the RWP will be fitted with a shoe to ease and direct the flow of rainwater.

Most gutters and rainwater pipe systems installed in new domestic properties and for replacement tend to be plastic, although sheet metal ones such as aluminium are also seen. In the past materials such as cast iron, lead, copper and asbestos cement were used. You may even come across solid timber section gutters known as spouting.

When working on a maintenance job, if you suspect that the gutters or down pipes are made of asbestos cement you should not attempt any repair and your supervisor should be called immediately for advice as to the correct procedures to follow.

Gutters

These are specified by the shape of their cross-section, see Figure 12.39. Also shown in the figure is a range of fittings used to join, terminate and support gutters. Gutters are available in 2 m and 4 m lengths and range from 75 mm to 150 mm in width. Conversion fittings are normally available to connect new plastic lengths of guttering with existing cast-iron sections. Gutters are normally laid at a fall of 1 : 600; that is 1 mm out of level for every 600 mm run.

Rainwater pipes

Rainwater pipes are either square or round in section and are available in lengths of 2.5 m, 4 m and 5.5 m. For most domestic replacements, either 65 mm square or 68 mm diameter are used.

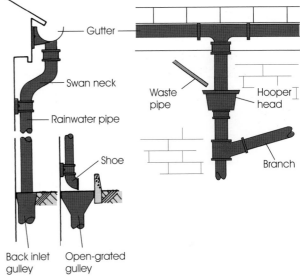

Gutter

Swan neck

Rainwater pipe

Shoe

Back inlet gulley

Open-grated gulley

Waste pipe

Hooper head

Branch

▲ **Figure 12.38** Rainwater pipes and connections

▼ **Figure 12.39** Gutter profiles and associated fittings

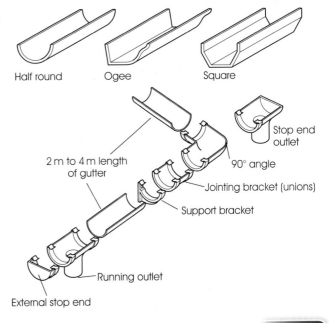

Half round

Ogee

Square

2 m to 4 m length of gutter

Stop end outlet

90° angle

Jointing bracket (unions)

Support bracket

Running outlet

External stop end

Related information

Also see Chapter 10 in Book 1, which deals with working platforms.

Heritage Link

Traditionally rainwater was discharged from the roof via an extended spout to shed rainwater clear of the building itself and away from the foundations. These spouts are termed **Gargoyles** (meaning throat or gullet); some are just plain undecorated channels, others are elaborately decorated in the form of carvings of people, animals or grotesque figures.

Gargoyles are a common feature in the more expensive range of buildings from the medieval times and beyond, particularly in churches and cathedrals. The name is also loosely applied to any grotesque carving or figure on a medieval building.

Replacing gutters and rainwater pipes

A proper working platform is required when replacing gutters; working off a ladder is not considered appropriate.

The following general procedure can be used when replacing guttering and down pipes (see Figure 12.40).

1. Old cast-iron gutters are very heavy and difficult to manage and, in addition, the bolts used to join sections will probably have rusted up. You will have to cut off the bolts with a hacksaw and remove and lower the gutter to the ground length by length, before any replacement can start.

2. Use a plumb line centred over the gulley to position the outlet. Fix the outlet to the fascia board; the gap between the top edge of the gutter and the bottom of the roof tile should be a maximum of 30 mm and, in addition, any tile underfelt should extend into the gutter.

3. Work out the fall over the length of the gutter and fix the last bracket. This amount above the outlet for example a 6 m run of guttering, will require a fall of at least 10 mm.

4. Fix a string line between outlet and end bracket; use the line to position either plain brackets or union brackets at a maximum of 1 m centres.

5. Once support brackets are in place, lengths may be cut to suit and snapped into position. Make sure when cutting that you allow for thermal movement about 3 mm per 1 m run. Most plastic unions have an 'insert to here' mark on the inside. Cutting too long will result in expansion, buckling and strange noises in hot weather. Cutting too short may result in the joint coming apart or leaking in cold weather due to contraction.

6. The final length of gutter will require a stop end; this should be positioned about 50 mm beyond the roof tiling, to aid the collection of wind-driven rain.

7. A sawn neck or offset is normally required at the upper up of the rainwater pipe to clear the width of the eaves soffit and allow the pipe to be clipped back to the surface of the wall. These are in two pieces that require cutting to the appropriate length. About 6 mm should be allowed for expansion.

8. Once the swan neck is in place, lengths of down pipe can be positioned and clipped in place with pipe brackets at around 1 m intervals. Check with a spirit level for plumb before drilling brackets. Again, about 6 mm should be allowed between the end of one pipe and the inside shoulder of the fitting for expansion.

9. Finally, either connect the down pipe to the gully or fit a shoe.

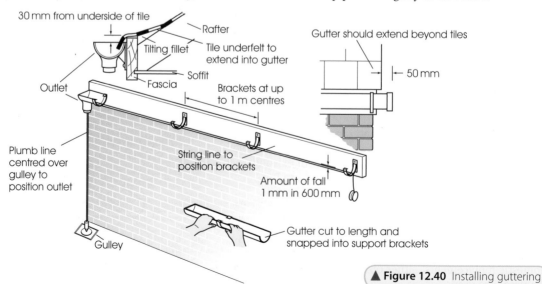

▲ **Figure 12.40** Installing guttering

Index